AF684332

Quality and Safety in Radiotherapy: Learning the New Approaches in Task Group 100 and Beyond

Medical Physics Monograph No. 36

Quality and Safety in Radiotherapy: Learning the New Approaches in Task Group 100 and Beyond

Bruce Thomadsen, Peter Dunscombe,
Eric Ford, Saiful Huq,
Todd Pawlicki, and Steven Sutlief

Editors

American Association of Physicists in Medicine
2013 Summer School Proceedings
Colorado College
Colorado Springs, Colorado
June 16–20, 2013

Published for the
American Association of Physicists in Medicine
by Medical Physics Publishing, Inc.

Corrected Title Page

To order American Association of Physicists in Medicine (AAPM) publications, contact:

Medical Physics Publishing, Inc.
4513 Vernon Boulevard
Madison, WI 53705-4964
Phone: (800) 442-5778 or (608) 262-4021
Fax: (608) 262-2121
E-mail: mpp@medicalphysics.org
Web: www.medicalphysics.org

© 2013 by the American Association of Physicists in Medicine. All rights reserved. No part of this publication may be reproduced, stored in a retrieval system, or transmitted in any form or by any means (electronic, mechanical, photocopying, recording, or otherwise) without the prior written consent of the publisher.

The CD attached to this book contains large tables, figures, and many of the text's images in color.

Published by:
Medical Physics Publishing, Inc.
Madison, Wisconsin

Published for:
American Association of Physicists in Medicine (AAPM)
One Physics Ellipse
College Park, MD 20740-3846
Phone: (301) 209-3350
Fax: (301) 209-0862

Library of Congress Control Number: 2013938355

ISBN book: 978-1-888340-49-5
ISBN CD: 978-1-936366-23-1

Printed in the United State of America on acid-free paper

Contents

Preface

Radiotherapy physics has had a very long history of performing quality assurance. Physicists working with the very earliest radiotherapy units developed before the beginning of the last century realized the need to verify the calibration and proper operation of the equipment. The International Commission on Radiation Units and Measures and the National Council for Radiation Protection and Measurement initiated standardization of quality assurance protocols through a series of recommendations. This was followed by a large collection of professional guidance documents, particularly from the American Association of Physicists in Medicine (AAPM). Through these efforts, radiotherapy became one of the first medical disciplines to establish standardized quality assurance.

In light of the increasing complexity of radiotherapy treatments and technology and the diversity of methods used by facilities for various clinical procedures, it is no longer practical to rely solely on generic, prescriptive lists of comprehensive quality assurance (QA) steps to ensure quality and safety for patients. This was the conclusion that guided the AAPM's Task Group 100 away from writing such guidelines for intensity-modulated radiotherapy and toward the examination of quality and safety techniques that have proven so effective first in industry and, more recently, in many medical disciplines. These techniques consider not only QA for equipment, but also for procedures as a whole in the context of the facility. These approach and methodologies may be new and somewhat daunting to many practicing radiotherapy medical physicists. The goal of this book is to make these new techniques accessible to the broader medical physics community.

The editors thank Todd Hanson of Medical Physics Publishing for tirelessly working on the manuscripts for this book and Karen McFarland of the AAPM for assistance in coordinating this project. The editors particularly thank the authors, who have worked to give this volume a smooth-flowing, integrated discussion of the topic.

Bruce Thomadsen
Peter Dunscombe
Eric Ford
Saiful Huq
Todd Pawlicki
Steven Sutlief

June 17, 2013
Colorado Springs, Colorado

List of Contributors

Jean-Pierre Bissonnette, MCCPM
Department or Radiation Oncology
University of Toronto
Toronto, Ontario
and
Department of Radiation Physics
Princess Margaret Cancer Centre
Toronto, Ontario
Jean-Pierre.Bissonnette@rmp.uhn.on.ca

Derek Brown, Ph.D., FCCPM
Adjunct Associate Professor
Departments of Oncology and Physics and Astronomy
University of Calgary
Calgary, Alberta
CanadaDerek.Brown@albertahealthservices.ca

Barrett S. Caldwell, PhD, CHFP, FHFES
Professor, School of Industrial Engineering (and Aeronautics by courtesy)
University Faculty Scholar
Purdue University
West Lafayette, Indiana
bscaldwell@purdue.edu

Peter Dunscombe, Ph.D., FCCPM, FAAPM, FCOMP
Director, Medical Physics
Tom Baker Cancer Centre
Calgary, Alberta
and
Professor, Department of Oncology
Adjunct Professor, Department of Physics and Astronomy
University of Calgary
Calgary, Alberta
Peter.Dunscombe@albertahealthservices.ca

Suzanne Evans, MD, MPH
Assistant Professor
Department of Therapeutic Radiology
Yale University School of Medicine
New Haven, Connecticut
suzanne.evans@yale.edu

Luis E. Fong de los Santos, Ph.D.
Department of Radiation Oncology
Mayo Clinic
Rochester, Minnesota
fongdelossantos.luis@mayo.edu

Eric Ford, Ph.D.
Associate Professor
Department of Radiation Oncology
University of Washington Medical Center
Seattle, Washington
eford@uw.edu

M. Saiful Huq, Ph.D., FAAPM, FInstP
Professor and Director
Department of Radiation Oncology
Medical Physics Division
University of Pittsburgh Cancer Institute
and
UPMC Cancer Center
Pittsburgh, Pennsylvania
huqs@upmc.edu

Ibbott, Geoffrey S., Ph.D.
Professor & Chairman
Radiation Physics
University of Texas MD Anderson Cancer Center
Houston, Texas
gibbott@mdanderson.org

Jennifer L. Johnson, M.S.
Senior Medical Physicist
Department of Radiation Physics
University of Texas MD Anderson Cancer Center
Houston, Texas
Ms_JL_Johnson@yahoo.com

Arno J. Mundt, M.D., FACRO, FASTRO
Professor and Chair
Department of Radiation Medicine and Applied Sciences
University of California – San Diego
La Jolla, California
amundt@ucsd.edu

Sasa Mutic Ph.D.
Professor and Co-director
Division of Medical Physics
Department of Radiation Oncology
Washington University School of Medicine
St. Louis, Missouri
smutic@radonc.wustl.edu

Zoubir Ouhib, M.S.
Chief Medical Physicist
Lynn Cancer Institute of Boca Raton Regional Hospital
Delray Beach, Florida
and
Research Assistant Professor
Florida Atlantic University
Boca Raton, Florida
Zouhib@brrh.com

Jatinder R. Palta Ph.D., FAAPM, FASTRO
Professor and Chair of Medical Physics
Virginia Commonwealth University Department of Radiation Oncology
Richmond, Virginia
jpalta@mcvh-vcu.edu
and
Chief Physicist, National Radiation Oncology Program
Veterans Health Administration
Richmond, Virginia
Jatinder.Palta@va.gov

Todd Pawlicki, Ph.D., FAAPM
Professor and Vice Chair of Medical Physics and Technology
Department of Radiation Medicine and Applied Sciences
University of California – San Diego
La Jolla, California
tpaw@ucsd.edu

Frank Rath, MSIE
Department of Engineering Professional Development
University of Wisconsin–Madison
Madison, Wisconsin
rath@engr.wisc.edu

George W. Sherouse, Ph.D., DABR, FAAPM
Landauer Medical Physics
Glenwood, Illinois
GWS@GWSherouse.com

Steven Sutlief, Ph,D., DABR
Chief Medical Physicist
VA Puget Sound Health Care System
and
Affiliate Instructor, Department of Radiation Oncology
University of Washington Medical Center
Seattle, Washington
Steven.Sutlief@va.gov

Bruce Thomadsen, Ph.D.
Professor, Departments of Medical Physics, Industrial and Systems Engineering, Engineering Physics, and Biomedical Engineering
University of Wisconsin–Madison
Madison, WI
brthomad@wisc.edu

Jeffrey F. Williamson, Ph.D., FAAPM, FACR
Professor of Radiation Oncology
Virginia Commonwealth University
Richmond, Virginia
JWilliamson@mcvh-vcu.edu

Ellen Yorke, Ph.D., FAAPM, FACMP
Medical Physics Department
Memorial Sloan Kettering Cancer Center
New York, New York
yorkee@mskcc.org

Chapter 1

Introduction to Quality

Peter Dunscombe, Ph.D.[1], **Suzanne Evans, M.D.**[2], **and Jeff Williamson, Ph.D.**[3]

[1]Director of Medical Physics, Tom Baker Cancer Centre, Calgary, Alberta
[2]Assistant Professor, Department of Therapeutic Radiology, Yale University School of Medicine, New Haven, Connecticut
[3]Professor of Radiation Oncology, Department of Radiation Oncology Virginia Commonwealth University, Richmond, Virginia

1.1 Quality and Safety in Radiotherapy

Quality and safety of clinical care are the cornerstones of the radiation medicine professional's practice. The radiation medicine community has not been complacent in the face of advancing technologies and clinical strategies. Indeed, much effort by the American Association of Physicists in Medicine (AAPM) and other professional organizations across the globe has been directed toward maintaining and enhancing quality in an era of rapid change. However, it is also true that early reports of safety issues in medicine generally (Institute of Medicine 1999) and radiotherapy in particular (International Atomic Energy Agency 2000) have not received the attention they perhaps deserve.

A series of high-profile *New York Times* articles (Bogdanich 2010) and some well-publicized incidents in Europe (Scottish Ministers for The Ionising Radiation 2006) have forced the community to reassess how safe radiotherapy actually is. This belated but heightened interest in radiotherapy safety has spawned a flurry of activity, including the rapidly convened Miami meeting "Safety in Radiation Therapy: a Call to Action" (Hendee and Herman 2011), a series of white papers from

ASTRO (Moran et al. 2011; Solberg et al. 2012), and this AAPM summer school. We now acknowledge that we need to work harder if we want to make radiotherapy as safe as those industries in which the public has a high level of trust (Ford and Terezakis 2010).

In addition to the question of safety, we need to reflect on the quality of the treatments we provide to radiotherapy patients. It is a tenet of radiotherapy, and any other medical intervention, that quality impacts outcome. However, definitively establishing the relationship between quality and outcome is fraught with difficulty. Certainly it cannot be done in any sort of prospective randomized manner. Fortunately, recent retrospective analyses have shed light on this relationship and provide support to the notion that quality does indeed enhance outcome and, conversely, that sub-optimal quality compromises outcome. Mardon et al. have explored the relationship between general measures of quality as reflected in the safety culture of an organization and adverse events (Mardon et al. 2010). They found that of the 15 patient safety culture variables investigated, e.g., frequency of events reported and management of handoffs and transitions, nearly all were negatively correlated with adverse events as measured by the Patient Safety Indicators developed by the Agency for Healthcare Research and Quality (AHRQ). The more events reported and the more attention given to handoffs, the fewer the number of clinical adverse events observed.

Looking specifically at technical quality issues in radiotherapy, a comprehensive review of the clinical trials literature by Weber et al. demonstrated that protocol violations for Hodgkin's lymphoma, head and neck, and pancreas radiotherapy resulted in inferior outcomes for patients (Weber et al. 2012). Relatively high rates of protocol violations were observed. Their conclusions are all the more worrisome as the data were collected from institutions participating in multi-institutional clinical trials, and such organizations tend to be larger and commit more resources to quality patient care in all its dimensions.

An interesting and relevant question is: should we view quality and safety as two separate issues? If the answer is yes, then the measures adopted to enhance these two aspects of patient care will be different with minimal overlap. If, on the other hand, quality and safety are closely interrelated, then a measure implemented to enhance one could reasonably be expected to benefit the other.

To answer this question, we need to adopt working definitions of safety and quality, both of which are inherently dependent on context. Conventionally, a radiotherapy system was considered safe if there were no obvious injuries to patients. As both the efficacy and toxicity of radiotherapy only become fully apparent after a period of time, we need to broaden the definition of safety to "the absence of an unacceptable risk of harm, where harm is excess morbidity or sub-optimal tumor control."

Quality is an equally difficult concept to define. It depends on perspective. For a radiation oncologist, quality would include the opportunity to cure cancer at the rate expected for the stage of disease in the most modern fashion. For a physicist, quality would probably include being confident in the output of a linear accelerator to within 2%. And for a therapist, quality might include being able to spend adequate time with each patient. Patients, of course, have their own perspective, and this is discussed in some detail later in this chapter. With a bias toward the provider's perspective, the Institute of Medicine has defined quality of care as the degree to which health services for individuals and populations increase the likelihood of desired health outcomes and are consistent with current professional knowledge (Institute of Medicine 1999).

With a few simple, but admittedly questionable, assumptions, we can make a quick estimate of the number of patients who *may* have their outcome compromised by poor quality radiotherapy. Let's assume all radiation prescriptions reflect the height of medical knowledge and appropriate use of radiation. In addition, assume that the distribution of deviations from the prescription, i.e., current professional knowledge, is gaussian. Taking a conservative view of the literature on accident rates in radiotherapy, we might postulate a probability of 10^{-5} clinically adverse events per course (British Institute of Radiology 2007). If, in terms of dose, an adverse event occurs beyond a deviation of 20%, then we can calculate the standard deviation of the distribution of deviations to be 4.5%. If we then say that, conservatively again, a sub-optimum outcome is represented by a dose deviation of 10%, then given a U.S. population of 3×10^8, a cancer incidence rate of 5×10^{-3}, and a radiotherapy utilization rate of 50%, approximately 20,000 patients per year have their outcome compromised to some unknown extent by poor-quality radiotherapy.

This is an astonishingly large figure. The derivation warrants some discussion. First of all, is the distribution of deviations from the prescription really normal? In the center of the distribution there is evidence to suggest that it is with a standard deviation not greatly different from the 4.5% estimated above (Fiorino et al. 2000). The issue, of course, is whether the assumed normality of the distribution out into the tails? This we do not know—it's pure conjecture.

The 20% dose deviation for a clinically significant adverse event is used by regulators, and that's the main justification for its use here. The 10% dose deviation threshold for a sub-optimal clinical outcome is the third questionable assumption in this derivation. There are clinical dose-response curves that are very steep, suggesting that 10% is not unreasonable for an estimate of this kind (Mijnheer, Battermann, and Wambersie 1987).

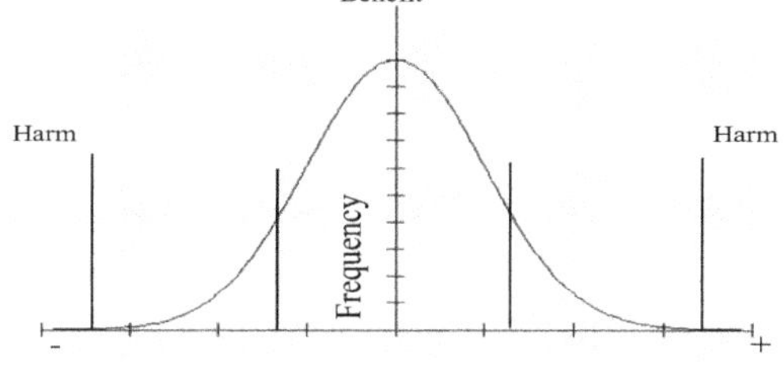

Figure 1–1. A possible distribution of deviations from the prescription.

What about the result of the calculation? Twenty thousand patients is a large number, but with probably around 750,000 first courses of radiation delivered per year in the U.S., it represents just 2.5% of the radiotherapy population. In the context of the Radiological Physics Center's experience that 30% of head and neck verification phantom tests cannot reach agreement within 7% and 4 mm between measurement and calculation, the 2.5% figure does not seem excessive (Ibbott et al. 2008).

A largely unknown contributor to sub-optimal radiotherapy is variability in contouring. There are numerous studies that have quantified this variability for many treatment sites, although the impact of contouring variability on outcome is only now becoming clearer (Weber et al. 2012). Professional organizations such as the American Society of Radiation Oncology (ASTRO) and the European Society for Therapeutic Radiology and Oncology (ESTRO) recognize contouring variability, along with field design and prescription, as quality issues, and both organizations are developing and actively promoting training experiences directed toward reducing this variability.

If we accept that the distribution of departures from quality is gaussian, we can suggest a relationship between quality improvement and safety improvement. Improving safety, defined as "reducing the risk of harm," corresponds to shrinking the area under the tails of the distribution which—if the quality-eroding mechanisms in operation around the peak and in the tails are similar—will also narrow the peak, i.e., reduce the standard deviation of the distribution. Likewise, quality improvement can be interpreted graphically as narrowing the central part of the distribution, which brings the delivered treatments closer to those prescribed. Narrowing the distribution, if it is a gaussian and with the caveat of similar quality-eroding mechanisms, reduces the area under the tails, thus leading to greater safety. If we accept that there is an intimate relationship between quality and safety as argued above, then it is hardly surprising that most of the recommendations made in the recent burgeoning literature on safety in radiotherapy could equally be interpreted as enhancing quality (Dunscombe 2012).

If the discussion above has convinced the reader that there is an issue in radiotherapy that goes well beyond the regrettable but rare reports in the popular press of major radiotherapy accidents, the question becomes what to do about it. Feasible solutions must be set in the context of healthcare funding. Total healthcare spending in the U.S. is predicted to rise to 20% of GDP by 2020. Spending on cancer care was $27 billion in 1990, and it is predicted to rise to $157 billion by 2020 (Sullivan et al. 2011). While much of the increase in the cancer care budget can be attributed to the introduction of expensive new drugs, the technological component of radiotherapy has been estimated to be rising at 5% per year prior to the widespread implementation of IMRT and the recent expansion of heavy particle therapy (Ploquin and Dunscombe 2008). Cancer care as we know it today may be financially sustainable into the future. The growth of the cancer care budget as we have known it over the last decades is almost certainly not. The dilemma is how to improve radiotherapy quality and safety within budgetary constraints. The topics in this summer school point to some of the answers.

The complexity and cost of quality assurance (QA) and quality control (QC) activities have grown as the technological complexity of radiotherapy has grown. Some of this growth is undoubtedly inevitable, but some of it may have occurred almost subconsciously with little reflection on the real value of adding additional tests at frequencies that could seriously drive up resource commitments.

Along with the complexity of the equipment used in modern radiotherapy, there is a concomitant increase in the complexity of clinical processes. One of the less obvious implications of this increasing complexity is the decoupling of the treatment delivery therapist's intuition of what constitutes correct treatment—traditionally the last line of defense—and the technical correctness of the treatment. Just a few years ago, an experienced therapist could tell if the calculated monitor units were at least in the right ballpark. With the advent of intensity-modulated radiation treatment (IMRT), that informal line of defense has disappeared.

Traditionally, the radiation oncology physics community has had a largely device-centric perspective. As discussed in more detail in Section 1.4, this has changed in recent years with the recognition that many of the safety and quality issues have significant human factors content. The AAPM's Task Group 100 is working to bring objectivity to quality management programs, covering both equipment and people, with the example being IMRT (Huq et al. 2008). The methodology employed by TG 100 is to develop a process map for the modality of interest, and then to perform failure modes and effects analyses (FMEA) on the steps in the process map. The FMEA identifies the weakest points in the process and, if desired, a fault tree analysis can then be conducted to select the locations of QA and QC interventions. A discussion of TG–100's efforts is a central feature of this summer school. To augment the insights gained through the formal FMEA, the accepted techniques of error management—such as incident learning systems and root cause analysis—are extensively discussed, as are the more familiar topics of quality assurance and control. From the discussion above, it will be apparent that although terms such as *incident learning* and *root cause* emphasize the safety aspect of these interventions, their presence as practical tools in the clinic will have a beneficial effect for radiotherapy quality in the broader sense.

Optimizing the care pathway, a.k.a. the process map, is another opportunity for enhancing not only quality and safety, but also efficiency. Some relevant techniques are discussed throughout the summer school.

As the discussion of these issues evolves during the summer school, it is worth placing the proposed interventions and measures in the context of the financial reality of today's health care. While a full-blown FMEA or root cause analysis (RCA) could be expensive to perform, it is not hard to devise shortcuts and expedited approaches, e.g., do a simple analysis on your own without a team, which will bring benefits. The first hurdle is to develop a familiarity and comfort level with these error management techniques that are foreign to most of us.

Recent documents, such as "Safety is No Accident," suggest measures for enhancing safety and quality that have minimal resource requirements (Zietman, Palta, and Steinberg 2012). For example, "no interruption zones," which are widely considered to be effective, require some leadership but little else.

The all-important patient's perspective is discussed in some detail in the following section of this chapter. Addressing many of the patient's quality indicators requires a good attitude coupled with emotional intelligence to gauge the patient's response to the interaction. Resource requirements are minimal to meet many of the patient needs discussed later.

It is unfortunate that it took the well-publicized suffering of a few patients who were victims of catastrophic errors to sharpen our focus on quality and safety in radiotherapy. However, the radiation treatment community has made a commitment to and considerable progress in addressing the key issues. This summer school is further evidence of our determination to provide cancer patients with high-quality, safe radiation treatment.

1.2 Quality in Radiotherapy: Patient's Perspective

The patient's perspective on quality is important to understand, as ultimately it is the patient who consumes health care and is the end user. There have been several studies of patients' perceptions of quality, including a study by Nijman et al. (2012), upon which the following discussion is based. Seven main themes emerged from the focus groups and concept mapping conducted by Nijman and colleagues: the provision of information, a patient-centered approach, professional competence, minimized planning and waiting times, accessibility, cooperation and communication, and follow-up care.

The provision of information encompasses a complete understanding by the patient and his or her family of the rationale for radiation therapy, the potential consequences of receiving or forgoing radiation, and the prognosis of the cancer. Additionally, there is no one universal approach for information provision that works for all patients. Some patients want all the information up front; some patients prefer it piecemeal. Some patients want specific prognostic information (percentage survival at five years) whereas others want generalizations (there is a low likelihood of long-term survival). All patients in the focus groups expressed a preference for frank discussions about the potential side effects of treatment.

A patient-centered approach is also highly desired. This is characterized by a kind, open, and friendly attitude by the treatment team, with personal attention. This also included respect for privacy, as well as frequent checks by the therapists and treatment team on a daily basis for how the individual is doing.

Professional competence was highly valued, although this was poorly and variably defined. It certainly includes a thorough knowledge of the individual patient's medical history. It was also described as a decisive manner, inspiring confidence, and giving the perception of intelligence and expertise. Patients also stated that they desired a level of confidence in their radiation therapists.

While planning and waiting times are considered best when minimized, the concept of accessibility requires more explanation. The concept of accessibility includes handicapped-accessible facilities, ample and easy parking, easy transport to the facility, and the presence of restrooms close to waiting areas (particularly for pelvic malignancies).

Table 1–1. Top 10 mean importance score of quality aspects of radiation care

Ranked Importance	Statement of Quality Aspects	Mean Importance Score
1	Radiation oncologist is familiar with my medical file	4.36
2	Confidence in the expertise of the healthcare professionals	4.25
3	Intake consultation at which the radiation oncologist explains everything in detail	4.21
4	Cooperation between doctors from different departments or hospitals	4.19
5	Healthcare professionals who take me seriously	4.07
6	Healthcare professionals who take enough time for me	4.04
7	Healthcare professionals who give honest answers to my questions	4.04
8	Treating physician is informed about the radiotherapy treatment	3.93
9	Healthcare professionals who give me the opportunity to ask questions	3.93
10	Healthcare professionals who do their work conscientiously	3.89

Cooperation and communication between healthcare providers was highly valued. In particular, communication between oncologists and surgeons was valued. However, communication between therapists as they change linear accelerators was also considered essential. Follow-up care was also important, both in having a contact person to turn to with problems or questions, as well as to have continued monitoring during treatment.

It is clear that the patient perspective of quality is concerned much more with the entire facility experience than just the quality of treatment. It is notable that patients take for granted that treatment will be safe, appropriate, and accurate. The physician perspective is much more concerned with the multifaceted components of quality, whereas the patient perspective strives for a quality atmosphere as well as a quality treatment.

1.3 Quality in Radiotherapy: A Physician's Perspective

Quality of care is defined as the degree to which health services for individuals and populations increase the likelihood of desired health outcomes and are consistent with current professional knowledge. For the field of radiation oncology, such quality has many facets, which can be loosely grouped into high-quality decision-making, high-quality performance, and high-quality resources.

High-quality decision-making encompasses all of the aspects of the patient's diagnosis, staging, and treatment. This is dependent on high-quality processes and healthcare interactions throughout the entire diagnostic workup. The accuracy of treatment is dependent on the accuracy of staging. The accuracy of staging is dependent on proper interpretation of diagnostic imaging, pathologic data, and surgical procedures. Proper multidisciplinary management involves systemic therapy decision-making. It is clear that the radiation oncologist is only as strong as the weakest link of the team.

Even assuming that staging and diagnosis are made correctly, the next phase of decision-making comes with the judicious use of local and systemic therapies, including chemotherapy, surgery, and radiation. This requires a wide breadth of knowledge of an ever-changing body of literature. In the case of radiation, the modality is highly dependent on technique. The proper modality must then be chosen (for example, brachytherapy, external beam, or stereotactic techniques). The quality of the treatment preparation activities is a major determinant of overall treatment quality: the method of immobilization, the use of motion management techniques, the degree of bladder/gastric filling, and the use of contrast or anatomic markers. All of these decisions affect the quality of treatment delivered. The proper fractionation, total dose, and regions to treat must then be chosen in order to encompass the anatomic regions at risk of harboring gross or microscopic disease. There is considerable variation in normal tissue and target volume segmentation (Feng et al. 2012). Once the fields are arranged or the target volumes are segmented accurately, the quality issues continue. The development of a quality radiation plan also includes inclusive and accurate segmentation of organs at risk. However, those practitioners trained in the two-dimensional era might argue that their experience would allow them to visually review a field configuration and determine—based on this inspection and without rigorous review of the dose volume histogram (DVH)—what constitutes an acceptable plan. Once the plan is designed, there is no universal standard to guide practitioners to accept or reject the inevitable trade-offs involved in a treatment plan (Marks et al. 2011). For example, how many cc's of mandible will you accept over 60 Gy? What dose inhomogeneity is considered acceptable on a breast plan? Does one stop a lung case at 60 Gy to keep the lung V20 to less than 30%? The answers to these questions are often institutionally biased and subjective. Although peer review is often held up as a method to improve quality, smaller institutions with fewer peers at their disposal must rely on services such as Chartrounds.com.

The next grouping is of high quality performance. This is includes:

- the quality of images taken at simulation;
- safe radiotherapy delivery and accurate delivery of the dose prescribed, which is discussed elsewhere in this chapter;
- the quality of the therapist, nursing, and patient interactions;
- the presence and quality of weekly treatment visits and nursing education; and
- peer review at chart rounds, weekly chart checks, and ongoing machine QA procedures.

Of interest to the performance discussion is the work done in the quality and accountability study by Keroack et al. (2007). This effort sought to quantify what made high-performing organizations high performing. The final result of this study found that a shared sense of purpose, a certain leadership style, an accountability system, a focus on results, and a culture of collaboration were vital to high performance. High-performing organizations seemed to have a very distinct mission of "patients first," which they articulated often when looking at gaps between any present unsatisfactory state and the desired future state. The leaders in these high-performing organizations were found to have a hands-on style, were intimately involved with the staff, exhibited a great deal of passion about achieving excellence, and sought grassroots input from the team. With respect to the accountability system, the essential elements were the acceptance of responsibility for safety from the chairperson, accompanied by the development of measures and goal setting at every level. The focus on results incorporated a strong focus on human behavior and the redesign of work to enhance productivity and safety. The culture of collaboration included frequent recognition of employee efforts to perform and collaboration between all staff as vital team members. As physicians are most often the radiation department chairs, we would do well to heed this advice and work to emulate these values.

Along a similar vein, another aspect of high-quality performance is seen in the relationship between a hospital's patient safety culture and its number of adverse events (Mardon et al. 2010). This study examined several patient safety indicators (including iatrogenic pneumothorax, selected infections due to medical care, post-operative hemorrhage, etc.), and found them to be more common in institutions scoring poorly on the indicators of safety culture tested by the Hospital Survey on Patient Safety Culture. The safety culture indicators that met statistical significance were frequency of events reported, good handoffs and transitions, management support for patient safety, non-punitive responses to errors, organizational learning and continuous improvement, overall good perceptions of patient safety, adequate staffing, management expectations and actions, teamwork across units, and teamwork within units. It is increasingly clear that high-quality performance requires an excellent safety culture and excellent organizational culture.

The final component of quality is high-quality resources. While not every clinic needs the latest radiotherapy gadget, there are certain capabilities that can add greatly to the capabilities of a given center. The availability of a stereotactic system for intracranial and extracranial applications adds to the quality of a center. The ability to perform motion management, image guidance, IMRT, and contrast administration all add to the quality of radiation delivered. The use of all of these tools presupposes that the center has adequate expertise to use these technologies in a proper manner.

1.4 Quality in Radiotherapy: A Physicist's Perspective

1.4.1 Evolution of Quality Management in Radiation Therapy

From the earliest days of radiation therapy, physicists and clinicians recognized the need to accurately quantify treatment intensity in terms of well-defined physical quantities (Rutherford and Chadwick 1911). This effort continues to be an important area of medical physics research and clinical practice to the present day. In the 1960s and 1970s, physicists recognized that the steepness of dose-response curves for tumor control and normal-tissue complications limited the level of dose-delivery uncertainty consistent with optimal clinical outcomes to about 5% (Herring and Compton 1971). The 1976 *ICRU Report 24* was one of the first formal efforts to outline tolerances for dosimetric quality-assurance (QA) tests based upon this insight (International Commission on Radiation Units and Measurements 1976). It soon became evident that field placement and other geometric delivery uncertainties also influenced radiation therapy dose delivery accuracy and outcomes, and that these factors also needed to be addressed by the QA program (Marks and Haus 1976). These analyses assumed that delivery errors were relatively frequent, normally distributed, and relatively small, enabling the problem of maintaining quality to be formulated as one of minimizing the variability of delivered doses about the prescribed dose which, in turn, was selected to optimize the trade-off between tumor control and toxicity. This type of error is referred to as *quality-compromising* or *quality-eroding* error. However, this perspective was enlarged by the Riverside Methodist Hospital tragedy, in which the output of a ^{60}Co teletherapy unit was incorrectly decayed over a two-year period (1974–1976). Because output checks were not performed, the output error was not detected, resulting in 10% to 40% overdoses to nearly 400 patients (Rubin 1978). This incident, and the regulatory response that followed it, demonstrated the need for quality assurance to address mitigation of low-likelihood, but *catastrophic errors*, i.e., unintended radiation exposures capable of causing deterministic harm (U.S. Nuclear Regulatory Commission 1979).

During the last 30 years, the physicist's understanding of the scope and goals of the radiation therapy Quality Management Program (QMP), along with the associated QA and QC tools, have evolved significantly. Currently, the physicist's concept of QMP is on the cusp of another dramatic transition due to the rapid increase in technical complexity and conformality of modern image-guided radiation therapy (Williamson et al. 2008). Because this transition is not yet complete, this contribution will present an historical overview of this progression, as documented by published QA guidance.

1.4.2 A Physicist's View of the Radiation Therapy Process

The QMP scope is dictated by the totality of error mechanisms of failure modes that it is to guard against. Enumerating the totality of possible failure modes, in turn, requires a model of the process under consideration. To this end, Figure 1–2 presents one such understanding of the radiation therapy process, generalized to include iterative image-based planning and some level of image guidance embedded in the delivery subprocess.

The radiation therapy process, as conceptualized by the authors, consists of three coupled subprocesses.

1. Clinical assessment and evaluation (red pathways and processes)

The clinical assessment process is initiated by referral of the patient to radiation oncology. The radiation oncologist first reviews the patient's clinical record—including prior imaging studies, laboratory tests, prior medical history, and prior consultations—and then sees the patient during consultation. The consultation fur-

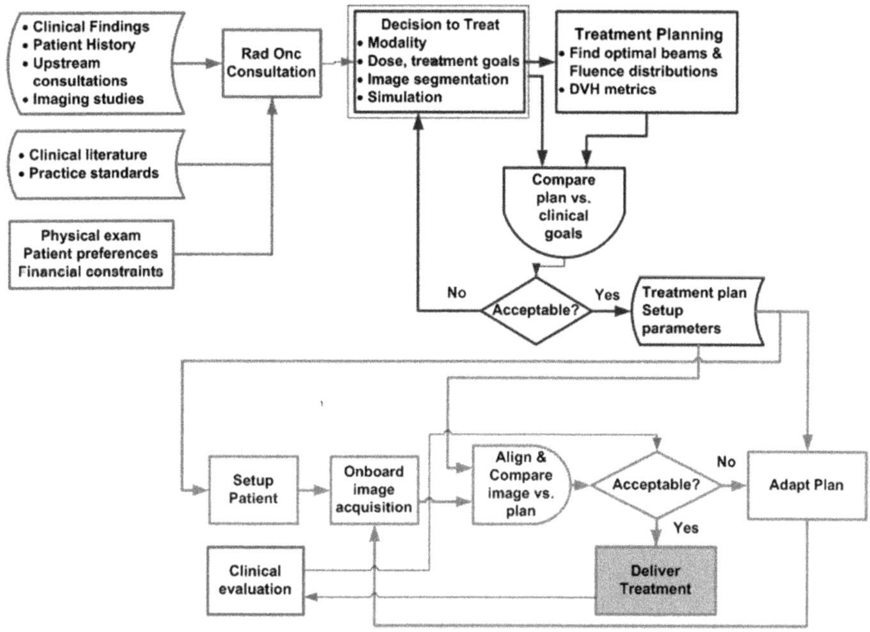

Figure 1–2. Overview of the radiation therapy process adapted to include image-guidance. The upper left, upper right, and bottom boxes and pathways denote physician-dominated subprocesses (clinical assessment, evaluation, and prescription), physics-dominated subprocesses (planning), and treatment delivery subprocesses, respectively. (Refer to the CD for the color version of this figure.)

ther enlarges this information domain through the patient's history and physical exam findings, additional communications among the patient's multi-disciplinary team physicians, or additional studies. The process includes synthesis of published studies and practice standards, culminating in a recommended course of treatment. The next step is "decision to treat," which includes developing a prescription, i.e., a specification of dose and volume to be treated. This can be as simple as drawing a field aperture on a 2D radiograph and specifying total dose and fraction size at mid-line. More likely, it involves contouring target volumes, margins, and organs at risk on a 3D imaging study and specifying dose-volume constraints for each of these contoured structures. The radiation oncologist controls downstream processes via the "clinical evaluation" function, in which the physician directly monitors acute patient responses to treatment and then exercises final authority over treatment planning and plan adaptation decisions.

2. Treatment planning (blue pathways and processes)

In this subprocess, the physics and dosimetry team takes the physician's prescription and transforms it into a set of detailed instructions and treatment parameters that both realizes the physician's clinical intent and provides a detailed and complete step-by-step guide that the therapist uses to deliver each daily fraction independently of the physicist and physician. This "treatment plan" can be as simple as instructions for patient positioning and isocenter alignment, along with field size, beam modifiers, and monitor units for each field. More modern IGRT and IMRT involve much more complex sets of parameters, including leaf sequences for modulating each field and detailed instructions for using onboard imaging to realize the setup. Planning is inherently an iterative process where each candidate "plan" is compared with the physician's prescribed treatment goals until a compromise acceptable to the physician is reached. Linkages to the red clinical evaluation pathways indicate that physician review of candidate treatment plans and interaction with the physics team to refine the treatment goals and their priorities are key steps in a properly functioning planning process.

3. Treatment delivery (green pathways and processes)

Finally, the approved plan is downloaded into the computer-controlled treatment delivery device (a linac interfaced with the department's information management system), which initiates the treatment delivery process. Contemporary external beam radiation therapy is a potentially iterative process in which pretreatment imaging (EPID, CBCT, or kV radiography) is used to verify accurate positioning of the delivered dose distribution relative to bony, soft-tissue, or implanted fiducial landmarks. This process of adapting the treatment from the simulation to treatment position anatomy includes physician review of images comparing the treatment and simulation anatomies and approval of any plan adaptations needed to resolve discrepancies. In its simplest form, EPID portal images are used to adjust the plan isocenter location on a weekly basis. Increasingly, more sophisticated technologies, e.g., adaptive radiation therapy (ART) or image-guided radiation therapy (IGRT), are used in which the on-line setup corrections are made on a daily basis and more complex off-line adaptive replanning is used to compensate for systematic time-trending changes associated with weight loss, edema, or tumor regression. The IGRT/ART paradigm clearly blurs the distinction and temporal independence of the clinical assessment, treatment planning, and treatment delivery processes.

1.4.3 The Current Approach to Radiation Therapy Quality Management

The first effort to comprehensively formalize QA practice that included both geometric and dosimetric uncertainties was AAPM's *Report 13* published in 1984 (Baily et al. 1984). Its basic approach to determining QC test tolerances continues to be referenced (if ignored in practice) by current device-oriented QA guidance found in the reports of TG 40 and TG 42 (Kutcher et al. 1994; Klein et al. 2009). *Report 13* defined quality assurance as "...those procedures that ensure a consistent and safe fulfillment of the dose prescription to the target volume, with minimal dose to normal tissues and minimal exposure to personnel." According to this definition, the domain of medical physics is limited to ensuring that the delivery of radiation to the patient, in terms of absorbed dose level and correct anatomic location, is consistent with the radiation oncologist's prescription.

Report 13 posited an overall maximum dose delivery uncertainty of 5% and positional uncertainty of 10 mm. Based on combining contributing errors in quadrature, tolerances on individual measurements (calibration, relative dose measurement, dose-calculation accuracy) and linac performance metrics (field size, flatness, symmetry, SSD, etc.) for corresponding QA and QC checks were assigned. The result of these efforts is a table of device performance endpoints and tolerances (the maximum allowed discrepancy between the QA test outcome and the expected value). While patient positioning error, human error, etc. are acknowledged as potential sources of uncertainty, *Report 13* is almost exclusively focused on ensuring consistent and accurate functioning of delivery devices and the accuracy of measurements. The only recommended approaches for mitigating process-related errors were portal film imaging and *in-vivo* dosimetry.

The report of Task Group 40 (Kutcher et al. 1994), published a decade later in 1994, significantly broadened the definition of quality: "The 'quality' of radiation oncology can be defined as the totality of features or characteristics of the radiation oncology service that bear on its ability to satisfy the stated or implied goal of effective patient care." TG 40 outlined a vision of quality assurance called a "comprehensive quality assurance program" that described radiation therapy as a sequence of coupled processes involving clinical and nursing activities, as well as physics-dominated subprocesses of treatment planning, dosimetry, and assuring correct device function. TG 40 rightly emphasized the importance of a well-organized and integrated team of professionals—including radiation oncologists, nursing personnel, physicists, dosimetrists, and therapists—to deliver clinically effective radiation therapy. It recommended that a mulidisciplinary QA committee have overall responsibility for the quality of the RT process, using formal quality improvement tools such as audits to measure performance and quality.

TG 40 expanded upon *Report 13* recommendations for individual patient treatment QA, recommending a comprehensive new treatment chart review, weekly portal film imaging, and physical review of new treatment setups as well as *in vivo* QC checks. TG 40 included recommendations on clinician-dominated aspects of the planning and delivery processes, including detailed recommendations on the content and organization of charts and physician peer review of all new cases, plus a weekly "film review"/chart rounds conference.

TG 40 significantly expanded the physicist's role in the patient-specific treatment-delivery process by codifying a detailed and comprehensive process for checking new monitor unit (MU) calculations and treatment plans prior to delivery of the third fraction. The goal of this check is to confirm that treatment parameters (machine settings, e.g., MU and field sizes, patient setup instructions, OAR dose tabulations) are consistent with the physician's initial prescription, simulation instructions, and simulator films. A major emphasis of this check for most physicists is independent verification of graphic treatment or MU calculations. For most medical physicists, this "new start" checking procedure—updated to reflect the many technical innovations introduced after the early 1990s like electronic charting, EMR, 3D CRT, and IMRT—remains the physicist's primary mode of interaction with the clinical process (American College of Radiology 2010). TG 40 also outlined weekly and close-out physics chart review procedures that continue to be the standard of practice to the present day. (American College of Radiology 2009, 2011)

While TG 40 updated the device QA recommendations in *Report 13* to address then state-of-the-art multimodality linacs and reduce the overall spatial accuracy target to 5 mm, the basic approach remained unchanged: detailed menus of QC/QA tests to be performed at fixed intervals (initial acceptance testing and commissioning, and thereafter at daily, monthly, and annual intervals) were given along with test outcome tolerances adapted from *Report 13*. In addition to accelerators, the devices covered included brachytherapy sources and applicators, 2D treatment planning systems, simulators, and QA test equipment. These periodic QA protocols are all "one size fits all" with no allowance for individual device stability, sensitivity of the target clinical process to device performance specifications, or the device features exploited by its specific clinical implementation of each clinical practice. Prescriptive QA guidance for brachytherapy was also published (Nath et al. 1997).

TG 40 and TG 56 were based on radiation therapy as practiced in the pre-3D CRT era, and hence considered only dose-calculation algorithm accuracy and 2D planning issues. A major advance was AAPM's Task Group 53, which outlined comprehensive testing methodologies for many 3D (image-based) planning performance endpoints—accuracy of image identification and display, 3D image reconstruction from 2D slices, manual segmentation, construction of mesh- or voxel-based 3D anatomic structures from manual contour stacks, and ray tracing (Fraass et al. 1998).

The QA guidance described above is typically formulated under the sponsorship of a scientific body, most commonly the AAPM. Task Group reports are created by convening a small group of expert practitioners who, with rare exceptions, are all medical physicists, to review practice patterns and develop consensus standards, usually in the form of prescriptive QC protocols that comprehensively test the system's compliance with pre-specified performance standards. One set of tests, tolerances, and frequencies, designed to intercept all imaginable error paths, is utilized for all clinical implementations of a given device.

Despite its simplicity, this approach has contributed to maintaining high-quality performance of many classes of devices and for many specialized devices, and for clinical procedures as well (Kutcher et al. 1994). These include linear accelerator safety and performance (Purdy et al. 1993; Nath et al. 1994); ionometric dosimetry (Almond et al. 1999; Ma et al. 2001); treatment planning software (Fraass et al. 1998; International Atomic Energy Agency 2004); CT simulation (Mutic et al. 2003); mega-voltage portal imaging (Herman et al. 2001); brachytherapy devices (remote afterloading systems, sources, applicators, and non-image-based planning systems) (Nath et al. 1997); and brachytherapy dosimetry and calibration practices (DeWerd et al. 2004; Rivard et al. 2004; Williamson et al. 2005). This has led to codification and widespread community acceptance of practice standards. This codification has, and in some cases, directly stimulated significant scientific advances,

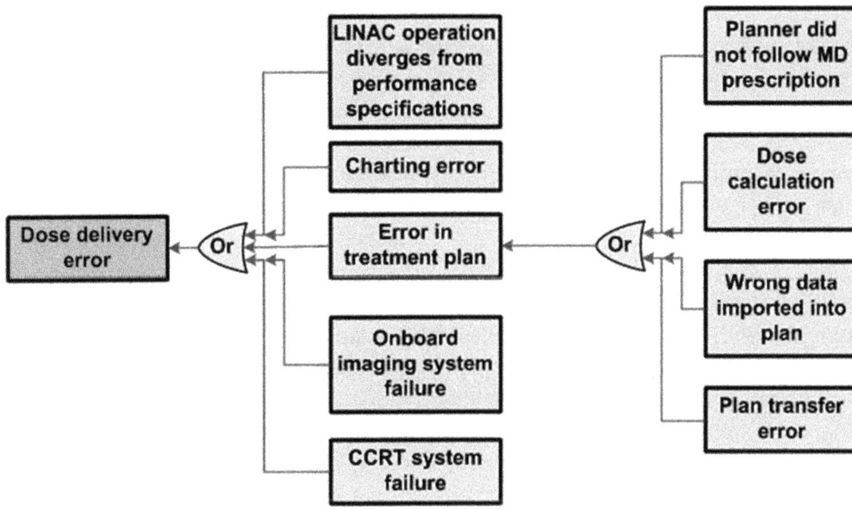

Figure 1–3. A physicist's view of "possible" errors that are (a) capable of causing quality to diminish and (b) are in the scope of practice of conventional medical physics QA.

e.g., the evolution from air-kerma-based calibration protocols (Schulz et al. 1983), absorbed dose-based calibration protocols (Almond et al. 1999), low-energy brachytherapy calibration (Seltzer and Lamperti 1999), and dosimetry standards (Rivard et al. 2004).

However, modern radiation therapy techniques—such as image-guided and image-based brachytherapy, 3D CRT, IMRT, and image-guided radiation therapy—approach a level of complexity and institution-to-institution variability that was not anticipated by the conventional QA guidance consensus panels. For example, MLC-based IMRT demands a higher standard of MLC leaf speed and positional accuracy than 3D CRT (Palta, Liu, and Li 2008). Also, specific tolerances for leaf positioning accuracy, speed, and gap calibration depend significantly on both MLC design and the selected treatment approach, e.g., dynamic vs. step-and-shoot and minimum segment width allowed. A more problematic example for clinical physicists is planning system QA. TG 53 and TRS 430 outline well over 100 different dosimetric and non-dosimetric tests for comprehensively testing 3D CRT planning systems (Fraass et al. 1998; International Atomic Energy Agency 2004). While TG 53 acknowledges that performing all of these tests would be prohibitively costly, and that each clinic must prioritize its effort according to its own needs, it offers no guidance on how to customize its recommendations. Image-based RT QA guidance, where it exists for commercially available systems, is widely perceived by clinical practitioners as being so complex and time-consuming as to be useless (Williamson et al. 2008). In general, prescriptive rule-driven QA/QC overwhelms physicists by the volume of QC testing that must be done. There is a clear need for risk-informed and possibly more customized approaches to prioritizing QA resources to address high-risk vulnerabilities specific to the individual clinic, the patient clinical setting, and the specific modes of operation and integration extant in the clinic.

Another significant problem is that QA guidance lags significantly behind the market penetration of new and technically complex products. Perhaps the most important omission is the rise of highly integrated "treatment management systems" (TMS) which integrate treatment planning, imaging, and delivery activities with patient tracking, charting, billing, and scheduling functions using networked hardware and software systems. These systems are evolving into computer-controlled, image-guided, highly conformal dose-delivery systems sometimes equipped with sophisticated automated plan adaptation features. They are almost unrecognizable by their "record/verify" antecedents of a decade ago. Systems of such complexity not only have substantially more potential failure modes, but they have more implementation variability: the relevance, likelihood of occurrence, and clinical impact of a given error pathway depends on the specific commercial products selected, the institution's clinical process in which the software is embedded, and the tumor site. Hence, a prescriptive "one size fits all" QA protocol that comprehensively intercepts all possible pathways is not even, in principle, feasible (Williamson et al. 2008; Fraass 2008; Fraass 2008).

The current QA approach focuses too narrowly on device operation and dosimetric accuracy. Figure 1–3 illustrates the kinds of error pathways that can be intercepted and corrected by the current "physics-centric" approach to device QA and chart checking. Numerous classes of hardware failure can be detected and corrected via equipment QC/QA testing. In addition to detecting dose-calculation errors, the chart check is sensitive to prescription and treatment plan discrepancies and to certain classes of data transfer errors. However, it is unlikely to detect import of incorrect imaging studies, deviation of individual physician prescriptions from standards of practice, differences between setup and final plan isocenter, and structure contouring errors. As illustrated by Figure 1–4, large CTV segmentation errors can be caused by inadequate image quality, poorly designed or commissioned contouring software, poor training of the contouring operator, or failure to check contours. Achieving uniformly accurate contours for a patient cohort requires engineering the entire contouring process with the goal of achieving maximum quality, including those process steps dominated by physicians and dosimetrists.

Published radiation therapy treatment-delivery error analyses illustrate both the benefits and weaknesses of the current QA paradigm (see Table 1–2). These analyses support the contention that the vast majority of serious errors are caused, at least in part, by human errors, data transfer errors, or process-related errors other than direct failure of a device or failure to properly execute a device-related QA task. In an analysis of 100 serious "unintended" radiotherapy exposure incidents collected by the IAEA, equipment or software failure was an "initiating event," i.e., proximate cause, for only 12 events, while another 15 events were initiated by failures of commissioning or device QC, e.g., calibration failure (Boadu and Rehani 2009). The remaining 73 incidents were due to human error or other forms of process error, the most common of which was basing planning or dose calculation on incor-

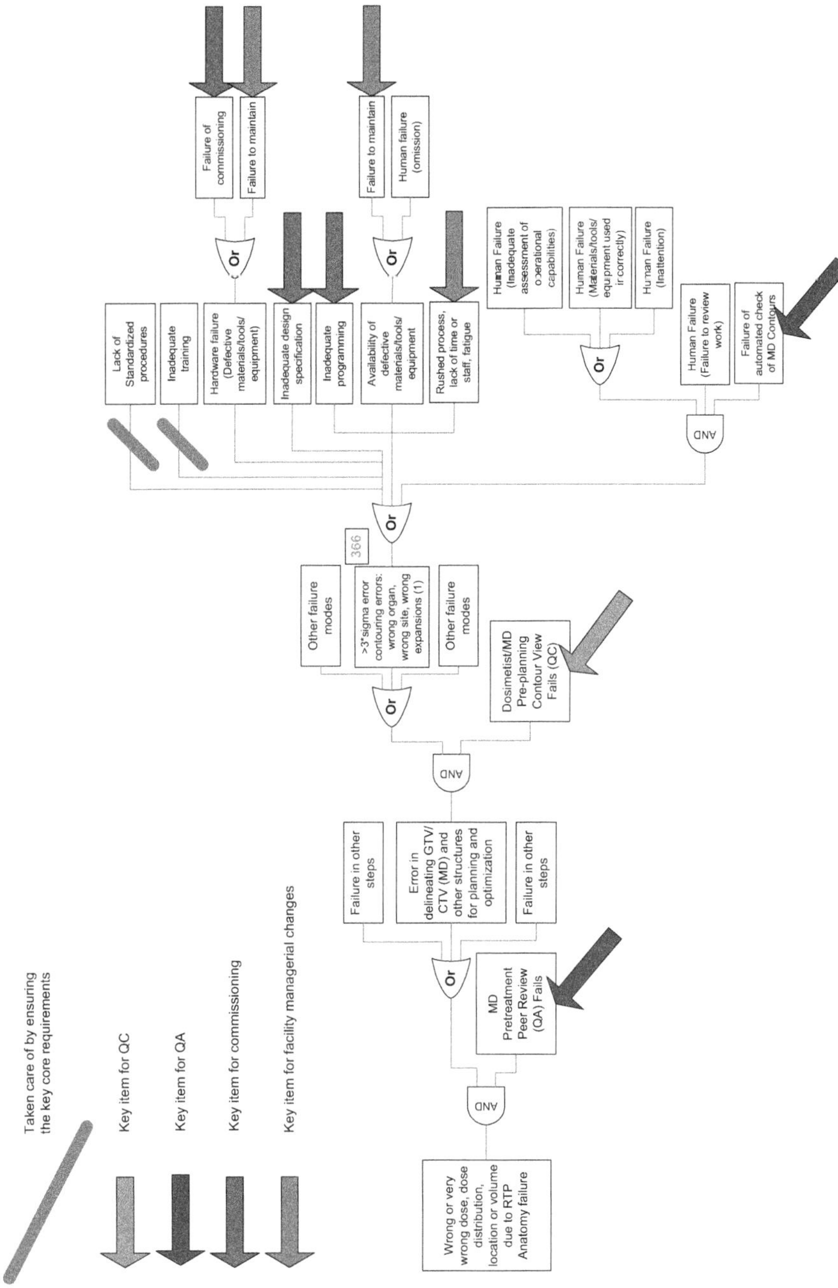

Figure 1–4. An example of a more global multidisciplinary fault-tree analysis of physician contouring error-induced treatment delivery errors. (From TG 100 Part 2 report)

rect patient data. The most frequent "contributing factor" was an error in executing protocols or procedures (33 events). In a qualitative and descriptive analysis of nearly 8000 patient delivery errors and near misses, the same pattern emerged (Shafiq et al. 2009). However, a small number of the most catastrophic random incidents involved device or software failures. In addition, several other systematic failures that afflicted large numbers of patients with large dose-delivery errors involved failures of software commissioning and basic linac QC procedures. An analysis of the Radiation Oncology Safety Information System (ROSIS) error reporting registry—a voluntary reporting system in which 101 mostly well-staffed and well-equipped clinics participate—found that fully 96% of the reported events involved clinical process failures and that only 2% were detected by device QC checks (Cunningham et al. 2010). In an analysis of 109 reported accidental brachytherapy exposures and misadministrations reported by Thomadsen et al., only 2 of 109 events were initiated by device failure (two instances of an HDR source transport mechanism failing to advance the source) (Thomadsen et al. 2003). Interestingly, 48% of the HDR events were caused by manual afterloading programming errors, a class of errors that can be largely eliminated by electronic transfer of treatment parameters from planning to delivery system. On the other hand, the Radiological Physics Center's experience with end-to-end IMRT testing by means of anthropomorphic phantoms attests to the fact that even in developed countries, failure to provide adequate device QC and QA and sufficiently accurate and comprehensive device commissioning were surprisingly common in the case of IMRT (Ibbott et al. 2008). In many cases, these errors were large enough to compromise the clinical effectiveness of IMRT. However, collectively, these studies support the conclusion that the perceived quality and safety of radiation therapy is dominated by error pathways that are not addressed by the current physics-centric QA paradigm.

The current device-centric quality management process outlines commissioning, acceptance testing, and periodic QC procedures that are designed to assure that integrated imaging, planning, and treatment devices are able to accurately deliver dose distributions with relatively low uncertainty of ±5%. This approach assumes that variations in device specification parameters are normally distributed about their target values, and that these stochastic properties are relatively stable through time. While this approach successfully addresses relatively predictable, non-catastrophic "quality eroding" errors, there is growing recognition that this model must be supplemented by broader strategies to mitigate low-probability "catastrophic" delivery errors that jeopardize the safety of smaller numbers of patients. Even though the safety record of radiation oncology, defined as the probability of deterministic injury due to avoidable errors, is superior to other medical and surgical oncologic interventions (Shafiq et al. 2009), such horrific events are often broadly reported (Bogdanich 2009, 2010), undermining patient confidence in radiation therapy and causing potential patients to seek alternatives that are less medically optimal. Even where the serious delivery errors summarized in Table 1–2 do involve device function and quality control, many of the errors arise through user misun-

Table 1–2. Selective review of compilations of radiation therapy delivery errors and near misses

Reference Era	Number of Events	Results	Comments

Cunningham et al. 2010
 	
2003–2008
	1074 incidents from 101 clinics.
	"Any incorrect delivery of radiation."
		• 96.4% process
		• 1.9% involved device QC
		• 24% found prior to treatment
		• 43% found at treatment
		   • 9.6% portal imaging
		   • 0.5% *in-vivo* dosimetry
		• 15.5% pretreatment chart check
		• 15.6% weekly chart check
		• 51% incorrect delivery
			Only 14% of departments fully networked

Shafiq et al. 2009 1976–2007	3125 major incidents with harm. 4616 Near misses. Each patient = 1 incident	*Major incidents:* • 55% treatment planning • 25% commissioning • 10% delivery: • 9% info transfer *Near misses:* • 38% information transfer • 18% delivery • 9% planning	Risk of mild-to-moderate injury is $1500/10^6$ (Williamson et al. 2008)
Holmberg and McClean 2002 1998–2000	15,386 2D computer plan and manual calculation checks. Errors and near misses: >1%–2% errors	Primary plan check yield: 3–4.5% Secondary plan check yield: 0.5% Actual delivery errors: 0.25% Detected computer plan errors: 4.3% • 22% RTP input and operation • 66% downstream calculations • 12% chart documentation	
Ibbott et al. 2008 2001–2005	250 irradiations of anthropomorphic head and neck phantom via a standard IMRT plan prepared on institution's inverse planning system.	28% of irradiations had discrepancies between institutional plan and QA phantom measurements giving a 7%/4 mm gamma index <0.7	Mostly inadequate commissioning and MLC QA
Boadu and Rahani 2009 1974–2006	100 serious incidents of unintended 'substantial' exposures collected by IAEA (IAEA 2000). The 12 most serious affected 2000 patients and caused 36 deaths.	*Initiating events:* 28: incorrect input data or dose calculation 21: inaccurate setup or incorrect treatment procedures 12: equipment/software failure 15: incorrect/no QA or commissioning 24: other process errors	Most common contributing factor: errors in executing procedures or protocols (33%)

Thomadsen et al. 2003
 	
1985–2001
	90 brachytherapy misadministrations reported to U.S. NRC and 23 brachytherapy events from IAEA (IAEA 2000).	• Planning errors: 30% HDR, 25% LDR
		• Delivery errors: 68% HDR, 75% LDR
		   • applicator positioning 8%
		   • source loading: 26% LDR; wrong channel: 7% HDR
		   • source positioning, treatment time: 42% LDR; manual programming: 48% HDR
		   • device failure: 0% LDR; 5% HDR
		• Wrong patient: 2%
			Human performance factors were major: failure to detect abnormal conditions and ineffective verification procedures

dertanding of device behaviors or response of the device to an input condition not previously encountered. Hence, an effective QM process for intercepting "outlier" failure modes must focus more broadly on the organization of the entire clinical workflow, including all participants and human-machine interactions.

1.4.4 Contemporary Approaches to Quality Management

1.4.4.1 Emerging Developments in Process-centric Quality Management

The first major efforts to systematically address error mitigation in a radiation therapy process were AAPM's guidance documents for clinical brachytherapy published in the late 1990s. The 1997 TG–56 report *Code of Practice for Brachytherapy Physics* (Nath et al. 1997) laid out a detailed process-centered QA approach for temporary low dose-rate brachytherapy procedures using "2-D" treatment planning. Meanwhile, the 1998 TG–59 Report *High dose-rate brachytherapy treatment delivery* (Kubo et al. 1998) serves as an extended example of applying TG–56 principles to the HDR brachytherapy domain. Both reports accepted that low-probability human errors—including measurement errors, communication failures, and transcription errors—must be detected and corrected to avoid catastrophic treatment delivery errors. This approach was designed to complement the prescriptive QA program outlined by TG 56 for HDR and LDR brachytherapy devices. These reports attempted to lay out a general QM system design process that could be adapted to many different kinds of clinical procedures. Reports 56 and 59 proposed that the QA program was not a separate activity imposed upon the clinical workflow, but that such processes should be prospectively designed from the ground up with the goal of making them robust to error propagation by building QC and QA checks into their basic structure according to the following recipe:

1. Define the anticipated or actual procedure flow, including preparation, implant design, applicator insertion, imaging, planning, etc. At each step, identify the critical activities to be performed, the required personnel, and the information to be captured. Figure 1–5 illustrates this step for HDR brachytherapy.
2. Develop forms for capturing and documenting all critical information, including the implant drawing, type of applicators, catheter numbering system, target localization data, and written prescription.
3. Identify vulnerable steps in the treatment delivery process where mistakes, misjudgments, or inaccurate transmission of data can jeopardize the procedure outcome.
4. For each such step, develop redundant checks, specifying who performs the check and the actions to be taken in the event of an unacceptable outcome. Both severity and likelihood of the target error can be taken into account in deciding how to distribute available QA resources. (See Table 1–3 for an example.)
5. Develop a written procedure that outlines the procedure chronology, team member functions, QA checks, etc.

TG 59 allocated tasks to ensure that no team member would be responsible both for execution and verification of a mission-critical process step. To identify "vulnerable" decision points, TG 59 advises, "Each key step and essential piece of information should be independently verified. How exhaustive, logically complete, and specific the QA test is depends on the physicist's estimate on the incidence of the targeted error and how serious the consequences are to the patient and staff." In practice, such "risk-informed" QA program design amounted to achieving consensus among the TG–59 authors (all highly experienced HDR physicists) as to which vulnerabilities posed sufficiently high risk to merit attention. To the extent that procedure flow and risk profiles are relatively constant across practice environments, this approach is feasible. TG 59 emphasized independent review by a physicist (or physician as appropriate) of the written directive, the implant geometry, the treatment planning input/output, patient setup, and console programming. Although manual dose-calculation accuracy checks were recommended, verification of key geometric (treatment lengths, catheter correspondence, implant location, etc.) and dosimetric (source strength, dose-specification criterion) parameters received the most emphasis. Both TG 56 and the IAEA outline less detailed, process-specific guidelines for conventional LDR temporary implants (International Atomic Energy Agency 1998).

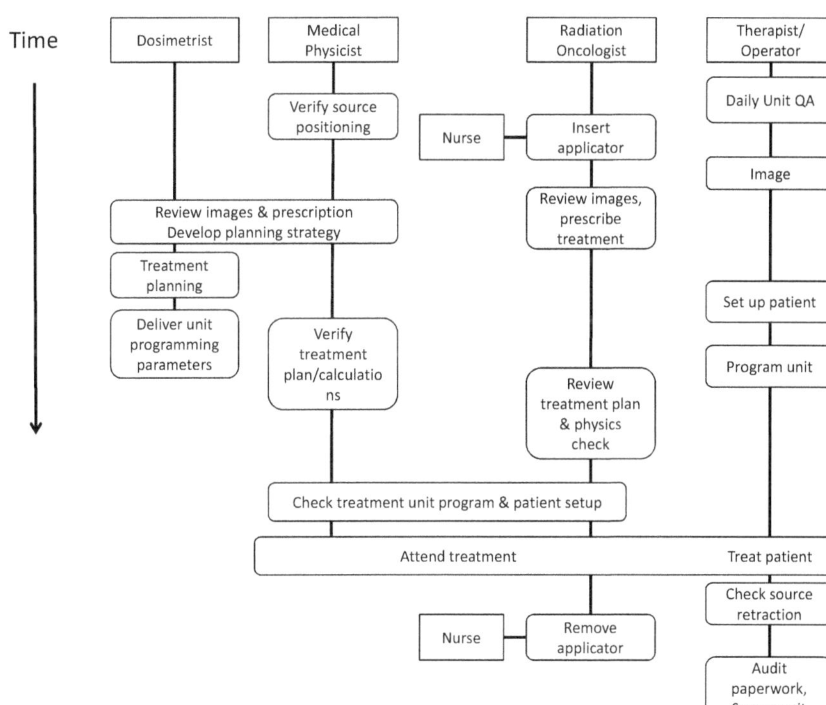

Figure 1–5. TG–59 process flow diagram of the one-physicist HDR brachytherapy workflow model.

Table 1–3. Essential elements of the treatment prescription process. The individual listed first is responsible for executing the activity while the second individual is responsible for verifying the procedure.

Responsible Individual	Activity	Methodology
Radiation oncologist	Complete, sign, and date prescription	Physicist/Dosimetrist: Question deviations from treatment policies or customary practice
Physicist	Review radiographs and mark treatment and/or target volume and sign	Physicist/Dosimetrist: Question deviations from treatment policies or customary practice
Radiation oncologist and physicist	Define dose prescription and optimization criteria Define treatment planning constraints (maximum and minimum doses, normal tissue tolerances, etc.)	If procedure is infrequent, customized, or otherwise unusual, physician and physicist should work together to address these questions If high-frequency procedure or clinical indication for treatment, written treatment policies which describe the dose specification and optimization processes should be available
Physicist and dosimetrist	Active dwell positions selected and documented	If standard pattern used (e.g., intracavitary implant), procedure type and protocol identified If identified and marked radiographically, confirm that all relevant data defining target volume is consistent with proposed dwell position distribution

Both TG 56 and TG 59 significantly advanced QMP as a multi-disciplinary effort involving all team members. While earlier guidance documents recognized the importance of team participation, these two reports provided a detailed, step-by-step clinical realization of this concept applicable to a broad class of radiation oncology procedures. Other groups, e.g., ASTRO and ACR, have long recognized that extending the QM program to encompass physician-dominated components of the radiotherapy process is essential to overall quality. One of the most explicit and detailed sets of guidelines is the ACR Accreditation Guidelines (American College of Radiology 2012). The ACR recommends a comprehensive physician peer-review system, including independent physician and physicist reviews of all new fields and starts, ongoing periodic review of patients under treatment via chart rounds, and a morbidity/mortality conference to assess all charts and patient treatments for patients exhibiting unexpected acute or late complications.

A more radical break with conventional QA guidance development is the AAPM Task Group 100 "Method for Evaluating QA Needs in Radiation Therapy," which was formed in 2003 and whose final report is currently under review (Huq et al. 2008). TG 100 adapted Juran's definition of quality in radiotherapy as follows:

- Those features which meet the needs of the patient, including rational medical, psychological, and economic goals while also taking into account the professional and economic needs of the caregivers and the institution.
- A clinical process that is designed to realize cancer treatments that conform with nationally accepted standards of practice and specifications; and
- Freedom from errors and mistakes." (Juran et al. 1999)

This definition incorporated the concepts of cost-benefit trade-off and clinical quality defined as adhering to standards of practice that are collectively believed to support the best clinical outcomes. The starting points of the TG–100 approach were that:

- Modern image-based or image-guided radiation therapy modalities are so complex and variable in their clinical realizations that the process QA design paradigm of TG 56 is not workable.
- Effective quality management must consider not only devices, but the entire clinical process as a multidisciplinary activity prone to human and communication errors.
- Formal risk assessment tools derived from industrial and human factor engineering provide general solutions to the problems of customizing broad QMP principles and goals to match individual clinic risk profiles and to prioritize deployment of limited QA resources.

The approach taken was failure modes and effect analysis (FMEA) which will only be briefly described here, but which is discussed in detail in Chapter 4 and in Stamatis (Stamatis 1995). To illustrate the application of this prospective design tool, the TG–100 membership developed a generic IMRT planning and delivery process. First they created a process tree, a graphical representation showing the temporal order of all-important steps in IMRT starting with the physician seeing the patient in consultation. For each process step, the group developed a list of "failure modes" (FM) or the unintended or erroneous step outcomes which, if not detected and corrected, could compromise the treatment's clinical outcome. Each group member independently assessed the clinical risk of an uncorrected FM by assigning to it probabilities of occurrence, its downstream detection potential, and a severity score. Each of these metrics was assigned a number from 1 to 10 according to the scoring system developed by the task group. The product of these metrics was the *risk priority number* or RPN, a measure of the risk this process step error poses to the patient. The failure modes can then be arranged in descending order of overall risk, allowing the highest risk process steps to be identified and addressed by QA/QC tests. Perhaps not surprisingly, the highest RPN scores were given to clinician-dominated steps (such as patient selection and anatomy delineation), along with basic planning system and linac QA. The bulk of the report consists of a model QM system developed by the group using FMEA and derived fault-tree analyses. FMEA, if used properly, is a powerful tool for the treatment team to work together to prospectively redesign their clinical processes for robustness against individual step failure modes in a way that takes into account clinic-specific environmental factors and implementation choices. At least two groups have described their experience with FMEA in various radiation therapy procedures (Ford et al. 2009; Perks et al. 2012).

A more reactive QMP formulation tool (although it can be applied "proactively" as discussed below) is root cause analysis (RCA). RCA is a formalized process for collecting and analyzing errors that have actually occurred in one's practice (Mutic et al. 2010). An important aspect of event analysis is the application of an error taxonomy which serves to assign a set of more general descriptors to each event. The radiotherapy event classification scheme developed by Ekaette et al. classifies each event as to subprocess, prescription element (dose or volume), source (process or infrastructure), and occurrence (systematic or random) (Ekaette et al. 2006; Dunscombe et al. 2008; Mutic et al. 2010). Such a taxonomy permits similar incidents to be grouped in the same category, enabling the analyst to look for patterns indicative of an underlying systematic fault. The taxonomic classification also helps focus remedial action by identifying subprocess in which the error occurred. There are many other error taxonomies. For example, the Madison Medical Error Taxonomy is a more complex four-tiered system intended to identify underlying causes rather than the describe events (Thomadsen and Lin 2008). The final result of this tool is a list of underlying cognitive mechanisms of human malfunction, organizational problems, or environmental factors that contributed to the event which can be used to more effectively design QM interventions. An extension of this approach is an event-reporting database in which participating institutions voluntarily report medical errors and near misses in an anonymized format. By providing sufficiently detailed descriptions and an appropriate taxonomy-driven analysis, other institutions will be able to proactively apply the lessons learned from these events to their own practices, thereby reducing the likelihood of similar errors. This enhancement allows other practices to proactively apply the lessons learned from RCAs performed in other institutions. Two examples are the well-established, ESTRO-sponsored Radiation Oncology Safety Information System (ROSIS) (Cunningham et al. 2010) and the IAEA-sponsored Safety in Radiation Oncology (SAFRON) event reporting systems (Abdel-Wahab et al. 2011).

1.4.4.2 Emerging Developments in Device-centered Quality Management

The faculty of the 2007 ASTRO-sponsored symposium "Quality Assurance of Radiation Therapy: Challenges of Advanced Technology" developed a set of recommendations that addressed resolved and unresolved issues in QA of radiotherapy devices (Williamson et al. 2008). Areas where adequate guidance and practices exist included linac safety and performance, ionometric dosimetry, CT simulators, brachytherapy delivery systems, brachytherapy dosimetry, and electronic portal imaging systems. This expert panel concluded that MLC-based IMRT delivery systems, dosimetric aspects of IMRT planning system commissioning, and stereotactic radiotherapy delivery systems could likely be handled by extensions of the current consensus panel-based QA guidance formulation approach. One suggested extension (Williamson et al. 2008) was menu-based prescriptive QA and QC protocols in which different device performance tolerances and constellations of tests are recommended for different applications, e.g., different suites of MLC tests and tolerances for 3D CRT, dynamic IMRT, and sequential static segment-based IMRT (Williamson et al. 2008). This approach has been adopted by the most recent AAPM guidance on linac QA in Task Group 142 (Klein et al. 2009).

Formulating straightforward guidance for performing cost-effective acceptance testing and commissioning of IMRT planning systems and other forms of image-based planning remains an unsolved problem. Perhaps the most complete guidance for commissioning IMRT planning systems is the 2003 AAPM IMRT Subcommittee Report, which outlines general issues and strategies but a not complete "cook book" approach (Ezzell et al. 2003). The ASTRO 2007 Symposium recommended that more emphasis be given to higher-level "end-to-end" testing to ensure that systems continue to perform as commissioned through time and to ensure that any systematic infrastructure failures, e.g., importing of images in the correct coordinate system, are detected (Williamson et al. 2008). A recent ASTRO white paper on IMRT safety expands upon the "end-to-end" testing concept and provides somewhat more detailed guidance on devising such tests (Moran et al. 2011). Although end-to-end testing is not a substitute for careful and comprehensive commissioning, the RPC's experience with their IMRT head and neck phantom illustrates its value in detecting non-catastrophic but clinically significant quality-eroding commissioning and QA errors (Ibbott et al. 2008). At a more disturbing level, this auditing experience illustrates that inadequate system QA and commissioning was widespread, at least in the 2001–2005 period.

Another emerging approach that has potential to rationalize device QC tolerances is the study of sensitivity and error propagation. TG 40 and TG 142 recommended that tolerances are supposed to be based upon an error propagation analysis that ensures that the collective effect of all uncertainties is 5% total uncertainty in dose delivery. In fact, error propagation analysis as described in AAPM *Report 13* (Baily et al. 1984) was not performed by TG 142 in the course of developing its recommendations (Klein et al. 2009). During the 2007 ASTRO Symposium, several speakers drew attention to the need to better understand how devices operating at their performance tolerances, (e.g., standard deviations of individual seed strengths, MLC leaf positioning, and gap accuracies) influence dose delivery accuracy (Palta, Liu, and Li 2008; Bayouth 2008; Williamson 2008). For example, Kim and colleagues have devises a generalized model that predicts voxel-by-voxel dose-delivery uncertainty for clinical IMRT treatments given *a priori* machine delivery uncertainties derived from simple single-field measurements (Jin et al. 2008; Jin et al. 2010). Rangel, Dunscombe, and colleagues have used generalized equivalent uniform dose (gEUD) as a surrogate for assessing sensitivity of clinical outcomes to setup and device-performance uncertainties (Ploquin et al. 2005; Rangel et al. 2007; Rangel and Dunscombe 2009). For example, they conclude that systematic MLC leaf-bank positioning errors must be limited to 0.3 mm to limit gEUD delivery errors to 2% for clinical dynamic MLC IMRT treatments. Additional experience applying this approach to a broader array of device performance endpoints and clinical cases is needed.

1.4.5 Conclusions

This review demonstrates that over the years, the medical physics concept of quality management has significantly evolved. In the late 20th century, the main contributions of medical physics to quality were narrowly focused but essential efforts to ensure accurate and reproducible dosimetric documentation of treatment and reliable and consistent function of devices. Starting with brachytherapy, the physicist concept of promoting safe and effective treatment broadened to include prevention of catastrophic as well as quality-compromising failures by leading a multidisciplinary effort to design clinical processes so as to maximize robustness both to device malfunctions and human errors.

As treatments have become more complex and clinical implementations more variable, the quality management paradigm is once again faced with the need for radical adaptation and change. The hallmarks of the emerging paradigm include enlarging the QMP scope to encompass physician-dominated clinical assessment and treatment response subprocesses and the use of formal process engineering tools to adapt non-prescriptive, general QM programs to specific clinical processes.

1.5 Summary

- Quality and safety are closely related concepts. Many of the published recommendations to make radiotherapy safer will also enhance quality.
- Clinical evidence from trials data is now supporting more speculative opinions that the clinical outcomes of a significant proportion of radiotherapy patients are compromised by sub-optimal quality.
- Constraints on resources are forcing us to critically reappraise our quality management programs for efficacy and efficiency.
- There is growing recognition in the medical physics community that a "one size fits all" prescriptive approach to maintaining radiotherapy quality is not viable, given growing emphasis on low-probability catastrophic errors and increased technological complexity and variability of modern radiotherapy processes.
- The emerging approach to quality management is marked by increased emphasis on team-based interdisciplinary cooperation, more focus on mitigating process-related errors, and the use of formal risk analysis tools.

References

Abdel-Wahab, M., E. Rosenblatt, O. Holmberg, and A. Meghzifene. (2011). "Safety in radiation oncology: the role of international initiatives by the International Atomic Energy Agency." *J Am Coll Radiol* 8(11):789–94.

Almond, P. R., P.J. Biggs, B.M. Coursey, W.F. Hanson, M.S. Huq, R. Nath, and D.W. Rogers. (1999). "AAPM's TG–51 protocol for clinical reference dosimetry of high-energy photon and electron beams." *Med Phys* 26(9):1847–70.

American College of Radiology (ACR). (2009). "ACR technical standard for medical physics performance monitoring of image-guided external beam radiation therapy (IGRT)." In *Practice Guidelines and Technical Standards*. Reston, VA: American College of Radiology.

American College of Radiology (ACR). (2010). "ACR technical standard for the performance of radiation oncology physics for external beam therapy." In *Practice Guidelines and Technical Standards*. Reston, VA: American College of Radiology.

American College of Radiology (ACR). (2011). "ACR–ASTRO practice guideline for 3d external beam radiation planning and conformal therapy." In *Practice Guidelines and Technical Standards*. Reston, VA: American College of Radiology.

American College of Radiology (ACR). (2012). "Radiation oncology accreditation program requirements." Reston, VA: American College of Radiology.

Baily, N., R. Loevinger, R. Morton, R. Moyer, J. Purdy, R. Shalek, P. Wootton, and K. Wright. (1984). "AAPM Report No. 13: Physical aspects of quality assurance in radiation therapy: Report of Task Group 24 of the American Association of Physicists in Medicine." New York: American Institute of Physics.

Bayouth, J. E. (2008). "Siemens multileaf collimator characterization and quality assurance approaches for intensity-modulated radiotherapy." *Int J Radiat Oncol Biol Phys* 71(1 Suppl):S93–7.

Boadu, M. and M.M. Rehani. (2009). "Unintended exposure in radiotherapy: identification of prominent causes." *Radiother Oncol* 93(3):609–17.

Bogdanich, W. (2009). "At V.A. Hospital, a Rogue Cancer Unit." *New York Times*, June 20, 2009.

Bogdanich, W. (2010). "Case studies: when medical radiation goes awry." *New York Times*, January 27, 2010.

British Institute of Radiology, Institute of Physics and Engineering in Medicine, National Patient Safety Agency, Society and College of Radiographers, and The Royal College of Radiologists. (2008). *Towards Safer Radiotherapy*. London: The Royal College of Radiologists.

Cunningham, J., M. Coffey, T. Knöös, and O. Holmberg. (2010). "Radiation Oncology Safety Information System (ROSIS) – Profiles of participants and the first 1074 incident reports." *Radiotherapy and Oncology* 97(3):601–607.

DeWerd, L.A., M.S. Huq, I.J. Das, G.S. Ibbott, W.F. Hanson, T.W. Slowey, J.F. Williamson, B.M. Coursey, and Medicine Calibration Laboratory Accreditation Subcommittee of the AAPM. (2004). "Procedures for establishing and maintaining consistent air-kerma strength standards for low-energy, photon-emitting brachytherapy sources: recommendations of the Calibration Laboratory Accreditation Subcommittee of the American Association of Physicists in Medicine." *Med Phys* 31(3):675–81.

Dunscombe, P. (2012). "Recommendations for safer radiotherapy: what's the message?" *Front Oncol* 2:129.

Dunscombe, P.B., E.U. Ekaette, R.C. Lee, and D.L. Cooke. (2008). "Taxonometric applications in radiotherapy incident analysis." *Int J Radiat Oncol Biol Phys* 71(1 Suppl):S200–3.

Ekaette, E.U., R.C. Lee, D.L. Cooke, K.L. Kelly, and P.B. Dunscombe. (2006). "Risk analysis in radiation treatment: application of a new taxonomic structure." *Radiother Oncol* 80(3):282–7.

Ezzell, G.A., J.M. Galvin, D.Low, J.R. Palta, I. Rosen, M.B. Sharpe, P. Xia, Y. Xiao, L. Xing, C. X. Yu, IMRT Subcommitte, and AAPM Radiation Therapy Committee. (2003). "Guidance document on delivery, treatment planning, and clinical implementation of IMRT: report of the IMRT Subcommittee of the AAPM Radiation Therapy Committee." *Med Phys* 30(8):2089–115.

Feng, M., C. Demiroz, K.A. Vineberg, A. Eisbruch, and J.M. Balter. (2012). "Normal tissue anatomy for oropharyngeal cancer: contouring variability and its impact on optimization." *Int J Radiat Oncol Biol Phys* 84(2):e245–9.

Fiorino, C., D. Corletto, P. Mangili, S. Broggi, A. Bonini, G.M. Cattaneo, R. Parisi, A. Rosso, P. Signorotto, E. Villa, and R. Calandrino. (2000). "Quality assurance by systematic in vivo dosimetry: results on a large cohort of patients." *Radiother Oncol* 56(1):85–95.

Ford, E.C. and S. Terezakis. (2010). "How safe is safe? Risk in radiotherapy." *Int J Radiat Oncol Biol Phys* 78(2):321–2.

Ford, E.C., R. Gaudette, L. Myers, B. Vanderver, L. Engineer, R. Zellars, D.Y. Song, J. Wong, and T.L. Deweese. (2009). "Evaluation of safety in a radiation oncology setting using failure mode and effects analysis." *Int J Radiat Oncol Biol Phys* 74(3):852–8.

Fraass, B., K. Doppke, M. Hunt, G. Kutcher, G. Starkschall, R. Stern, and J. Van Dyke. (1998). "AAMP Radiation Therapy Committee Task Group 53: quality assurance for clinical radiotherapy treatment planning." *Med Phys* 25(10):1773–829.

Fraass, B.A. (2008). "QA issues for computer-controlled treatment delivery: this is not your old R/V system any more!" *Int J Radiat Oncol Biol Phys* 71(1 Suppl):S98–S102.

Hendee, W.R. and M.G Herman. (2011). "Improving patient safety in radiation oncology." *Prac Rad Oncol* 1:16–21.

Herman, M.G., J.M. Balter, D.A. Jaffray, K.P. McGee, P. Munro, S. Shalev, M. Van Herk, and J.W. Wong. (2001). "Clinical use of electronic portal imaging: report of AAPM Radiation Therapy Committee Task Group 58." *Med Phys* 28:712–737.

Herring, D.F. and D.M. Compton. (1971). "The degree of precision required in the radiation dose delivered in cancer therapy." *Br J of Radiol* Supplement 5:51–58.

Huq, M.S., B.A. Fraass, P.B. Dunscombe, J.P. Gibbons Jr., G.S. Ibbott, P.M. Medin, A. Mundt, S. Mutic, J.R. Palta, B.R. Thomadsen, J.F. Williamson, and E.D. Yorke. (2008). "A method for evaluating quality assurance needs in radiation therapy." *Int J Radiat Oncol Biol Phys* 71(1 Suppl):S170–3.

Ibbott, G.S., D.S. Followill, H.A. Molineu, J.R. Lowenstein, P.E. Alvarez, and J.E. Roll. (2008). "Challenges in credentialing institutions and participants in advanced technology multi-institutional clinical trials." *Int J Radiat Oncol Biol Phys* 71(1 Suppl):S71–5.

Institute of Medicine. (1999). *To Err is Human: Building a Safer Health System.* Washington, DC: National Academy Press.

International Atomic Energy Agency (IAEA). (1998). "Design and Implementation of a Radiotherapy Programme: Clinical, Medical Physics, Radiation Protection and Safety Aspects." Vienna: IAEA.

International Atomic Energy Agency (IAEA). (2000). "Lessons Learned from Accidental Exposures in Radiotherapy." In Safety Report Series. Vienna: IAEA. International Atomic Energy Agency (IAEA). (2004). "Commissioning and Quality Assurance of Computerized Planning Systems for Radiation Treatment of Cancer." Vienna: IAEA.

International Commission on Radiation Units and Measurements (ICRU). (1976). "Determination of absorbed dose in a patient irradiated by beams of X or gamma rays in radiotherapy procedures." Bethesda, MD: ICRU.

Jin, H., J. Palta, T.S. Suh, and S. Kim. (2008). "A generalized a priori dose uncertainty model of IMRT delivery." *Med Phys* 35(3):982–96.

Jin, H., J.R. Palta, Y.H. Kim, and S. Kim. (2010). "Application of a novel dose-uncertainty model for dose-uncertainty analysis in prostate intensity-modulated radiotherapy." *Int J Radiat Oncol Biol Phys* 78(3):920–8.

Juran, J.M., A.B. Godfrey, R.E. Hoogstoel, and E.G. Schilling. (1999). *Juran's Quality Handbook.* 5th ed. New York: McGraw Hill.

Keroack, M.A., B.J. Youngberg, J.L. Cerese, C. Krsek, L.W. Prellwitz, and E.W. Trevelyan. (2007). "Organizational factors associated with high performance in quality and safety in academic medical centers." *Acad Med* 82(12):1178–86.

Klein, E.E., J. Hanley, J. Bayouth, F.F. Yin, W. Simon, S. Dresser, C. Serago, F. Aguirre, L. Ma, B. Arjomandy, C. Liu, C. Sandin, T. Holmes, and AAPM Task Group 142. (2009). "Task Group 142 report: quality assurance of medical accelerators." *Med Phys* 36(9):4197–212.

Kubo, H.D., G.P. Glasgow, T.D. Pethel, B.R. Thomadsen, and J.F. Williamson. (1998). "High dose-rate brachytherapy treatment delivery: report of the AAPM Radiation Therapy Committee Task Group No. 59." *Med Phys* 25(4):375–403.

Kutcher, G.J., L. Coia, M. Gillin, W.F. Hanson, S. Leibel, R.J. Morton, J.R. Palta, J.A. Purdy, L.E. Reinstein, G.K. Svensson, et al. (1994). "Comprehensive QA for radiation oncology: report of AAPM Radiation Therapy Committee Task Group 40." *Med Phys* 21(4):581–618.

Ma, C.M., C.W. Coffey, L.A. DeWerd, C. Liu, R. Nath, S.M. Seltzer, J.P. Seuntjens, and American Association of Physicists in Medicine. (2001). "AAPM protocol for 40–300 kV x-ray beam dosimetry in radiotherapy and radiobiology." *Med Phys* 28(6):868–93.

Mardon, R.E., K. Khanna, J. Sorra, N. Dyer, and T. Famolaro. (2010). "Exploring relationships between hospital patient safety culture and adverse events." *J Patient Saf* 6(4):226–232.

Marks, J.E. and A.G. Haus. (1976). "The effect of immobilisation on localisation error in the radiotherapy of head and neck cancer." *Clin Radiol* 27(2):175–77.

Marks, L.B., C.M. Rose, J.A. Hayman, and T.R. Williams. (2011). "The need for physician leadership in creating a culture of safety." *Int J Radiat Oncol Biol Phys* 79(5):1287–9.

Mijnheer, B.J., J.J. Battermann, and A. Wambersie. (1987). "What degree of accuracy is required and can be achieved in photon and neutron therapy?" *Radiother Oncol* 8(3):237–52.

Moran, J.M., M. Dempsey, A. Eisbruch, B.A. Fraass, J.M. Galvin, G.S. Ibbott, and L.B. Marks. (2011). "Safety considerations for IMRT: executive summary." *Prac Rad Oncol* 1:190–195.
Mutic, S., J.R. Palta, E.K. Butker, I.J. Das, M.S. Huq, L.N. Loo, B.J. Salter, C.H. McCollough, J. Van Dyk, and AAPM Radiation Therapy Committee Task Group No. 66. (2003). "Quality assurance for computed-tomography simulators and the computed-tomography-simulation process: report of the AAPM Radiation Therapy Committee Task Group No. 66." *Med Phys* 30(10):2762–92.
Mutic, S., R.S. Brame, S. Oddiraju, P. Parikh, M.A. Westfall, M.L. Hopkins, A.D. Medina, J.C. Danieley, J.M. Michalski, I.M. El Naqa, D.A. Low, and B. Wu. (2010). "Event (error and near-miss) reporting and learning system for process improvement in radiation oncology." *Med Phys* 37(9):5027–36.
Nath, R., L.L. Anderson, J.A. Meli, A.J. Olch, J.A. Stitt, and J.F. Williamson. (1997). "Code of practice for brachytherapy physics: report of the AAPM Radiation Therapy Committee Task Group No. 56." American Association of Physicists in Medicine. *Med Phys* 24(10):1557–98.
Nath, R., P.J. Biggs, F.J. Bova, C.C. Ling, J.A. Purdy, J. van de Geijn, and M.S. Weinhous. (1994). "AAPM code of practice for radiotherapy accelerators: report of AAPM Radiation Therapy Task Group No. 45." *Med Phys* 21(7):1093–121.
Nijman, J.L., H. Sixma, B. van Triest, R.B. Keus, and M. Hendriks. (2012). "The quality of radiation care: the results of focus group interviews and concept mapping to explore the patient's perspective." *Radiother Oncol* 102(1):154–60.
Palta, J.R., C. Liu, and J.G. Li. (2008). "Quality assurance of intensity-modulated radiation therapy." *Int J Radiat Oncol Biol Phys* 71(1 Suppl):S108–12.
Perks, J.R., S. Stanic, R.L. Stern, B. Henk, M.S. Nelson, R.D. Harse, M. Mathai, J.A. Purdy, R.K. Valicenti, A.D. Siefkin, and A.M. Chen. (2012). "Failure mode and effect analysis for delivery of lung stereotactic body radiation therapy." *Int J Radiat Oncol Biol Phys* 83(4):1324–9.
Ploquin, N., W. Song, H. Lau, and P. Dunscombe. (2005). "Intensity modulated radiation therapy for oropharyngeal cancer: the sensitivity of plan objectives and constraints to set-up uncertainty." *Phys Med Biol* 50(15):3515–33.
Ploquin, N.P. and P.B. Dunscombe. (2008). "The cost of radiation therapy." *Radiother Oncol* 86 (2):217–23.
Purdy, J.A., P.J. Biggs, C. Bowers, E. Dally, W. Downs, B.A. Fraass, C.J. Karzmark, F. Khan, P. Morgan, R. Morton, et al. (1993). "Medical accelerator safety considerations: report of AAPM Radiation Therapy Committee Task Group No. 35." *Med Phys* 20(4):1261–75.
Rangel, A. and P. Dunscombe. (2009). "Tolerances on MLC leaf position accuracy for IMRT delivery with a dynamic MLC." *Med Phys* 36(7):3304–9.
Rangel, A., N. Ploquin, I. Kay, and P. Dunscombe. (2007). "Towards an objective evaluation of tolerances for beam modeling in a treatment planning system." *Phys Med Biol* 52(19):6011–25.
Rivard, M.J., B.M. Coursey, L.A. DeWerd, W.F. Hanson, M.S. Huq, G.S. Ibbott, M.G. Mitch, R. Nath, and J.F. Williamson. (2004). "Update of AAPM Task Group No. 43 Report: A revised AAPM protocol for brachytherapy dose calculations." *Med Phys* 31(3):633–74.
Rubin, L.S. (1978). "The Riverside Radiation Tragedy." *Columbus Monthly*, April, 1978.
Rutherford, E. and J. Chadwick. (1911). "A balance method for comparison of quantities of radium and some of its applications." 24:141–150.
Schulz, R.J., P.R. Almond, J.R. Cunningham, J.G. Holt, R. Loevinger, N. Suntharalingam, K.A. Wright, R. Nath, and G.D. Lempert. (1983). "A protocol for the determination of absorbed dose from high-energy photon and electron beams." 10:741–771.
Scottish Ministers for The Ionising Radiation. (2006). "Unintended overexposure of patient Lisa Norris during radiotherapy treatment at the Beatson Oncology Centre, Glasgow in January 2006." Report of an investigation by the Inspector appointed by the Scottish Ministers for The Ionising Radiation (Medical Exposures) Regulations 2000.
Seltzer, S.M. and Paul J. Lamperti. (1999). "Status of NIST Primary Standards for 125I and 103Pd Therapy Seeds Based on the Wide-Angle Free-Air Chamber (WAFAC): Proceedings of a CIRMS Workshop." Gaithersburg, MD: National Institute of Standards and Technology.

Shafiq, J., M. Barton, D. Noble, C. Lemer, and L.J. Donaldson. (2009). "An international review of patient safety measures in radiotherapy practice." *Radiother Oncol* 92(1):15–21.

Solberg, T.D, J.M. Balter, S.H. Benedict, B.A. Fraass, B. Kavanagh, C. Miyamoto, T. Pawlicki, L. Potters, and Y. Yamada. (2012). "Quality and safety considerations in stereotactic radiosurgery and stereotactic body radiation therapy: Executive summary." *Prac Rad Oncol* 2:2–9.

Stamatis, D.H. (1995). *Failure Mode and Effect Analysis: FMEA from Theory to Execution.* Milwaukee, WI: ASQC Quality Press.

Sullivan, R., J. Peppercorn, K. Sikora, J. Zalcberg, N.J. Meropol, E. Amir, D. Khayat, P. Boyle, P. Autier, I.F. Tannock, T. Fojo, J. Siderov, S. Williamson, S. Camporesi, J.G. McVie, A.D. Purushotham, P. Naredi, A. Eggermont, M.F. Brennan, M.L. Steinberg, M. De Ridder, S.A. McCloskey, D. Verellen, T. Roberts, G. Storme, R.J. Hicks, P.J. Ell, B.R. Hirsch, D.P. Carbone, K.A. Schulman, P. Catchpole, D. Taylor, J. Geissler, N.G. Brinker, D. Meltzer, D. Kerr, and M. Aapro. (2011). "Delivering affordable cancer care in high-income countries." *Lancet Oncol* 12(10):933–80.

Thomadsen, B. and S.W. Lin. (2008). "Taxonometric guidance for developing quality assurance." *Int J Radiat Oncol Biol Phys* 71(1 Suppl):S204–9.

Thomadsen, B., S.W. Lin, P. Laemmrich, T. Waller, A. Cheng, B. Caldwell, R. Rankin, and J. Stitt. (2003). "Analysis of treatment delivery errors in brachytherapy using formal risk analysis techniques." *Int J Radiat Oncol Biol Phys* 57(5):1492–508.

U.S. Nuclear Regulatory Commission (NRC). (1979). "Regulation of the Medical Uses of Radioisotopes." *Federal Register* 44:8242.

Weber, D.C., M. Tomsej, C. Melidis, and C.W. Hurkmans. (2012). "QA makes a clinical trial stronger: evidence-based medicine in radiation therapy." *Radiother Oncol* 105(1):4–8.

Williamson, J.F. (2008). "Current brachytherapy quality assurance guidance: does it meet the challenges of emerging image-guided technologies?" *Int J Radiat Oncol Biol Phys* 71(1 Suppl):S18–22.

Williamson, J.F., P.B. Dunscombe, M.B. Sharpe, B.R. Thomadsen, J.A. Purdy, and J.A. Deye. (2008). "Quality assurance needs for modern image-based radiotherapy: recommendations from 2007 interorganizational symposium on Quality Assurance of Radiation Therapy: Challenges of Advanced Technology." *Int J Radiat Oncol Biol Phys* 71(1 Suppl):S2–12.

Williamson, J.F., W. Butler, L.A. Dewerd, M.S. Huq, G.S. Ibbott, M.G. Mitch, R. Nath, M.J. Rivard, D. Todor, and American Association of Physicists in Medicine. (2005). "Recommendations of the American Association of Physicists in Medicine regarding the impact of implementing the 2004 Task Group 43 Report on dose specification for 103Pd and 125I interstitial brachytherapy." *Med Phys* 32(5):1424–39.

Zietman, A.L., J.R. Palta, and M.L. Steinberg. (2012). *Safety is No Accident.* Fairfax, VA: American Society for Radiation Oncology.

Chapter 2

Work System Design and Culture

Jennifer L. Johnson, M.S.[1], Barrett Caldwell, Ph.D.[2], and Suzanne Evans, M.D.[3]

[1]Senior Medical Physicist, Department of Radiation Physics, University of Texas MD Anderson Cancer Center Houston, Texas

[2]Professor, School of Industrial Engineering, Purdue University, West Lafayette, Indiana

[3]Assistant Professor, Department of Therapeutic Radiology, Yale University School of Medicine, New Haven, Connecticut

2.1 Work System Design

In addition to providing services to patients, healthcare environments are work systems. Nurses, physicians, radiation physicists, and others are expected to combine their skills, training, and motivation to perform effectively in the task environment (Carayon 2007; Reed et al. 2005). However, any worker's ability to perform effectively can be affected, limited, or even prevented by aspects of the physical environment or technology designs, regardless of skill or intention (Proctor and van Zandt

1994; Sanders and McCormick 1993). Human factors and ergonomics (HFE) professionals have developed a large body of research and practical references to help determine the capabilities and limits of workers in a variety of task settings (Kroemer et al. 1994; Stanton et al. 2005; Wickens et al. 2004; Wilson and Corlett 1990).

Given the advances in robotics, software-controlled linear accelerator systems, and electronic medical records systems, many of the tasks in a clinical oncology treatment facility can be examined as computer-intensive tasks. Thus, examining and improving task performance in these work environments may focus on the cognitive, sense-making, and socio-cultural aspects of computer-based work, which have been studied in HFE for decades (see, for example, Bailey 1989; Bradley 1989; Granjean 1987). This chapter provides a few guidelines that address these and other major elements of how the work environment can affect task performance.

2.1.1 Information Processing and Skilled Tasks

Information processing can be seen as the basis of how humans make sense of the world and perform any cognitive tasks (Bailey 1996; Proctor and van Zandt 1994; Rasmussen 1986; also see section 3.4). There are four major stages in human information processing:

- detection of incoming signals from the environment using our sensory organs and attention processes,
- interpretation of those signals using prior experience available from long-term memory and signal classifications,
- encoding of signals into short-term (*working*) memory through rehearsal and active processing, and
- transfer and storage of short-term memory into long-term memory using a variety of cues and access links.

Each of these stages is subject to interference or disruption from the environment, which is why workplace design can be critical to effective information processing performance. Signal detection is limited by our ability to see or hear a relevant signal against background noise; if we are distracted from paying attention to that signal, or if glare or echoes prevent us from detecting the signal, then it is not available for further processing. Ambiguous signals that have multiple or confusing meanings are difficult to effectively interpret, and are easily misread or forgotten. Encoding of the signal is limited by working memory limits—if there are too many items to process—or if we are distracted from that encoding task, then items are lost from short-term memory before they can be cued and stored into long-term memory.

Most stages of information processing are both time- and resource-limited. Temporary input signals stored in visual, auditory, or touch *sensory registers* are usually limited to 0.5 to 2.0 seconds of availability. If not detected (or in some cases, not interpreted for meaning), they are lost (Procter and van Zandt 1994;

Wickens et al. 2004). The encoding and working memory stages of processing are also limited by the time without active processing (up to 30 seconds) and the number of simultaneous units being processed (often described as "7 ± 2", but can be potentially larger in skilled and trained operators performing familiar tasks). While there are no known limits to the amount of information that can be stored in human memory, functional limits to memory seem to be due to loss of access cues, similar to losing keys to a safe deposit box or to misplacing the information that indicates which coded key opens which box; the material in the box may still exist, but it is no longer available.

For the purposes of this discussion, we will assume that healthcare professionals are already competently trained in the health specialty domains for which they are licensed—internal medicine, clinical nurse practice, radiation physics, etc. Domain knowledge of this type is gained from long exposure to clinical subject matter, which is learned and committed to memory using the cognitive tools of information processing. The application of prior experience and memory to information processing of new signals in a given task is, of course, limited—we cannot retain and recover everything we have ever tried to learn. More importantly, domain knowledge is not the only aspect of skilled performance that a healthcare provider develops to apply to a given work task. Task context, technology interface tools, and interactions with other team members (also discussed later in this chapter) can be seen as additional *dimensions of expertise* to consider in skilled task performance (Garrett et al. 2009). Some authors describe the development of expertise and effectiveness in a task as a progression from "knowledge-based" (novice, effortful) to "rule-based" (procedure-following) to "skill-based" (expert, automatic) stages of performance (Rassmussen 1986). However, it is better to consider complex, problem-solving, knowledge-based tasks; structured, procedural rule-based tasks; and physically coordinated skill-based tasks to be different domains in which performance proceeds from a novice, effortful, error-prone stage to an expert, implicit stage (Caldwell 1997).

2.1.2 Situation Awareness and System Monitoring

Computer systems assisting in treatment planning and delivery are examples of *supervisory control* human-machine interactions. These supervisory control systems link information and other flows between a human-interactive function (displays and controls interfaces) and a task-interactive function (e.g., beam optimization algorithms, leaf collimation, and radiation intensity calculation and modulation) to perform complex system tasks (Sheridan 1992).

To effectively monitor and manage such supervisory control systems, healthcare providers are required to develop and maintain *situation awareness* (SA) of the current state of the task and the information being provided by the computer system (Endsley et al. 2003). SA is a dynamic, task-specific process of information processing and sense-making using both experience with the task and technology support. There are three major elements to SA, which are affected by the signals presented to the provider and by their own cognitive skills:

- perception of task-related objects in the environment,
- comprehension of the current task situation and system states, and
- projection of those task situations and systems state conditions into the future.

To improve SA, work systems need to make information about the current task and computer system interfaces as clear and task-relevant as possible. Error codes and system status alerts, for example, should provide clearly interpreted updates about the current state of the patient or treatment, not just documentation of the software or equipment diagnostics. In addition, task procedures and instructions provided to providers should be appropriate for the range of task conditions (both normal and abnormal conditions) as well as sensitive to time pressures and constraints affecting the provider's ability to understand the effect of current actions in the future.

2.1.3 Function Allocation and Team Coordination

Function allocation is, at its simplest, a clear determination of who or what is responsible for completing which tasks in a complex work setting. With the continued development of increasingly automatic and computer-driven hardware and software systems, appropriate allocations of functions between humans and computer systems are recognized as an important aspect of systems design (Sheridan 1992). Such allocation strategies can fall along a continuum. At one end of this continuum is a purely tele-operated system, where the human makes all placement, movement, and activation decisions—the machine executes commands only when directed and in the manner directed by the provider. For example, when the technician pushes a button to activate the x-ray beam, that is an example of this *human-intensive* level of function allocation. At the other end of this continuum is a purely automated, fully autonomous machine. An axial tomography machine that automatically moves the patient and modulates the source beam according to an optimization program represents this type of *machine-driven* function allocation.

Rather than remaining a fixed aspect of a task design, function allocation may be modified depending on situation changes or other task-related conditions. These aspects of *dynamic function allocation* introduce even more work design challenges to appropriately determine who should be doing what and at which specific steps in the task. As computer systems become more capable of autonomous behavior and complex, task-sensitive performance, it is possible to consider a computer-controlled system (such as a linear accelerator, treatment planner, or functional imaging machine) as a member of the healthcare provider team requiring effective information and task coordination with nurses, physicians, physicists, and other healthcare providers (Cuevas et al. 2007). However, the need for effective function allocation between humans and computer systems highlights the limited understanding of how to ensure effective function allocation between humans in a team performance setting (Helmreich and Merritt 2000; McGrath and Hollingshead 1994; Stanton 1996). Healthcare provider teams require complex task coordination because their various domains and dimensions of expertise are applied to sophisticated, high-consequence tasks (Caldwell et al. 2010; Sundstrom et al. 1990; Swezey and Salas 1992).

2.1.4 Work Processes: Documentation, Layout and Procedures

One of the most basic elements of an engineering or HFE analysis of any work system is a process or work flow diagram showing required tasks and the flows of information or material between them. Work flow diagrams date back to the work of Frederick Taylor and the Gilbreths in the early years of the 20th Century (Gilbreth 1921). These diagrams provide members of the work team with a clear and easily communicated visualization of what tasks they are accomplishing and how those tasks relate to the larger work system goals. Although a clear understanding of the interplay and flow of information and material in one's healthcare work unit is critical to documenting a work process, members of a work team frequently lack such an understanding, especially as their units change or new systems are added (Caldwell and Anderson 1996; Garrett et al. 2010).

Thus, it is vital that complex work systems, such as clinical oncology processes in healthcare delivery, have clear documentation of work flows, layouts, and procedures that affect the flows of patients, information, artifacts (such as blood samples for chemistry panels, or medications), and even expert knowledge across the treatment facility. This documentation should reflect not only overall structures, but also timings, sources of delays, and the need for particular resources (and when they are needed) to support resource foraging and responses to critical patient events (Caldwell and Garrett 2011; Garrett et al. 2010). Checklists are seen as a valuable source of information support and documentation for procedure-based healthcare tasks, although their development and use across medical specialties is still subject to inconsistencies and limited rigor in implementation or validation (Hales et al. 2008).

2.1.5 Error Pathways and Risk Perception

Despite common misconceptions and cultural perspectives, errors in health care or other high-risk, high-intensity environments are rarely due to single-point failures or unique personality or motivation defects by "bad actors" (Helmreich and Merritt 1998). One of the most respected perspectives of human error across multiple work domains highlights that many adverse events and degraded system outcomes are due to multiple pathways and vulnerabilities in the overall work system (Reason 1990). Commonly known as the "Swiss cheese" model, the work system involves interactions between components with the potential for information breakdowns, inaccurate or unreliable procedures, fatigued workers, and other latent system vulnerabilities. Effects from a variety of performance shaping factors (PSFs)—including work environment, organizational structure, technology implementation, and procedural elements—affect the overall reliability of the work system. Some system reliability and safety analysis measures provide various mechanisms for identifying various PSFs and root cause contributions to adverse events; however, a more comprehensive approach that incorporates multiple PSFs and contributing error pathways and root cause factors should be considered (Caldwell 2008).

Risk perception and the attribution of error sources can also be substantially affected by organizational culture and social factors in the workplace as empirical, analytical results. Helmreich and Merritt's approach to "threat-error management" indicates that even the perspective of determining blame can be a cultural bias of risk perception that can degrade effective system performance and response to weakened or dangerous conditions (Helmreich and Merritt 1998). Further discussion of the effects of work culture on risk perception management and mitigation is provided in the following sections.

2.2 Safety Culture

2.2.1 Definition/Background

A definition of *culture* in the *Oxford Dictionary* includes "the ideas, customs, and social behaviour of a particular people or society." Culture with a modifier, such as *organizational culture* or *safety culture*, becomes the attitudes and behaviour characteristic of a particular social group. Although safety culture does not have a universal definition (Reason 1998; Halligan and Zecevic 2011), it may be generally described as the interaction of shared values and beliefs with the organization's structures and controls (Uttal 1983; Bate 1992; Thompson et al. 1996). (The distinct term *safety climate* is sometimes also used to describe the safety culture.) Safety culture became its own concept in the early twentieth century owing to accidents in high-risk industries such as nuclear power plants and commercial aviation. Safety culture became relevant in health care with the Institute of Medicine's report, "To Err is Human" (Institute of Medicine 1999).

Part of the ambiguity in defining safety culture is due to the various theories that describe it. Some of the most frequently cited theories include High Reliability Organization (HRO) Theory (Roberts 1999); the Safety Culture Maturity Model by The Keil Centre (HSE 2000); Donabedian's Process-Structure-Outcome Model (Donabedian 1980); Organizational Theory (Barnard 1968, Halligan and Zecevic 2011), where the goal is to control and manipulate workers and their environment for stability; and Systems Theory (Scott 1981, Halligan and Zecevic 2011), where organization components are (nonlinearly) interrelated and interact with and adapt to their environment.

2.2.2 Assessing Safety Culture

Halligan and Zecevic completed a literature review to determine the common set of definitions, dimensions, and measures of assessing a safety culture (Halligan and Zecevic 2011). They analyzed 139 peer-reviewed, English-language studies on safety culture published between 1980 and 2009. Those studies most commonly used or cited quantitative surveys with various measures, lengths, and reliability levels. Some of the most common surveys included:

- the Agency for Healthcare Research and Quality's (AHRQ) Hospital Survey on Patient Safety Culture (HSOPS) (AHRQ 2004),
- the Safety Attitudes Questionnaire (Sexton et al. 2006),
- the Patient Safety Culture in Healthcare Organizations Survey (PSCHO) (Singer et al. 2007), and
- the Modified Stanford Patient Safety Culture Survey Instrument (MSI) (Ginsburg et al. 2009).

While each of these metrics has its subtleties, it is worthwhile to note the sort of culture elements noted in such survey. The HSOPS survey focuses on 14 main variables and produces a composite score. The 14 variables of safety culture measured are: communication openness, feedback and communication about error, frequency of events reported, handoffs and transitions, management support for patient safety, nonpunitive response to error, organizational learning and continuous improvement, overall perceptions of patient safety, adequate staffing, management expectations and actions promoting safety, teamwork across units, teamwork within units, patient safety grade given by workers of A or B, and the number of events reported. As a low HSOPS score has been linked to more frequent adverse events (Mardon 2010) among a study of nearly 200 hospitals, it is clear that patient safety culture surveys are able to gauge the impact of safety culture on patient outcome.

Assessing one's safety culture identifies areas for improvement, even in the best of organizations. Various interventions on the work system design for a single area of improvement are created, executed, and evaluated for effectiveness. They may include changes in organizational controls, structures, and work processes, as well as changes in employee attitudes and behaviors. Some interventions include team training, leadership walkarounds, and patient safety education to raise staff awareness. Remediation and quality management will be discussed in their respective chapters. However, it is worth stressing the importance department leadership has on nurturing a safety culture that maximizes patient safety.

2.2.3 Interactions of People, Systems, and Processes

People have inherent attitudes, beliefs, and values about safety that are difficult to directly change. However, as noted from work system design, people interact with systems of the organization as they complete work processes. Their attitudes, beliefs, and values are shaped through these systems of controls, structures, practices, and policies. Thus, by changing its systems and processes, an organization can create an effective safety culture (Reason 1998). Work systems should be designed with safety in mind, as detailed previously. With ongoing changes in technology and health care delivery, developing an effective safety culture is itself a continuous process.

2.2.4 Traits of Successful Safety Cultures

A full understanding of operational risks, as well as fear of those risks, defines the success of a safety culture; failures in understanding these risks to varying degrees weaken the protection system (Reason 1998). An organization that does not provide appropriate front-line support—such as appropriate training, communication, and procedure design—increases the probability it will experience higher front-line errors and encourage non-compliance with safe practices. This will be particularly true if production and efficiency outweigh safety goals. In other words, if a department insists on keeping 15-minute time slots for complicated image-guided radiotherapy and intensity-modulated radiation therapy cases rather than expanding hours and creatively scheduling therapist time so as not to exceed 8-hour shifts, it is inviting frontline errors. An organization that does not provide adequate maintenance, quality assurance, or appropriate resources increases the probability it will experience more latent error conditions and longer-lasting safety gaps. This is particularly troublesome since latent conditions are harder to safeguard and lead to organizational accidents resulting with more widespread consequences. However, an organization that does not address and correct known issues has the least effective safety culture.

Successful safety cultures have similar traits: they are informed of their accidents, have an effective reporting structure, and have a "just culture" (Reason 1998), which will be explained in detail. For a safety culture to exist, an organization has to be informed of its incidents, which includes both events and near misses. These incidents can then be examined to determine the causes of the incident: which contributing factors were due to the organization's culture (its beliefs and values interacting with its structures, controls, and processes) and which were due to an individual? For instance, did a failure to recognize an extended source-to-skin distance setup occur in part because the physician rushed a non-emergent patient (waiting in the clinic after simulation) into an early start, thereby rushing the second check, thereby rushing the therapist's chart check? Or did the incident happen because a rattled therapist had a minor car accident in the garage 10 minutes prior and was preoccupied? Accidents are primarily due to active failures of both organizational and local workplace factors, combined with unsafe acts committed by individuals, and to failures along latent condition pathways (Reason 2004). A *latent failure* is one that is the result of a decision, action, or practice by the organization or the work system design that predated the incident; it is called latent because it exists for some time before any adverse consequences are realized. In contrast, an *active failure* is the result of an interaction between an individual and a larger part of the system (such as a linear accelerator or a patient) and is immediately apparent (Reason, 2004). The improper commissioning of electrons on a linear accelerator or a poor dose calculation algorithm is a latent failure; a therapist's failure to engage the gating system prior to treating a breath hold patient is an active failure.

An informed culture is built upon an effective reporting culture. A reporting culture has the following features:

- reporter confidentiality or de-identification,
- report filing ease,
- a reporting analysis group that is separated from disciplinary authority,
- rapid and useful feedback, and
- trust.

Incidents need to be reported in a timely manner, preferably as soon after their occurrence as possible, but certainly as soon as they are discovered, to be the most beneficial. Utilizing a reporting system helps an organization systematically track events and variables that may be analyzed for trends; near misses should ideally also be recorded in the system. Root cause analysis (RCA), discussed in Section 7.3, and its simplified variations are commonly used to investigate incidents and develop effective interventions. A reporting culture is predicated on a just culture, where the distinction between acceptable and unacceptable behavior is both clear and understood (Reason 1998).

2.2.5 Concepts of "Just Culture"

A "just culture" is fundamental to creating a reporting culture. Blame and punishment can range from highly punitive to completely exonerating (a "no blame" culture); in between these extremes is a just culture. As mentioned previously, in a just culture, acceptable and unacceptable behaviors are clearly identified, with unacceptable behaviors resulting in appropriate disciplinary action. The focus here is on the individual's underlying behavior at the time of the error rather than the magnitude of the error or its consequences. Instead, an unintended error or violation needs to be evaluated in terms of what was reasonable to do in the circumstances at the time; some use the substitution test (Johnston 1995) to help distinguish the fine lines by asking, "Would a new individual behave any differently or a peer commit a similar unsafe act?" A just culture seeks to distinguish between human error and the less forgivable reckless conduct, negligent conduct, or intentional rule violations. "Mistakes are a fact of life. It is the response to the error that counts," says Giovanni (Giovanni 1970).

2.2.6 Lessons from Other Industries

Lessons learned from other industries are applicable to health care in general and to radiation oncology in particular. There are many examples of serious incidents from other industries, including the National Aeronautics and Space Administration's Challenger and Columbia shuttle incidents and the Chernobyl explosion in 1986.

"An error doesn't become a mistake until you refuse to correct it" (Battista 1970). Systematic or recurrent errors can be found in the nuclear power industry, the aircraft industry, the oil exploration industry, and the production industry. These recurrent errors have three elements (Reason 1998): (1) universals, or obvious, ever-present hazards; (2) local traps, or workplace or task characteristics which lead to unsafe acts when combined with human error or violations; and (3) drivers, or motivating forces that drive individuals into erroneous behavior. One of the most common drivers is the constant conflict between safety and productivity.

Protection through safety controls affects the production of health care. Trying to strike a balance between the two can inadvertently lead to any one of the following problems identified by Reason (2004): "trading off improved [safety] defenses for increased production; keeping your eye on the wrong ball; attempting too much with too little: the 'can-do' syndrome; and believing that past non-events predict future non-events (forgetting to be afraid)." These problems are regularly experienced in a radiation oncology department:

- shortening a patient's allotted treatment time to "work in" another patient;
- assuming a standard order of fields for treating a total skin irradiation treatment;
- radiation therapists treating alone during the lunch hour, the early morning, or the late evening; and
- treating a patient at the wrong isocenter because shifting the couch from outside the room on a second isocenter without going into the room to visualize the isocenter on the patient is more efficient, and it hadn't been a problem from prior treatments.

Although other industries include highly reliable organizations, they too are prone to errors, and lessons learned in both their incidents and solutions can be applied to health care in general and radiation oncology specifically. Improvements in patient safety can be made in any industry when safety culture influences work system design.

2.3 Safety Culture Influences on Work System Design

2.3.1 Biases and Complacency

Human beings regularly use biases in judgment and decision-making. *Cognitive* biases are used for information processing to make decisions or judgments, but *motivational* biases such as wishful thinking may also factor in. Numerous biases fall into these two categories, and some fall into both categories. Since it would be impossible to discuss all the biases humans are prone to have, a relevant subset will be presented.

Biases are clearly present in a radiation oncology environment. When a major incident is shared within the community, the uninvolved may wonder how something so obvious could be missed. Balcetis and Dunning (2006) have demonstrated in five studies that participants' motivational states influence their visual perception, i.e., that individuals see what they want to see. Even physicians can be subject to pro-innovation bias, where they fail to identify limitations and weaknesses when introducing new technology into the clinic. Other biases include:

- anchoring, where the starting point has a strong influence on the final result;
- tunnel vision, where the narrow focus on a limited possibility range eliminates other possibilities; and
- confirmation bias, where the information selected confirms a person's theory and ignores contradicting information.

After an incident, some biases may adversely affect retrospective analyses, such as root cause analysis and error reporting systems, and these biases should be minimized as much as possible. The following investigative biases are common in safety culture (Reason 2004): hindsight bias, outcome bias, symmetry bias, and counterfactual fallacy. Knowing the outcome affects the causality lines of how and why an event occurred, i.e., events may seem more foreseeable than they really were at the time leading up to the event. This is called hindsight bias (Reason 2001), because it exaggerates the ability to predict the event before it occurs. Henriksen and Kaplan (2003) distinguish this from outcome bias, where the knowledge of the outcome influences the evaluation of the decision quality (e.g., a bad outcome is assumed to be associated with bad decision processes.) Considering other possible outcomes helps minimize these biases. Symmetry bias occurs when there is presumed symmetry of magnitude between a cause and its consequences (Reason 2004). Counterfactual fallacy thinking involves thoughts of what might have been had the situation been different (Epstude and Roese 2008). An example would be to think that an incident would not have happened if certain organizational factors were different, even when such factors were not causal to the incident. Roese and Maniar (1997) have noted that counterfactual fallacy actually increases hindsight bias and can be minimized by generating causal explanations.

Complacency may be generally described as a feeling of uncritical satisfaction with something (Oxford 2010). Complacency has been identified as a factor in safety culture (Harvey et al. 2002). A study relevant to radiation oncology completed by Parasuraman, Molloy, and Singh found evidence of performance consequences of automation-induced complacency (Parasuraman 1993). A system under automatic control would provide immediate notification of a system malfunction; participants were asked to track the notifications to monitor the performance of the system (system-monitor). When participants were required to exclusively system-monitor only what was under automatic control, they were efficient at detecting the system malfunctions (automated reliability) regardless of the variation of frequency; tracking results were worse if they had to complete additional tasks, even if the automated reliability was held constant. As a result, Singh (1993) developed a complacency-potential rating scale to measure four factors: confidence, reliance, trust, and safety. A similar tool could be used to measure the potential for complacency regarding new treatment delivery systems due to the interrelations between automation and operator trust in and reliance on automation.

Work systems and processes need to be designed to minimize bias and complacency as much as possible in the work environment. Creating an environment that proactively seeks out additional information and possibilities should limit biases. Creating a safety culture and designing systems to counter complacency should improve patient safety.

2.3.2 Disruptive Individuals

Disruptive individuals can have a direct, detrimental effect on observers' job satisfaction and morale. The results of the 2004 Physician Behavior Survey from the American College of Physician Executives indicate that more than 95% of physician executives have encountered disruptive behaviors—which include primarily disrespect, refusal to complete tasks or duties, yelling, and insults—on a regular basis (Weber 2004), with 70.3% respondents indicating the source of the disruptive behaviors is nearly always the same physician(s). Just as unacceptable behavior needs to be defined for a just culture, disruptive behavior also needs to be defined and distinguished from impaired and incompetent behavior. Impaired behavior addresses personal issues, including physiological or psychological illness, while incompetent behavior relates to meeting standards, guidelines, and specific professional norms (Martin 2008).

Disruptive individuals can have a direct, detrimental effect on patient safety. Rosenstein and O'Daniel (2008) surveyed member hospitals in four VHA regions using the instrument Nurse-Physician: Impact of Disruptive Behavior on Patient Care. Their results found that 77% and 65% of respondents had observed disruptive behavior by physicians and nurses, respectively, and that 67% of respondents agreed that such behaviors were linked with adverse events, including 71% of the medical errors and 27% of patient mortality. Taking steps to prevent disruptive behavior occurrences is important, but so is dealing with events as they happen to prevent harm and to complete post-event analysis and actions. Rosenstein and O'Daniel recommend that organizations improve awareness, education, communication, and incident reporting to help address these behaviors. Policies and procedures need to be in place and enforced, and they should include the American Bar Association recommendations for a five-step process from verbal warning to governance action (Martin 2008).

2.3.3 Normalization of Deviance

Standards of performance and processes are included in the work system design for reasons that include reproducibility, efficiency, and safety. As discussed previously, a protection-production tension exists because safety goals compete with other goals, like efficiency. Human behavior tends to drift from the standard, to take the shortcut. This drift, when repeated over time without any adverse consequences, may become an acceptable norm that is no longer perceived as risky or reckless. "Normalization of deviance," which describes this phenomenon (Vaughn 2005; King 2010), was coined as a result of studying the Challenger space shuttle launch decision to describe the social acceptance of risk. It is important to note that early (and regular) warning signs are misinterpreted or ignored (as acceptable risk) until it is too late. Therefore, having an effective safety culture to investigate such deviations and provide good work system designs can help prevent or minimize drift.

2.3.4 Checklist Systems

The aviation industry uses checklists in its airline safety practices. However, there are human factor system design elements—such as a checklist's "functions, format, design, length and usage" (Degani and Wiener 1993)—that encourage proper rather than improper checklist use, or even worse, nonuse. Checklists are not used in a vacuum, however, and the existing safety culture influences the use of these tools.

Checklist systems have been introduced into health care to enhance patient safety, not only to ensure failures are reduced by completing the procedural steps, but also to reduce failures in communication among team members. Provonost, Needham, Berenholtz et al. introduced five evidenced-based procedure interventions that included a checklist, and by doing so they decreased the rate of catheter-related bloodstream infections in Michigan intensive care units from 2.7 infections per 1000 catheter-days at baseline to 0 after the study intervention (Provonost et al. 2006). The experience of Haynes, Weiser, Berry, et al. (2009) demonstrated that introducing and using a 19-item surgical safety checklist statistically reduced surgery complications and deaths associated with noncardiac surgery in eight hospitals around the globe. In radiation oncology, Potters and Kapur (2011) developed their checklist system (quality check list) with proactive stopping rules to prevent rushing or compromising patient safety; treatment initiation required reconciliation of stopping events. This study provides an example of a necessary safety culture that implements "a checklist, process-driven, and variance-reducing culture" while mitigating risk.

Checklist design for non-linear procedures can be challenging, and uncertainty about their usefulness may discourage adoption. However, if designed well and used properly, a checklist system can help enhance communication, reduce cognitive load, and produce measurable results (this topic is discussed further in Section 5.6.5).

2.3.5 Crew Resource Management

With its origins in the aviation industry, crew resource management (CRM) is formal team training for pilots and aircrew to foster better communication and teamwork (Musson and Helmreich 2004). Elements of CRM that are relevant to health care include briefings, training in acceptable methods of challenging authority, verifying the actions of another, and discussing causes of human performance decline. While CRM in aviation has developed for over 25 years since its origins in the 1980s by NASA, the Institute of Medicine's 2000 report "To Err is Human" and the Agency for Healthcare Research and Quality project funding has prompted a demand in health care that exceeds resources, with private, commercial training filling the gaps. As such, CRM in medicine and health care needs to be developed with its own behaviors, practices, and medical experiences to enhance patient safety. Studies (Salas 2006; Pizzi 2001) have not been able to ascertain the effectiveness of CRM as it relates to the safety of an organization, which may be because of the evaluation framework (reactions, learning, behavior, and organizational impact developed by D.L. Kirkpatrick in 1976) or because of the safety metrics. The differences between organizations can be quite marked, as demonstrated in a study by J.B. Sexton, E.J. Thomas, and R.L. Helmreich (2000) that compared attitudes of teamwork between cockpit crew members and operating room personnel. The rejection of steep hierarchies was higher in the cockpit crew group (94% of crew members versus 55% of for attending surgeons) as was the recognition of performance degradation (74% of pilots versus 30% of attending surgeons). Perceptions of teamwork varied even between operating room members (10% to 73%). Although environments like the operating room and emergency room may benefit from CRM, issues of cost and measurable safety benefits may slow CRM development in health care.

2.4 Models of Clinical Responsibility

No discussion on work system designs and safety culture would be complete without acknowledging that responsible care is vital to both quality and safety. Although what clinical responsibility entails can be understood intuitively, there are still multiple lenses through which one can view clinical responsibility that need to be evaluated. There is a distinct vocabulary within this discipline that is not intuitive to the average practitioner. The clinical responsibility literature refers to loci, domains, procedures, and models of clinical responsibility (Emanuel and Emanuel 1996). This chapter section will explain the meaning and impact of these terms on patient safety and quality impressions.

Loci of accountability are the locations or parties that can be held accountable for specific clinical actions. It also includes parties that can hold others accountable. The loci of accountability, therefore, are truly the "who" of clinical responsibility. As such, loci include:

- individual patients who neglect palpable masses,
- individual physicians who fail to act on critical test results,
- nurses who do not perform proper hand hygiene prior to patient care,
- physicists who neglect appropriate QA measures,
- hospitals that engage in willful understaffing as a cost cutting measure, or
- managed care plans that fail to cover appropriate screening tests.

Among the loci holding others accountable are:

- professional organizations such as ASTRO or the AAPM that set a field's standards,
- employers who expect individual departments to conduct themselves in a conscientious and ethical manner,
- private payers (e.g., Blue Cross and Blue Shield) that expect to pay for quality healthcare services,
- the federal government (which can regulate radiation therapy standards), and
- investors and lenders who may front money for equipment such as a CyberKnife® unit, but expect it to be run in an efficient and proper manner.

Finally, the tort system is the system of last resort, which can be accessed by any of the aforementioned parties to hold other loci accountable.

Domains of accountability are the practices, activities, issues, or actions for which any of the loci can be rightfully held responsible. These domains may require justification; for instance, an insurance company's refusal to pay for routine screening mammograms for women under age 50 or colonoscopies in those under age 50. Accountability in health care consists of a minimum of five domains, or "whats." The first domain is professional competence. There are certainly many certifications that are relied upon to ensure professional competence, such as board certification from the American Board of Radiology for physicists and radiation oncologists, or medical dosimetry certification for dosimetrists. However, these certifications do not encompass all the practices or domains in which a clinician might be engaged. For instance, a physician may be board certified, but have very little experience in salvage brachytherapy for a base of tongue cancer or a vulvar Syed implant and, therefore, be board certified but incompetent in those particular practices. Likewise, a medical physicist may be exceptionally qualified but be incompetent with respect to the quality assurance and commissioning of a hyperthermia unit. As a result, board-certifying organizations are discussing and creating subspecialty certifications such as vascular and interventional radiology.

Sound legal and ethical conduct is another domain of accountability. This conduct includes proper billing by dosimetrists, physicists, and physicians to guard against fraud, as well as conducting a legal and ethical department. This domain incorporates a wide variety of medical ethics, professional ethics, and legal standards. This domain encompasses:

- not tolerating sexual harassment or racial discrimination among patients or employees,
- encouraging honest and frank communication with patients regarding prognosis to enable appropriate medical decision making,
- reporting of any quality assurance deviations in a facilities' linear accelerators, and
- properly disclosing medical errors as a matter of practice.

One can see how this would be of interest to all of the loci of accountability; who would care to be involved with, pay for, receive care from, or work for an organization that is neither legally compliant nor ethical?

Another domain is financial performance. Inefficiencies in care are always more expensive. Poor-quality radiotherapy has been defined variably, but it generally includes the potential for harm requiring additional care. So, a miscalculation or wrong setup resulting in overdose to a spine field may require closer monitoring and additional blood counts due to cytopenia, subsequent surgery down the line for a small bowel obstruction, or a myriad of costs caused by paraplegia from radiation myelopathy. Physicists and radiation oncologists are called upon to provide excellent and cost-effective care in less extreme situations as well. Should one choose intensity-modulated radiation therapy (IMRT) with multiple control points for breast cancer to reduce inhomogeneity? Or should one utilize the more economical electronic compensator, which can achieve much the same results but be considered a three-dimensional plan with a complex device? Financial performance can be skewed depending on the loci receiving the information. A hospital receiving the technical charges would advocate for IMRT, while an insurance payer would advocate the electronic compensator.

Community benefit is a domain that includes the community's ability to access quality health care. There are a number of locations in the United States and elsewhere in the world where access to radiotherapy is a vital issue. There is no doubt that daily radiotherapy is less likely to be used by patients who live a substantial distance from the treatment facility. However, adequacy of access to not only radiotherapy, but also to specialized services, must be considered. Access to stereotactic body radiosurgery for medically inoperable stage I lung cancer can mean the difference between a local control rate of >90% and a local control rate of 15% with conventionally fractionated radiation. Access to brachytherapy for cervical cancer is essential to local control. Therefore, the radiation community has a responsibility to provide not only facilities, but also all the appropriate services within its competence.

Public health promotion is a domain of accountability as well. This particular domain is something more widely ascribed to the medical community at large. However, radiation oncology departments are not excluded from taking measures to promote public health. These might include a walking trail surrounding a cancer center, showers to allow workers to bike to work, nonsmoking facilities, and healthy food choices in on-site cafeterias.

The final component of the accountability lingo is a discussion of the procedures of accountability. This is the meat of the discussion in that this is the "how" of the accountability process. The procedures of accountability encompass the evaluation and compliance with the criterion set forth by the loci of accountability. So if the codes of ethics of the American Medical Association or the AAPM define the standards for behavior, how is a physician or physicist judged? By whom are they judged? Is it a rigorous, well-defined evaluation by a supervisor? Is it an evaluation by a national or state organization? Is it only after multiple examples of glaring nonconformities that nonstandard behavior comes to light? Once a violation has been determined, the information must be distributed to the affected parties, and responses must be heard from those whose actions were questioned. It seems this happens too often within the tort system, although it can also take place through complaints to state medical boards or certifying organizations.

As one might surmise, there is a complex relationship between all of the loci and domains of accountability referred to as the matrix of responsibility. This means that each person may be responsible to any number of people or organizations for any number of activities. So, a physicist may be responsible to the chief physicist for proper work performance, to the patients for maintaining reliable linear accelerator calibration and performance, to the hospital for proper work behavior, and to the American Board of Radiology for professional ethics. Likewise, a patient is responsible for communicating any untoward effects of therapy, non-compliance with medication, or changes in general medical health.

2.4.1 Understanding Models of Clinical Responsibility

One can consider clinical responsibility through several lenses. The simplest of these is the *professional model.* In the professional model, the two main loci of accountability are the physician to the patient, and the physicians to their colleagues and organizations. Historically the primary concerns of these relationships have been competence, legal activity, and ethical conduct. This is one of the main functions of true, well-run chart rounds—that is, one that incorporates peer review. While there are many judgments to be made in medicine (in other words, situations with no clear-cut answers), far more situations have well-defined standards. Peer review provides both an opportunity for course correction and a way to assess when a colleague is repeatedly making the same poor choices and thus nearing incompetence. The physician-patient interaction is intended to improve the health of patients within the boundaries of what patients will allow and comply with, i.e., within their individual value systems. For instance, while twice-daily thoracic radiotherapy has been shown to be superior to daily treatment in small cell lung cancer, some patients are unwilling to commit to this regimen and prefer daily treatment. Likewise, some patients forego concurrent chemotherapy because they are concerned about toxicity, or even (rarely) alopecia. Patients are involved in the decision-making, and they are also recipients of care. In this model, health care is a professional service and not a marketplace commodity. If the patient is dissatisfied, the usual recourse is litigation.

The second model of accountability is the *economic model*, which is primarily concerned with the relationship between individual consumers and the providers of health care. In this model, the providers of health care include physicians, hospitals, and managed care systems. The consumer is the patient, and the product is the delivered health care. This model relies on an entity to ensure fair operation of the market, a role that has historically been filled by the government. The domains of accountability here are quality and price. This model holds that patients will leave poor providers and flock to good ones, and thereby the market will be regulated.

This presupposes that patients are able to determine who is a good provider. Certainly patients do not have the ability to judge the competence of each of their medical specialists, nor is there accurate, reliable information about the competence of individual providers. Of course, the problem with this model is that price (financial performance) among providers is heavily affected by the types of patients for whom they care. Consider a location with a relatively medically disenfranchised patient population, who present with late-stage head and neck cancer, have substantial treatment breaks, continue to drink and smoke after therapy, and are non-adherent to follow-up. Contrast that with a more affluent area where cancers are detected earlier, compliance with radiotherapy is impeccable, and health-threatening lifestyles are stopped. Consider also a radiation oncologist who treats breast cancer and treats many stage I or II patients with three-dimensional conformal radiation. Contrast her with a colleague who treats predominantly glioblastoma with intensity-modulated radiation with predictably poor outcomes. What does the economic model say about each of these situations? Clearly this model would call some clinics poor providers in an unjust manner.

The final model conceived of is the *political model*, in which the philosophies of the political sphere are applied to health care. This entails considering the realm of health care as a community, in which both physicians and patients are citizen members. There is input from all members of the healthcare community. The locus of accountability in this model is a governing board of representatives made up of all the participants involved in health care. Any changes to the healthcare interactions within this model would need to go through this board. This would not exclude the ability to bring litigation or to switch providers.

2.4.2 The Impact of the Model of Clinical Accountability on Quality and Safety

Each of the aforementioned models has its advantages. The political model has the most equality built into it, as its design focuses on a collaborative board that can evaluate safety and quality from all angles. However, this particular model also has the potential for inefficiencies, as it requires agreement among a large number of people. The economic model would seem to be the least likely to succeed. As discussed in Section 1.2 (Quality in Radiotherapy: Patient's Perspective), many of the metrics patients are concerned with when evaluating a physician are centered on how well the provider reviews the medical history or how well they interact on an interpersonal level. Additionally, there is often substantial resistance to leaving a provider, even when a patient is dissatisfied with him or her. People are often hesitant to leave because the rest of their medical care is given at that provider's hospital, or because they work with another provider they really value. Therefore, the loss of patients from a practice is unlikely to influence poor providers. Similarly, the professional model requires that colleagues are engaged enough to critically evaluate their colleagues, that those colleagues can receive that criticism in a constructive manner, and that leadership devotes enough time to this evaluation to make it meaningful. It also requires that organizations can define standards of practice in a meaningful way. Despite these challenges, it would seem that the best way to influence quality and safety daily and continually includes peer review, dedicated analysis of errors, and incident learning using many of the principles of the professional model.

2.5 Summary

In essence, work system design of processes and equipment, the safety culture, and the various models of clinical responsibility form the infrastructure of quality and safety in radiotherapy.

Work System Design

- Workplace design can be critical to effective information processing, where each stage is both time- and resource-limited and subject to interference or disruption.
- System monitoring requires situation awareness of the current task state; control system information provided should be as clear and task-relevant as possible.
- Appropriate function allocation—the determination of who (humans) or what (computer systems) is responsible for completing which tasks in a complex environment—is an important aspect of work systems design, particularly if function allocation is dynamic.
- Work processes such as documentation, layout, and procedures provide clear communication and documentation of the flows of patients, artifacts, information, and expert knowledge across the facility.
- Multiple error pathways and vulnerabilities in the overall work system cause adverse events and near misses; risk perception and attribution of error sources involve both cultural and technical factors.

Safety Culture

- A safety culture entails the interaction of shared values and beliefs with the organization's structures and controls; in successful safety cultures, organizations are informed of their incidents, have an effective reporting structure, and have a "just culture."
- Various surveys for assessing safety cultures may identify areas for improvement.
- Changing systems and processes through appropriate work system design will improve the safety culture.
- A "just culture" seeks to distinguish between acceptable behaviors (human error) and unacceptable behaviors, with unacceptable behaviors resulting in appropriate disciplinary action.

Safety Culture Influences on Work System Design

- Work systems and processes need to be designed to minimize bias and complacency as much as possible in the work environment.
- Disruptive individuals can have a direct, detrimental effect on observers' job satisfaction and morale, as well as on patient safety.
- Using an effective safety culture to investigate deviations in behavior, in conjunction with good work system designs to prevent them, minimizes normalization of deviance.
- If designed well and used properly, a checklist system can help enhance communication, reduce cognitive load, and be effective with measurable results.

Models of Clinical Responsibility

- The loci of accountability refer to the locations or parties that can be held accountable for clinical actions, which may include individuals, organizations, associations, and the legal system.
- The domains of accountability are the practices, activities, issues, and actions for which any of the loci can be rightfully held responsible. In health care, the five domains include professional competence, legal and ethical conduct, financial performance, community benefit, and public health promotion.
- The procedures of accountability encompass evaluation and compliance with the criterion set forth by the loci of accountability.
- Models of clinical responsibility include the professional model, the economic model, and the political model; each provides advantages but also challenges for quality and safety in radiotherapy.

Radiotherapy is a complex, multi-step process involving various professionals with particular expertise and skills. It also involves using advanced machinery, evolving technology, complicated software, and various human-machine and human-human interactions. This extremely complex work environment has a high-risk potential for error that may cause patients toxicity, injury, or death. Sound work systems and cultures are critical to providing safe, quality care.

References

Agency for Healthcare Research and Quality. Patient safety culture surveys. (2004). Available at: http://www.ahrq.gov/qual/patientsafetyculture/. Accessed December 30, 2012.

Bailey, R.W. *Human Performance Engineering: Using Human Factors/Ergonomics to Achieve Computer System Usability.* 2nd ed. Englewood Cliffs, NJ: Prentice Hall, 1989.

Balcetis, E. and D. Dunning. (2006). "See what you want to see: motivational influences on visual perception." *J of Personality and Social Psychology* 91:612–625.

Barnard, C.I. *The Functions of the Executive.* Cambridge: Harvard University Press, 1968.

Battista, O.A. Retrieved April 29, 2013, at http://www.brainyquote.com/quotes/quotes/o/orlandoab109834.html.

Bradley, G. *Computers and the Psychological Work Environment.* London: Taylor and Francis, 1989.

Caldwell, B.S. (1997). "Components of Information Flow to Support Coordinated Task Performance." *Intern J Cognitive Ergonomics* 1:25–41.

Caldwell, B.S. (2008). "Tools for Developing a Quality Management Program: Human Factors and Systems Engineering Tools." *Int J Radiat Oncol Biol Phys* 71(Suppl 1):S191–S194.

Caldwell, B.S. and S.K. Garrett. "Coordination of Event Detection and Task Management in Time-Critical Settings." In Mosier, K.L. and U.M. Fischer, eds. *Informed by Knowledge: Expert Performance in Complex Situations.* New York: Psychology Press, 2011.

Caldwell, B.S. and T.J. Anderson. (1996). "Design and Active Phases of Medical Process Improvement Systems." Paper presented at the Examining Errors in Health Care: Developing a Prevention, Education and Research Agenda. Rancho Mirage, CA.

Caldwell, B.S., S.K. Garrett, and K.C. Boustany. (2010). "Healthcare Team Performance in Time Critical Environments: Coordinating Events, Foraging, and System Processes." *J Healthcare Engineering* 1(2):255–276.

Carayon, P., ed. *Handbook of Human Factors and Ergonomics in Health Care and Patient Safety.* Mahwah, NJ: Lawrence Erlbaum Associates, 2007.

Cuevas, H.M., S.M. Fiore, B.S. Caldwell, and L. Strater. (2007). "Augmenting Team Cognition in Human-Automation Teams Performing in Complex Operational Environments." *Aviat Space Environ Med* 78(Suppl 5, Section II):B63–70.

Degani, A. and E.L.Wiener. (1993). "Cockpit checklists: concepts, design, and use." *Human Factors: The Journal of the Human Factors and Ergonomics Society* 35:345–359.

Donabedian, A. *Explorations in Quality Assessment and Monitoring: the Definition of Quality and Approaches to its Assessment.* Ann Arbor, MI: Health Administration Press, 1980.

Endsley, M.R., B. Bolté, and D.G. Jones. *Designing for Situation Awareness.* Boca Raton, FL: Taylor and Francis, 2003.

Epstude, K. and N.J. Roese. (2008). "The functional theory of counterfactual thinking." *Pers Soc Psychol Rev* 12:168–92.

Garrett, S.K., B.S. Caldwell, and P.R. Ebright. (2010). "Provider Information and Resource Foraging in Health Care Delivery." *Intern J Collaborative Enterprise* 1(3/4):381–393.

Garrett, S.K., B.S. Caldwell, E.C. Harris, and M.C. Gonzalez. (2009). "Six Dimensions of Expertise: A More Comprehensive Definition of Cognitive Expertise for Team Coordination." *Theoretical Issues in Ergonomics Science* 10:93–105.

Gilbreth, L.M. (1921). *The Psychology of Management.* Available from http://www.gutenberg.org/files/16256/16256-h/16256-h.htm.

Ginsburg L, D. Gilin, D. Tregunno, et al. (2009). "Advancing measurement of patient safety culture." *Health Serv Res* 44:205–224.

Giovanni, N. *Black Feeling, Black Talk, Black Judgment.* New York: W. Morrow, 1970.

Grandjean, E. *Ergonomics in Computerized Offices.* London: Taylor and Francis, 1987.

Hales, B., M. Terblance, R. Fowler, and W. Sibbald. (2008). "Development of Medical Checklists for Improved Quality of Patient Care." *Intern J Quality Health Care* 20:22–30.

Halligan, M. and A. Zecevic. (2011). "Safety culture in healthcare: a review of concepts, dimensions, measures and progress." *BMJ Qual Saf* 20:338–343.

Harvey, J., G. Erdos, H. Bolam, M.A. Cox, J.P. Kennedy, and D.T. Gregory, D. T. (2002). "An analysis of safety culture attitudes in a highly regulated environment." *Work and Stress* 16:18–36.

Haynes, A.B., T.G. Weiser, W.R. Berry, S.R. Lipsitz, A.H.S. Breizat, E.P. Dellinger, P.L. Kibatala, M.C. Lapitan, A.F. Merry, K. Moorthy, R.K. Reznick, B. Taylor, and A.A. Gawande. Safe Surgery Saves Lives Study Group (2009). "A surgical safety checklist to reduce morbidity and mortality in a global population." *N Eng J Med* 360:491–499.

Health and Safety Executive. *Safety culture maturity model: offshore technology report 2000/049.* Colegate, Norwich: HSE Books, 2001.

Helmreich, R.L. and A.C. Merritt. *Culture at Work in Aviation and Medicine: National, Organizational and Professional Influences.* Hampshire, UK: Ashgate Publishing LTD, 1998.

Henriksen, K. and H. Kaplan. (2003). "Hindsight bias, outcome knowledge and adaptive learning." *Qual Saf Health Care* 12(Supp II):ii46–ii50.

Institute of Medicine. *To Err is Human: Building a Safer Health System.* Washington, DC: National Academy Press, 1999.

Johnston, N. (1995). "Do blame and punishment have a role in organizational risk management?" *Flight Deck* 15:33–36.

King, C.A. (2010). "To Err is Human, to Drift is Normalization of Deviance." *AORN Journal* 91:284–286.

Kroemer, K., H. Kroemer, and K. Kroemer-Elbert. *Ergonomics: How to Design for Ease and Efficiency.* Englewood Cliffs, NJ: Prentice Hall, 1994.

Mardon, R.E., K. Khanna, J. Sorra, N. Dyer, and T. Famolaro. (2010). "Exploring Relationships Between Hospital Patient Safety Culture and Adverse Events." *J Patient Saf* 6:226–32.

Martin, W.F. (2008). "Is your hospital safe? disruptive behavior and workplace bullying." *Hosp Top* 86:21–28.

McGrath, J.E. and A.B. Hollingshead. *Groups Interacting With Technology.* Newbury Park, CA: Sage Publications, 1994.

Musson, D.M. and R.L. Helmreich. (2004). "Team training and resource management in health care: current issues and future directions." *Harvard Health Policy Review* 5:25–35.

Oxford Dictionaries. April 2010. Oxford University Press. http://oxforddictionaries.com/definition/english/ (accessed December 29, 2012).

Parasuraman, R., R. Molloy, and I.L. Singh. (1993). "Performance consequences of automation-induced complacency." *Internl Jf Aviation Psychology* 3:1–23.

Pizzi, L., N.I. Goldfarb, and D.B. Nash. (2001). "Crew resource management and its applications in medicine." In *Making Health Care Safer: a Critical Analysis of Patient Safety Practices.* Evidence Report/Technology Assessment, No. 43. AHRQ Publication No. 01–E058, July 2001. Agency for Healthcare Research and Quality, Rockville, MD. http://www.ahrq.gov/clinic/ptsafety/.

Potters, L. and A. Kapur. (2011). "Implementation of a 'No Fly' Safety Culture in a Multi-Center Radiation Medicine Department." *Int J Radiat Oncol Bio Phys* 81:S696.

Proctor, R.W. and T. VanZandt. *Human Factors in Complex Systems.* Boston: Allyn and Bacon, 1994.

Pronovost, P., D. Needham, S. Berenholtz, D. Sinopoli, H. Chu, S. Cosgrove, B. Sexton, R. Hyzy, R. Welsh, G. Roth, J. Bander, J. Kepros, and C. Goeschel. (2006). "An intervention to decrease catheter-related bloodstream infections in the ICU." *N Eng J Med* 355:2725–2732.

Rasmussen, J. *Information Processing and Human-Machine Interaction: an Approach to Cognitive Engineering.* New York: North–Holland, 1990.

Reason, J. (1998). "Achieving a safe culture: theory and practice." *Work and Stress* 12:293–306.

Reason, J. "Managing the risks of organizational accidents." Risk Management Conference V. Oct. 27, 2004. http://rmc.nasa.gov/archive/rmc_v/presentations/reason%20managing%20the%20risks%20of%20organizational%20accidents.pdf. (Accessed December 29, 2012).

Reason, J. *Human Error.* Cambridge: Cambridge University Press, 1990.

Reason, J.T., J. Carthey, and M.R. de Leval. (2001). "Diagnosing 'vulnerable system syndrome': an essential prerequisite to effective risk management." *Quality in Health Care* 10 (Suppl II):ii21–ii25.

Reed, P., W.D. Compton, J.H. Grossman, and G. Fanjiang, eds. *Building a Better Delivery System: a New Engineering/Health Care Partnership.* Washington, DC: The National Academies Press, 2005.

Roberts, K.H. (1999). "Some characteristics of one type of high reliability organization." *Organ Sci* 1:60–176.

Roese, N.J. and S.D. Maniar. (1997). "Perceptions of Purple: Counterfactual and Hindsight Judgments at Northwestern Wildcats Football Games." *Pers Soc Psychol Bull* 23:1245–1253.

Rosenstein, A.H. and M. O'Daniel. (2008). "A survey of the impact of disruptive behaviors and communication defects on patient safety." *Jt Comm J Qual Patient Saf* 34:464–471.

Salas, E., C.S. Burke, C.A. Bowers, and K.A. Wilson. (2001). "Team training in the skies: does crew resource management (CRM) training work?" *Human Factors: The Journal of the Human Factors and Ergonomics Society* 43:641–674.

Salas, E., K.A. Wilson, C.S. Burke, and D.C. Wightman. (2006). "Does crew resource management training work? An update, an extension, and some critical needs." *Hum Factors* 48:392–412.

Sanders, M.S. and E.J. McCormick. *Human Factors in Engineering and Design*, 7th ed. New York: McGraw Hill Book Co., 1993.

Scott, W.R. *Organizations: Rational, Natural, and Open Systems.* Englewood Cliffs, NJ: Prentice–Hall, 1981.

Sexton, J.B., E.J. Thomas, and R.L. Helmreich. (2000). "Error, stress, and teamwork in medicine and aviation: cross sectional surveys." *BMJ* 320:745–9.

Sexton, J.B., R.L. Helmreich, T.B. Neilands, K. Rowan, K. Vella, J. Boyden, P.R. Roberts, and E.J. Thomas. (2006). "The Safety Attitudes Questionnaire: psychometric properties, benchmarking data, and emerging research." *BMC Health Serv Res* 6:44.

Sheridan, T.B. *Telerobotics, Automation, and Human Supervisory Control.* Cambridge, MA: MIT Press, 1992.

Singer, S., M. Meterko, L. Baker, D. Gaba, A. Falwell, and A. Rosen. (2007). "Workforce perceptions of hospital safety culture: development and validation of the patient safety climate in healthcare organizations survey." *Health Serv Res* 42:1999–2021.

Singh, I.L., R. Molloy, and R. Parasuraman. (1993). "Automation-Induced 'Complacency': Development of the Complacency-Potential Rating Scale." *The International Journal of Aviation Psychology* 3:111–122.

Stanton, N.A. "Team performance: communication, co-ordination, co-operation and control." In N.A. Stanton, ed. *Human Factors in Nuclear Safety.* London: Taylor and Francis, 1996.

Stanton, N.A., P.M. Salmon, G.H. Walker, C. Baber, and D.P. Jenkins. *Human Factors Methods: A Practical Guide for Engineering and Design.* Hampshire, UK and Burlington, VT: Ashgate 2005.

Sundstrom, E., K.P. DeMeuse, and D. Futrell. (1990). "Work Teams: Applications and Effectiveness." *Am Psychologist* 45:120–133.

Swezey, R.W. and E. Salas, eds. *Teams: Their Training and Performance.* Norwood, NJ: ABLEX, 1992.

Vaughan, D. *The normalization of deviance: signals of danger, situated action, and risk. How Professionals Make Decisions.* Mahwah, NJ: Lawrence Erlbaum Associates, 2005.

Weber, D.O. (2004). "Poll results: doctors' disruptive behavior disturbs physician leaders." *Physician Executive* Sept–Oct:6–14.

Wickens, C.D., J.D. Lee, Y. Liu, and S.E. Gordon Becker. *An Introduction to Human Factors Engineering*. Upper Saddle River, NJ: Pearson Prentice Hall, 2004.

Wilson, J.R. and E.N. Corlett, eds. *Evaluation of Human Work: A practical ergonomics methodology.* London: Taylor and Francis, 1990.

Chapter 3

Actions, Failures, and Guidance

Bruce Thomadsen, Ph.D.[1], Derek Brown, Ph.D.[2], Peter Dunscombe, Ph.D.[3], and Zoubir Ouhib, M.S.[4]

[1]Professor, Departments of Medical Physics, Industrial and Systems Engineering, Engineering Physics, and Biomedical Engineering, University of Wisconsin, Madison, Wisconsin
[2]Adjunct Associate Professor, Departments of Oncology and Physics and Astronomy, University of Calgary, Calgary, Alberta
[3]Director of Medical Physics, Tom Baker Cancer Centre, Calgary, Alberta
[4]Chief Medical Physicist, Lynn Cancer Institute, Boca Raton Regional Hospital, Boca Raton, Florida

3.1 Introduction

Avoiding actions and situations that could degrade the quality and safety of patient treatments in radiotherapy requires an understanding of how humans take actions and in what ways they can fail. Before looking at the mechanisms of actions and failures, a review of the types of events that have occurred in radiotherapy can provide perspectives on the problem. Looking at recommendations made by several organizations on how to avoid errors and mistakes can establish the arena and set a starting point for the material to come.

3.2 Events in Radiotherapy

The errors reported in this chapter are public information and can be found on the Nuclear Regulatory Commission (NRC) website (USNRC 2013). Readers should keep in mind that errors are expected to happen, and the purpose of sharing these cases is simply to learn from them and, hopefully, avoid making similar errors in the future. In the radiation therapy community, no one is immune to these unfortunate situations.

3.2.1 Radiotherapy Incidents in 2012

The NRC website recorded more than 79 cases of medical events involving radioactive material in the United States in 2012 (see Table 3.1). The majority of these appear to have produced no serious health consequences, based on statements from the authorized users involved in these cases.

3.2.2 General Incidents

General errors are errors that are not related directly to radiation delivery. They could be related to a missed diagnosis, medication (incorrect dose), wrong site (left vs. right, lower vs. upper), or faulty communication between the physician and the patient, between physicians of different specialties, or simply between members of the radiation oncology team. For physicists, these types of errors could be due to erroneous data collected during commissioning, inaccurate data entry in the TPS, or errors made during linear accelerator calibration.

Case #1. A good example is the case of a facility that commissioned their SBRT/SRS machine using inappropriate ionization chambers. Incorrect data was used for planning, and the treatment resulted in serious medical events. Could an error like this one have been prevented? Several potential solutions might have avoided this unfortunate outcome, including peer review or simple data comparison.

Table 3–1. Brachytherapy events for 2012 from the database of the U.S. Nuclear Regulatory Commission

Body Site	Type of Brachytherapy	Number of Cases	Error Type*	Isotope
Prostate	Permanent	33	Dose coverage (D90<80%)	^{125}I, ^{103}Pd, ^{131}Cs
Prostate	HDR	1	Dose coverage (D90<80%)	^{192}Ir
Prostate	Permanent	2	Leaking sources	
GYN	LDR	2	Applicator not in target; source outside applicator	^{137}Cs
GYN	HDR	1	Software issue (under dose)	^{192}Ir
Breast (APBI)	HDR	5	Under dose	^{192}Ir
Thyroid	LDR	4	Dose to Embryo/fetus (1); Incorrect dose (2); Contamination (1); Inadequate negative pressure in patient room (1)	^{131}I
Liver (primary and metastatic disease)	Microbrachytherapy: Sir Spheres/ Theraspheres	21	Wrong lobe (2); Wrong patient (3); Under dose (13) Spill (1); Dose to other organs (2)	^{90}Y
Endovascular	LDR	2	Source stuck (2)	^{90}Sr
Lung	LDR	2	Seed migrate to abdomen (1); Under dose (1)	^{125}I
Non-Hodgkin's Lymphoma	Radiopharmaceutical	1	Incorrect record	^{90}Y
N/A	Missing sources	2	N/A	^{90}Y
N/A	Cremation	2	Patient with ^{125}I mesh implant (1) Patient with ^{238}Pu pacemaker (1)	^{125}I, ^{238}Pu
Gastric case	HDR	1	Incorrect target area	^{192}Ir
Brain	Gamma Knife	1	Overdose	^{60}Co
Unknown	HDR	2	Wrong plan executed (1); Incorrect dose (1)	^{192}Ir

* The numbers in parenthesis give the number of cases with that particular type of error.

Case #2. A female patient that was given ^{131}I treatment while pregnant. The baby, three years old at the time of the error reporting, was still on thyroid hormone therapy. This error appears to be caused by the lack of a simple protocol that would have required the clinician to ask the patient whether she was pregnant. This could also have been a communication issue between the patient and the individual asking the question.

3.2.3 Brachytherapy Errors

3.2.3.1 Gynecological Cases

Low Dose-rate Brachytherapy Case. On July 9, 2010, a cervical cancer brachytherapy treatment was being performed using ^{137}Cs sources. The physician failed to place the source in the afterloader, and the source fell onto the patient's buttocks. The source was later recovered from the trash before it left the facility. The licensee estimates the maximum dose received was 10.5 Gy. The patient's physician and the patient were notified of this event. No reddening of the skin was noticed.

Recommendations:

- Once the applicator and sources are removed from the patient, a survey of the patient and all linens would detect the presence of an enhanced radiation level. In brachytherapy cases, no trash is to be removed from a patient's room without a survey performed by a medical physicist or an authorized user.
- Once the lead pig and sources are returned to the storage room, a source count must be performed to ensure that all sources are accounted for. Records of all sources logged out and logged in must be documented as per regulations. In this case, this was not done until the following day.
- If the medical physicist is unable to be present for the removal of the sources, the radiation oncologist should be instructed and trained to perform these simple checks.

High Dose-rate Brachytherapy Case. A patient received four treatments to the cervix. After the fourth fraction, the physician noticed, during examination, reddening of the skin on the upper thigh. Upon reviewing the setup with the tandems, a constriction and blockage was identified in the catheters caused by wear debris. Corrosion at the entry of the tandem used in the Fletcher Suite device was discovered. Because the treatment unit was being used in the "search-end mode" to find the end of the applicator to set the starting position, it was presumed that this caused the system to conclude the source had reached the end of the catheter when it had not. The starting point of fraction treatment within the tandem was misplaced by 9 cm. Dose calculations revealed a skin dose of 12.5 Gy to the right thigh and 12.7 Gy to the left thigh per fraction. The dose to the prescribed point for that fraction was 1.94 Gy instead of 4.00 Gy. The total dose for all fractions to the prescribed treatment site was 13.9 Gy instead of the intended 16.0 Gy.

Recommendations:

- A thorough evaluation of applicators should be performed (visual and x-ray), with some tests performed before each treatment and others done periodically (Thomadsen 1999).
- All high dose-rate (HDR) units should have a safety net to avoid these errors and to alert the user that the source did not reach the correct first dwell point as planned. For each case, the distance to the fist dwell position should be compared with the range of normal values.

3.2.3.2 Prostate Case

An institution was surveying packing materials used to ship ^{125}I seeds for a procedure conducted earlier that day, when they noted elevated readings. Further surveys revealed that the elevated readings were not coming from the packing material associated with the procedure done that day, but from some packing material in the area that was used to ship ^{125}I sources from a previous procedure. A survey of the packing material revealed no loose contamination on the exterior or interior of the box, but elevated readings of 2,500–350,000 counts/minute and 2 mSv/h (0.2 mrem/h) were found. No loose seeds were found in any of the packing material. The ^{125}I seeds were accompanied by the manufacturer's leak test report that indicated no contamination. A thyroid count was performed on the patient involved with the leaking sources, and it confirmed an uptake of iodine by the thyroid. Based on the institution's calculations, the patient involved received a thyroid uptake of 3.7 MBq (0.1 mCi) of ^{125}I and a dose of to the thyroid of 3 Gy.

Recommendations:

- Perform a survey prior to each case by isolating packing material. Focus should be given for each needle label to avoid the use of the wrong needle for the wrong location.

3.2.3.3 Breast Case

A breast cancer patient was undergoing treatment with a high dose-rate (HDR) brachytherapy unit for accelerated partial breast irradiation. A Savi-8™ device (Cianna Medical, Aliso Viejo CA) was connected to the HDR unit. The source apparently punched through the catheter stem and moved along the skin surface during the treatment. The patient did not notice the source outside the tube. The source retracted normally to the shielded position. The physicist estimates that the patient's skin received 5 Gy if the source moved along the skin surface, but if the source stuck in one position, the patient's skin could have received 50 Gy. According to the licensee, the catheter is easily kinked, which can cause this failure. This failure has happened previously but without any exposure. In the past the licensee would repair the catheter with heat shrink tubing.

Recommendations:

- Use of a zoom camera on the area of treatment could be helpful to interrupt treatment if the same error were to occur again.
- The use of an extra check cable run through all the catheters before starting treatment (for units equipped with such option) might prove useful in protecting against such failures.

3.2.3.4 Liver Cases

Case #1. A medical event took place involving a ^{90}Y selective internal radiotherapy (SIR) microsphere therapy patient treated on 3/17/2011. On 3/18/11 the radiation oncologist covering the SIRS procedure discovered that the activity of ^{90}Y was not 105% above the prescribed amount, but actually 150%. Upon investigation with the medical physicist, it was deduced that the medical physicist had not read the patient's written directive prescription correctly. The intended dosage was 1.11 GBq (30 mCi for an intended dose of 30.72 Gy) and the administered dosage was 1.66 GBq (44.8 mCi, giving a delivered dose 45.93 Gy).

Recommendation:

- An independent check should be performed of the ^{90}Y order—or any calculation related to the order—prior to submission to the manufacturer by

Table 3–2. Doses to some critical organs from microsphere treatment in Liver Case #2 as determined on SPECT imaging

Organ	Fraction of Activity	Volume (cm^3)	Dose, Averaged by Volume (Gy)
Liver	0.839	1209	53
Fundus of stomach	0.058	101	44
Spleen	0.093	200	35
Portion of duodenum	0.010	41	35

the prescribing authorized user and again prior to delivery by the medical physicist.

Case #2. After infusion of radioactive ^{90}Y for liver treatment, it was discovered by follow-up radionuclide scanning (SPECT imaging) that some of the material was not in the liver as intended. Material (10% to 15%) appeared in vessels involving the spleen and digestive track (gastric fundus and duodenum) instead. The results (with 20% uncertainty) are shown in Table 3.2.

Recommendation:

- Interventional radiologists must consider all flow from the arteries into which microspheres will be injected during the pretreatment localization arteriogram to avoid such events. This case could be a good teaching example for all microsphere users.

3.2.3.5 Lung Cases

Permanent Implant. A patient was given an invasive procedure in the right lung using a mesh containing ^{125}I seeds. The implant contained 50 seeds, with 5 strands of 10 seeds per strand. The total activity was 0.6 GBq (16.4 mCi). After the procedure, the patient was discharged on June 9, 2012. The patient was readmitted to the hospital on July 4, 2012. On July 7, 2012, radiographs revealed an abscess in the right lung with approximately 38 seeds in place. The area of the right lung received no more than 76% of the intended dose. On July 9, 2012, following chest and abdomen films, 35 seeds were noted in the right lung and 3 in the abdomen. These numbers changed to 13 in the lung and 17 in the abdomen by July 18, 2012. Nine seeds were recovered during the patient's hospitalization. The patient stated that he was coughing up phlegm and subsequently swallowing it.

Recommendation:

- On discharge, such patients should be given a container and instructions to spit phlegm into the container.

HDR Case. A patient has persistent small cell lung cancer after chemotherapy in an area that has received prior external radiation therapy and was symptomatic. A course of HDR endobronchial brachytherapy was recommended to the left upper and left lower lobe bronchus where the tumor was present with a prescription dose of 20 Gy in 2 fractions. During treatment planning, the catheters were correctly identified, but the direction of the catheter reconstruction was mistakenly reversed, thereby changing the starting position of the source. The treatment was given to a location that was considerably more proximal to the tumor, and which included the larynx. The plan was checked per protocol, but the orientation error was missed. After the error was discovered one hour later, a 3D plan was generated and revealed that the neck region received about 15 to 20 Gy.

Recommendation:

- The independent check of the treatment plan should include verification of the direction of digitization.

3.2.3.6 ^{131}I Thyroid Ablation

A patient was administered a dose of ^{131}I that differed from the intended dose by greater than 20%. The intended prescription was 3.7 GBq (100 mCi). The administered dose was 0.93 GBq (25 mCi). The dose provided by the vendor was divided into three capsules: two containing 0.93 GBq and one with 1.86 GBq. The bottle sent by the manufacturer stated that there was one capsule of "309 mci." The labeling was a software error. When the dose was administered to the patient, one capsule was dispensed from the bottle, while the other two stuck in the bottom of the bottle. The presumed empty bottle was then repackaged and shipped back to the company. Upon receipt of the package, the company discovered that two capsules remained in the bottle. The returned capsules contained a total of 75mCi.

Recommendation:

- An appropriate survey should have been performed on the supposedly empty bottle without the bottle being inside the shielding.

3.2.4 External-beam Radiotherapy Errors

In a treatment with intensity-modulated radiotherapy (IMRT) a patient received a fatal overdose. The following is brief summary of the events.

In this case, the first fraction went fine. Based on the imaging, the physician wanted to close the jaws (not the MLC) very slightly, a change that would not affect the dosimetry of the plan at all. The plan was performed and cleared for treatment. Because the new plan would not change anything substantial, the physics staff felt they could postpone redoing the patient-specific measurements for a couple days since they just did it for fraction 1.

When downloading the program, the therapists, as always, hit download and went into set-up as always. But this time an antivirus scan was underway, a program added by the IT service for safety. Moving to set-up interrupted the download, so that the treatment computer had a file but the MLC segments did not make it, and the file was empty. Had this been the first treatment, the unit would not have let the treatment begin, knowing that the MLC file was missing. But that interlock, once cleared, would not stop the unit again because it thought it had the file (it was created at the first treatment, but it was now empty). In addition, the computer had just been upgraded to a very much faster one. When this happened, the vendor removed a software stop that prevented going to set-up during download because they thought it could not be a problem with the faster hardware, and, indeed, it would not have been, except for the anti-virus scan which was underway and slowing the computer.

The therapist noticed that the MLC did not move on the screen during the case. They knew the computer program was new and lamented that they no longer had this cartoon showing them the motion. The event was detected when, after the fourth treatment (third without MLC), the therapist noticed on a different IMRT patient that the cartoon still worked!

Recommendation:

- This case presents difficulties because at each step the staff believed they were following reasonable quality assurance protocols and standard procedures used in many facilities. The physics staff delaying the patient-specific measurements was based on the experience that the change in jaw settings would not have dosimetric ramifications. The therapists looked for MLC movement, but they built mental models—albeit incorrect ones—to explain the absence of movement on the screen. The IT service installed the antivirus software to protect patients, not realizing it could cause problems.

Two things that could have prevented this event would have been a guideline directing the therapists to stop the treatment if they noticed something unusual (lack of motion in the cartoon of the MLC) and the physics staff following their protocol to the letter, regardless of extenuating circumstances (such as, in this case, it seemed as if they had just performed such measurements).

3.2.5 Understanding the Nature of Failures of the Radiotherapy System

Errors are inevitable and therefore will continue to happen. The goal is to minimize the probability of occurrence and render the effects of errors harmless.

3.2.5.1 Frequent Problems Encountered in Radiotherapy

Several factors can be associated with system failures, such as the lack of adequate details, issues with upgrades, insufficient support, and products released before they have had full and adequate testing.

Lack of Dissemination of Information. Sometimes certain situations are never encountered during the testing stage for devices, so users encounter very unusual situations (for example, the SAVI case described earlier). However, information about problems, once detected, should be available to all other users rapidly to avoid similar situations at other institutions. No one should have to wait for a state or an NRC report to learn from an event. Warning messages from equipment must be clear, and in the case of serious problems, the equipment should not allow users to clear an interlock without authorization from knowledgeable persons, such as the physicist and an engineer. Requiring a call to a manufacture to obtain a code to clear a serious fault would not be unreasonable and would also alert the manufacturer to the occurrence.

Issues with Upgrades. It seems that with any major upgrade, there is always a possibility of witnessing an error (from minor to major). The user is urged to repeat some test cases to confirm that the latest upgrade has not affected the expected results. This is not 100% fool proof, but it is the appropriate and recommended thing to do.

Premature Product Release. Sometimes products (hardware and software) are released before complete evaluation. A rushed product may lead to an event.

Insufficient support. Support can be within an organization or from the manufacturer. The institution might not have adequate staffing to perform certain tests and checks on a major software upgrade, while the equipment vendor might not have enough qualified support staff to respond to the issues addressed by users in a timely manner.

Training. Human factors perhaps contribute to failures in an estimated 80% of events. Human performance begins with the competency of individuals performing certain specialized tasks that require adequate knowledge and training. Adequate training is a responsibility of both users and manufacturers, where relevant. There are instances where the manufacturer, in addition to charging the cost of a device, will request an additional fee for the operator training. The cost of devices should have training included and not be additional. Where appropriate, facilities should provide training for more than one person to provide for backup and independent verification. Additional training may be necessary if there are changes in a product or procedure, or as refreshers if the number of cases performed by the institution is too low to maintain their skills.

3.2.5.2 The Importance of Quality Management

Radiotherapy is an evolving and complex technology that makes it difficult for the best-trained staff (radiation oncologists and medical physicists, for example) to deliver error-free patient care. Errors will happen, and a facility needs to focus on designing systems to anticipate or prevent errors from occurring and absorb them without harm. Those who do not follow appropriate checklists, but who rely on memory, could eventually miss important steps. Omission of QA as directed in a facility's protocols often precedes events.

Radiotherapy as a discipline needs increasingly to involve other specialties in patient care (radiologists, thoracic surgeons, etc.), using their expertise to improve outcome and patient safety. The use of everyone's expertise will only improve patient care.

Institutions sometimes create their own software for treatment planning, quality assurance, calculation sheets, and even devices. While this is creative and often innovative, the product might not go through rigorous and appropriate evaluation to confirm its safety. Caution should be exercised when using or borrowing in-house software.

Radiation therapy facilities continue to operate in a world where medical devices such as linear accelerators are more software- and computer-dependent. These devices are connected to the facility's information technology (IT) systems. One cannot ignore the benefits of such collaboration between medical technology and IT, such as better workflow, better communication of information, automation, etc. However, an ineffective device–IT interface can lead to hazards, such as incompatibility between multi-vendor software applications and systems or the transfer of appropriate data from one system to another (PACS to treatment planning system). These types of issues can lead to undesirable results.

3.2.6 Quality Improvement

Each institution should have a robust quality improvement program with a priority on value-based health care delivery. The following items are some of the items that each quality improvement management team should implement and maintain on a regular basis:

- Rigorous practice standards
- Accredited training and program
- Adequate education
- Outcome measures
- Improving the culture of safety
- Empowerment of team members
- Improved communication at all levels
- Training on error identification and prevention
- Standardization and simplification of procedures (as much as possible)
- Peer review, such as chart rounds
- Identification of opportunities for improvement
- Participation in root-cause analyses of events and the design of systems to implement improvements
- Evaluations of current clinical guidelines and improvement according to current recommendations

All near misses should be reported to the management team and used as an opportunity for improvement.

Most of the time, the first person to identify a medical error is the patient. For this reason, all staff should pay close attention to patient feedback regarding their treatment from start to finish. They might just identify an unusual setup, treatment, or outcome. There is a great value to listening.

Ostrom et al. studied radiotherapy events observed, "Licensees' corrective actions were often narrow in focus, as they were directed towards preventing a similar misadministration" (Ostrom et al. 1996). While it is easy to make changes that would have prevented a given event, events happen in an environment that allows them to happen. Making the environment safer and maintaining the quality of treatments requires looking past the obvious fixes using risk assessment techniques. The rest of this book discusses the issues and tools necessary to carry out such assessments.

3.3 Safety Guidance for Radiotherapy

In the opening chapter of these proceedings, the argument was made that safety and quality are so closely interrelated that separating them is probably not helpful. Certainly, most measures that enhance safety also improve quality. Taking that viewpoint, advice on quality (encompassing many aspects of safety) has been provided in abundance by professional bodies such as the American Association of Physicists in Medicine (AAPM), the Institute for Physics and Engineering in Medicine (IPEM), and the European Society for Therapeutic Radiology and Oncology (ESTRO) over many years. However, in the last 15 years or so, there have been two significant developments that have influenced our safety and quality focus. The first of these was the report by the Institute of Medicine (IOM), *To Err is Human: Building a Better Health Care System* (Institute of Medicine 1999). This document contained the astonishing statistic that between 44,000 and 98,000 deaths occurred in the U.S. health care system each year due to medical errors. Reported experience in health care systems in other developed countries is not dissimilar (Baker et al. 2004). A year after the IoM report, in 2000, the International Atomic Energy Agency (IAEA) published its Safety Report Series 17 *Lessons Learned from Accidental Exposures in Radiotherapy* (International Atomic Energy Agency 2000). The IoM publication was widely reported in the popular literature, arousing considerable concern in the patient population, while the IAEA document was largely of academic interest, describing "freak" incidents that did not appear to seriously undermine patients' confidence in radiotherapy.

Providers of radiotherapy did not become seriously concerned about the state of the services they provided until significant misadministrations began to be described in the popular press. In the U.S. it was the New York State incident reported in the *New York Times* some five years after the event, which cast a deep shadow over our profession (Bogdanich 2010). In Europe, it was incidents in Glasgow, Scotland (Scottish Ministers for the Ionizing Radiation 2006) and Epinal, France that resulted in the most publicity (Peiffert 2007). (As an interesting aside, the conditions that led to the New York incident were identified several years previously. A paper by Thomadsen et al. (2003) contains the prescient statement, "Errors often follow violations in protocols, particularly failures to perform verification procedures, and indicators that things are not correct are often present yet ignored during events.") So one major development over the last decade or two has been the increased realization that radiotherapy may not be as safe as we had preferred to believe.

The second major development links back to the extensive quality literature, protocols, and recommendations generated by the professional organizations. These documents were, and still are, extremely valuable and responsible for much of the high clinical quality we routinely achieve today in radiotherapy. However, in hindsight, an omission in much of the published quality and safety literature in our field until recently has been the discussion of failures of our processes, as opposed to failures of equipment, which has been the focus of much of our attention. Many publications in recent years, which we will discuss below, are redressing this imbalance. Indeed, they are moving beyond analyzing equipment and what we are doing with it to analyzing the culture and structure of the environment in which our tasks are performed. This observation provides us with an approach to the review of the recent literature. The recommendations can be divided into those that address the infrastructure and culture of an organization, those that address equipment specifically, and those that address processes that take place within an organization. Safety and quality should be a core feature of organizational infrastructure, equipment, and processes.

3.3.1 Method of Literature Analysis

The discussion that follows is based on an analysis by Dunscombe (2012). The analysis was complete before the publication of *Safety is No Accident: A Framework for Quality Radiation Oncology and Care* (American Society of Radiation Oncology 2012), so this contribution to the literature will be discussed separately. Seven documents, which encompass major works from national and international bodies, were included in the analysis. As might be expected, there was considerable duplication within the 117 recommendations in these documents and, hence, the task became to distill them down to a manageable number receiving broad endorsement.

The mapping process used during the distillation is described elsewhere (Dunscombe 2012). It is, however, worth briefly remarking on two of the challenges encountered. First, several of the documents combine two or more recommendations into one paragraph. For example, in Hendee and Herman (2011) one recommendation includes both "Workstations should be clutter-free..." and "Therapists should not be interrupted while treatments are underway." In such cases, the recommendations were broken out. The second predictable difficulty was that of interpretation. For the purposes of this analysis, a fairly liberal interpretation of the central issue underlying the recommendation was adopted. For example, all the following recommendations were mapped onto *training*:

- "Training records should be created and maintained...and [be] specific to particular procedures"
- "Funding to support training should be available"
- "A consistent effort on education and training"
- "Resources should be allocated so as to avoid substituting proper training..."
- "Competency certification" and "Staff competency assessment"
- "Training/additional study/analysis" and "More training"
- "Patient safety should be a competency"
- "Adequate training of staff"

Clearly, a significant and variable element of judgment needed to be employed depending on the recommendations being mapped.

3.3.2 Analysis of Recommendations for Radiotherapy Safety

The analysis identified twelve topics in three or more of the documents as being pertinent to the improvement of patient safety in radiotherapy. They are:

- Training
- Staffing
- Documentation
- Incident learning
- Communication
- Checklists
- Quality control and preventive maintenance
- Dosimetric audit
- Accreditation
- Minimizing interruptions
- Prospective risk assessment
- Safety culture

Before one can make any recommendations, it is critical first to identify all existing pitfalls associated with a procedure. The *New York Times* article of January 24, 2010 (Bogdanich 2010) summarized 621 radiotherapy events between January 2001 and January 2009 reported to the New York State event database, NYPORTS (New York State Department of Health 2008). The distribution of the events' classifications is as follows:

Classes of errors

- Missed all/part of target 46%
- Wrong dose 41%
- Wrong patient 8%
- Other (e.g., technology) 5%

Causes of errors

- QA flawed 28%
- Data entry and calculation errors 20%
- Misidentification of patient or site 14%
- Setup error (blocks, wedges) 11%
- Patients physical setup wrong 8%
- Flawed treatment plan 6%
- Hardware malfunction 5%
- Software/data transfer, software overrides, communication 5%

The majority of errors are human with fewer related to hardware. These errors are also directly related to the twelve items in the previous list.

Considering our categorization into infrastructure, equipment, and process issues, we can see that the majority of the recommendations these days are directed more strongly toward the healthcare system infrastructure, with less emphasis on equipment. In fact, the traditional topics in the above list are quality control, preventive maintenance, and dosimetric audits. If a process issue is defined as an activity that can be identified with a particular patient, then the only topic in this category is checklists. The remaining nine topics address the infrastructure of the organization, including the culture. It is acknowledged that safety culture could actually encompass all of these issues as it is a very loosely defined notion. However, as it was explicitly identified in three of the seven source documents, it is included in the list.

After this analysis was complete, ASTRO independently produced *Safety is No Accident* (Zietman et al. 2012), a comprehensive addition to the literature that was endorsed by all the major U.S. professional therapeutic and diagnostic radiology organizations. It is interesting to look at the overlap between *Safety is No Accident* and the results of the analysis described above. The items below are the headings in Section 3.4.0 of *Safety is No Accident* entitled "Examples of Tools/Initiatives to Facilitate Safety and the Safety Culture." The difficulty of establishing a one-to-one correspondence is clear.

- Staffing/Schedules
- Communication/Facilities
- Workflow/Efficiency
- Standardization
- Hierarchy of Effectiveness
- Human Factors Engineering
- Incorporating QA Tools/functionality into Software
- Peer and Interdisciplinary Review
- Daily Morning Meetings
- Safety Rounds
- Routine Public Announcements/Updates
- Address Errors and Near-Misses
- Quality Assurance Committee
- Credentialing and Training

However, with some freedom of interpretation, we can identify the following parallels between this list and the 12 items identified by Dunscombe (2012).

- Staffing/schedules → staffing
- Communication/facilities → communication
- Standardization → documentation

	Always/ Strongly Agree	Most of the Time/Agree	Sometimes /Neutral	Rarely/ Disagree	Never/ Strongly Disagree	Don't Know/Not Applicable
Clinical staff submits written reports of near-miss incidents	☐	☐	☐	☐	☐	☐

Figure 3–1. Sample question and responses from the Safety Profile Assessment (SPA) on-line assessment tool.

- Address errors and near-misses → incident learning
- Credentialing and training → training

At first pass, the agreement does not seem that good—only five out of 14 topics appear on both lists. However, looking deeper into *Safety is No Accident*, we find that all the topics identified in Dunscombe's list appear somewhere in the ASTRO document.

Returning to our distinction between infrastructure, equipment, and process issues, we see again that the *Safety is No Accident* list emphasizes safety infrastructure. Elsewhere in the document, QA and QC receive detailed treatment, but not in this list.

The importance of an effective safety culture has not missed the attention of the regulators, either. The U.S. Nuclear Regulatory Commission (www.nrc.gov) has suggested a non-exhaustive list of traits characteristic of a safety culture. While the NRC's list of nine items was not developed for inspection purposes, its existence is additional confirmation that the culture of an organization does impact the safety of its operations.

3.3.3 Safety Assessment Profile

We continue to be bombarded with suggestions and recommendations for enhancing the safety and quality of radiotherapy in our clinics. Presumably, if we adopt all or, more likely, just some of these recommendations, how do we know if it is making a difference? This question is all the more important in an environment of stretched resources. The AAPM's Working Group on the Prevention of Errors is in the process of developing a tool that will hopefully answer this question at the local level. The Safety Profile Assessment (SPA) is an on-line assessment tool with 92 statements, against which a clinic can assess its performance. An example of the SPA's statements and their response options is shown in Figure 3–1.

To provide a logical format for its practical use, SPA has been divided into four major sections: 1) Institutional Culture, 2) Quality Management, 3) Managing Change and Innovation, and 4) Clinical Performance. The fourth section, Clinical Performance, is further subdivided into the major process steps as identified previously (Ford et al. 2012). A text comment box is associated with each of the 92 questions. It is known that inadequate documentation is a major source of incidents in radiotherapy (Clark et al. 2010). In the section on Clinical Performance, the assessor is asked, for each of the 50 statements, whether or not a formal policy exists at his or her facility.

The output of the tool is the facility's overall assessment of its performance in the 92 categories using the five-point Likert scale (plus don't know/not applicable) shown in the example above. Performance can be illustrated using a bar or pie chart, and performance can also be rated against the average performance of all the facilities participating in the on-line assessment. SPA will help answer the critical question posed above by allowing users to track their safety profile over time as quality and safety improvement strategies, based on the many recommendations available, are implemented.

In summary, there is a huge amount of safety guidance in the recently published literature. We have witnessed an expansion of recommendations beyond equipment-specific quality control into matters of clinical processes and organizational culture. We are still left with the practical problem of which of these many recommendations warrant the commitment of significant resources. It is possible that tools like the Safety Profile Assessment tool will, over time, help us to not only enhance quality and safety in our own clinics, but also collate our collective experience with the variety of safety initiatives recommended in the burgeoning literature on the subject.

3.4 Nature of Actions

3.4.1 Arenas of Human Actions

Prevention of human failures requires an understanding of how humans succeed. Failure comes when some part of the process of successful action does not happen as expected. Attempts to understand how humans function go back to ancient times. With the increased understanding of the psychology behind activity, modern models provide many frameworks to identify where actions go awry. One must always bear in mind that models explain some aspects of a process and may be useful in specific applications; models are not intended be true descriptions. Reason discusses models for human activities (Reason 1990). One model that is particularly useful in assessing human performance was developed by Rasmussen (1982). His model notes a series of actions that take place in order to achieve a result, with actions falling into the following arenas:

1. *Observation* that an action is needed or desired
2. *Identification* of the problem or situation to address
3. *Interpretation* of the situation and its environment
4. *Definition* of the task to address the situation
5. *Execution* of the task

To these, others would add *monitoring* of the execution, either during the performance or sometime after completion. While listed as "observation," the cue need not be visual, of course, just brought to the consciousness of the person performing the activity (the *agent.*) One limitation of this model is how reactive it appears, that is, actions only arise from some observational cue of the need for that activity. In medicine, this tends not to be a great limitation since almost all activities are intended to address some problem.

3.4.2 Levels of Actions

While these steps lead the agent through the activity, the nature of the activities themselves fall into three levels:

A. Skill-based, which constitute practiced actions. Humans acquire skill through repetition of the actions, such as learning the lines to a play or playing an instrument. Common actions, such as tightening a screw, are learned skills, as can be seen watching a young child trying to perform the action. With enough practice, the agent becomes confident in his or her ability to correctly execute the skill, often to the point where very little thought goes toward the performance. For most people, driving a car becomes such a skill (sometimes with too little attention paid to the execution). There is no substitute for practice when developing a skill.

B. Rule-based, where the actions follow sets of instructions or rules. Many actions may be common, but not so commonly done that one becomes skillful at performing them. For some activities that occur periodically but not frequently, the agent may have to look up the rule—for example, the instructions for setting the kitchen range's clock with a change to daylight savings time. For other situations, the instructions may be known, but the agent needs to concentrate while working though the steps.

C. Knowledge-based, where the required actions are not readily apparent, and the agent must think about the problem. During knowledge-based actions, the agent must sort through everything he or she knows and identify items relevant to the problem. From this background information, the actions to address the problem must be crafted.

Faced with a problem, humans prefer to perform skill-based activities, if appropriate. A skill, by its nature, produces predictable and usually positive results. If it were otherwise, no one would bother with that "skill." If a known skill does not seem to fit a problem encountered, the next choice usually is to try a rule-based approach. This requires either recalling the rule or knowing where to find the rule. In finding the rule, sorting through possible rules and selecting the correct one is often a challenge. Of course, the further from one's experience the rule lies, the less confidence one can have that the rule is correct. Often, the rule includes steps that entail a skill. An example may be measuring the output of an accelerator using the protocol of Task Group 51 (Almond 1999). Many medical physicists need to follow the protocol steps, which would be a rule-based activity, but when it comes to the parts of the protocol like making measurements with an ionization chamber, the actions become automatic, or skill based. On the other hand, a frequently applied rule can become a skill.

If searching for a rule comes up with nothing, the last resort becomes a knowledge-based approach. In this level, actions must be determined either from what the agent knows or can learn in time. Rules usually have been developed to provide

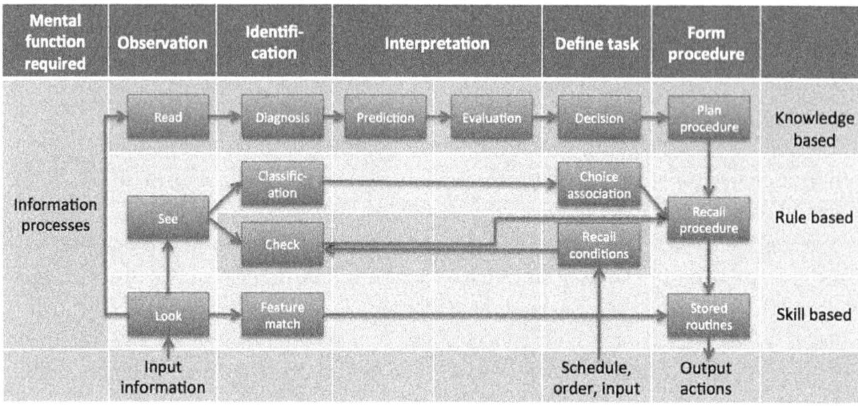

Figure 3–2. A diagram showing a model of the process by which humans perform actions. Based on the work of Rasmussen. (Rasmussen 1982)

some protection against mistakes. Actions designed on the spot seldom have such depth. The likelihood that the knowledge-based solution will adequately solve the problem falls much below that of a rule-based solution, and further still below one based on skill.

Both the arenas of actions required for achieving a result and the levels just discussed are different and independent ways of modeling what happens during an activity. Together, they form a matrix giving more insight into what is happening during some process. Figure 3–2 shows such a matrix, based on Rasmussen (1982).

3.4.3 Performance-shaping Factors

While an agent's ability and knowledge strongly affects the outcome of a process, other factors may also be involved. These performance-shaping factors (PSFs) influence the agent's actions or in other ways interact to affect the outcome. Examples include, but are definitely not limited to the following six items.

1. *Environmental factors.* The environment establishes the playing field for the agent. Lighting, for example, could limit what the agent can see, read, or distinguish; noise could prevent hearing important information; the size and position of a table might facilitate or prohibit resting materials near where they will be used.
2. *Resources.* Resources fall into physical resources and staffing resources. Having the right tools for a job improves one's probability of success and reduces the time required to do the job. Having adequate staff to allow an agent time to think through a process also increases the likelihood of success.
3. *Personal factors.* Events and situations in the agent's life can also affect performance. Concerns about family or finances, for example, can intrude

on the thought process during the performance of a task, creating a situation where attention wanders. Fatigue also compromises the agent's abilities (and this may be related to resources, or lack thereof).

4. *History.* The activities performed earlier can affect the job at hand for several reasons. Similarities between an earlier task and the current one become highlighted by their proximity in time and location. If a previous task failed when following the appropriate rule, the agent may avoid the rule for the current task, even though the rule may be the best approach most of the time.
5. *Equipment performance.* Equipment affects human performance. Equipment or instrumentation that fails to perform as it is supposed to increases the stress felt by the operator, compromising one's ability to make judgments or remember steps. At worst, instrumentation errors can lead the operator directly to errors, as seen in several high-profile radiotherapy events (see Section 3.2).
6. *Timing.* Timing plays a major factor in our lives. Like all creatures on the Earth, we have evolved such that the pattern of our lives follow, in some way, that of the sun. While some who work nighttime shifts adapt to the reversal of day and night, most humans function best working during the day and resting at night. Occasional changes often compromise abilities. The worst situation is a mixed schedule of day and night work, where the body and mind have little opportunity to adapt to a diurnal schedule. Similarly, while not evolutionary, members of many cultures have become accustomed to a workweek with a pattern of having particular days off. Breaking this pattern, for example missing a weekend, can leave workers tired and lethargic, and not at their best the following week. Toward the end of the week, anticipation of the weekend can lead to hurrying to wrap up business and to possibly caring less about performing tasks as well as one otherwise would. Similarly, the anticipation of a holiday or vacation can capture a worker's attention. Working on a holiday presents additional challenges because the supporting staff is likely to be thin.

3.5 Nature of Failures, Events, and Terminology

The purpose of this section is to 1) discuss terminology and define what constitutes a failure, incident, or error; 2) describe some of the terminology used to classify failures; and 3) examine the relevance and value of near-failures and hazardous conditions. It should be noted that the discussion presented here relies rather heavily on the work of Reason and Hobbs. Their work provides a good basis for understanding how systems fail, and their book is a valuable read (Reason and Hobbs 2003).

In order to discuss failures, incidents, and errors in a meaningful way, we need to define what we mean by these terms. This is more difficult than it should be, since different institutions use these terms to mean different things. Some definitions refer to actual failures and incidents only, while others include potential failures and incidents. Let's start by looking at some definitions used by well-known organizations. Where helpful, the distinction between definitions that include *actual or potential failures and incidents* are indicated in italicized brackets.

ESTRO's Radiation Oncology Safety Information System (ROSIS) simply defines an incident as the incorrect delivery of radiation (*actual only*) (ROSIS 2013). According to the IAEA safety standards, an incident is defined as: "Any unintended event, including operating errors, equipment failures, initiating events, accident precursors, near misses or other mishaps, or unauthorized act, malicious or non-malicious, the consequences or potential consequences of which are not negligible from the point of view of protection or safety (actual and potential)." (IAEA 2007)

The WHO World Alliance for Patient Safety general patient safety taxonomy contained within the International Classification for Patient Safety uses the following definitions (WHO 2009):

- A patient safety incident is an event or circumstance that could have resulted, or did result, in unnecessary harm to a patient (*actual and potential*).
- An adverse event is an incident that results in harm to a patient (*actual only*).
- An error is a failure to carry out a planned action as intended or application of an incorrect plan, and may be manifested by doing the wrong thing (an error of commission) or by failing to do the right thing (an error of omission), at either the planning or execution phase (*actual only*).

Cooke et al. adopted the following definition of an error: A failure to complete a planned action as it was intended or a situation in which an incorrect plan is used in an attempt to achieve a given aim (Cooke 2005).

Perhaps a more effective approach to defining failures, incidents, and errors is to first define what is meant by quality in health care. The Institute of Medicine defines quality in health care as follows—"quality care is safe, effective, patient centered, timely, efficient, and equitable" (Institute of Medicine 2001). In the context of this section then, we'll take the terms failure, incident, and error to mean any deviation from a process path that compromises, or has the potential to compromise, quality in health care, as defined above (*actual and potential*). This broad definition enables us to move beyond very specific definitions for these terms, allowing us to concentrate on failure categorizations. These categorizations are valuable because they provide insight that can help us devise mechanisms aimed at reducing the likelihood that the failure occurs or re-occurs.

Reasons and Hobbs define three different failure categories—errors, mistakes, and violations—and define an error as "…the failure of planned actions to achieve their desired goal, where this occurs without some unforeseeable or chance intervention" (Reasons and Hobbs 2003). The latter part of this definition is interesting, implying that the occurrence of an error is dependent upon there being some reasonable possibility of prevention. It also implies that errors are distinctly different from *sporadic*, chance occurrences. All of this is very positive, because it means that errors, by definition, are preventable, or at least that the likelihood of their occurrence can be reduced. This fits nicely with what we are attempting to do in this series—to learn about the various tools that will allow us to minimize the likelihood of errors.

This discussion also leads to another error classification scheme—sporadic versus systematic. Sporadic errors occur once and are not likely to occur again in the same place under the same circumstances, whereas systematic errors will re-occur in the same place under the same circumstances. Systematic errors have the potential to affect many patients.

If we try to reduce the likelihood of errors, it is useful for us to ask the question—what are the most common root causes for errors? Though there are infinitely many possible root causes, it is generally accepted that at least 60% of failures within a system are due to human factors (Woods 1991, Phillips 1994). Given this, let's take a closer look at human factors by returning to Rasmussen's categories of human performance that were introduced above. Once we've described Rasmussen's three categories of human performance, we'll then try to map them to Reason and Hobbs' failure categories (Reason and Hobbs 2003).

Below are some examples of failures based on Jens Rasmussen's three categories of human performance (see Section 3.4.2) (Rasmussen 1982):

Skill-based. Applies to straightforward, routine tasks that have been performed for some time. May or may not include checks along the way. Documentation rarely needs to be referred to and skill can be increased through repetition. Examples of predominantly skill-based activities for the experienced practitioner are:

- morning warm-up on a machine
- physics assistant monthly linac QC
- taking a general medical history

Rule-based. Applies to more complex or critical tasks that may be only occasionally performed. Documentation (procedures, instruction manuals, protocols) generally needs to be readily accessible, although rules can also be handed down through tradition. Examples of predominantly rule-based activities might include:

- adjusting lasers
- working up a patient for a clinical trial
- radiation safety

Knowledge-based. Applies to dealing with unfamiliar or unprepared-for tasks. The "rules" have to be made up and the "skills" developed during performance of the task. Prior specific documentation is not available. Uses more education than training. Examples of predominantly knowledge-based activities are:

- deciding what to do if the marks don't fit
- commissioning a new treatment technique for TBI
- contouring on 4D-CT

Reason and Hobbs define three primary failure types—errors, mistakes, and violations (Reason and Hobbs 2003). Having briefly described Rasmussen's categories of human performance, let's try to relate these to Reason and Hobbs' failure classifications.

Skill-based (Errors). There are three types of skill-based failures: 1) Recognition failures, which include misidentification (mistaking a mole for a tattoo) and non-detection (failure to observe a metastasis on a CT scan); 2) Slips, in which a step is missed in a frequently performed routine activity (failing to set the electrometer to zero between readings); and 3) Memory lapses, or forgetfulness (leaving the patient chart in the treatment room).

Rule-based (Errors/Mistakes/Violations). A rule-based error could be the misapplication of a good rule (doing an SSD calculation for an SAD patient). A rule-based mistake could be the application of a bad rule, maybe following a traditional, undocumented process (completing the prescription sheet after the oncologist has signed it). There are three types of rule-based violations: 1) Routine violations, which can include taking shortcuts that are not in the written procedure (not checking the patient's ID properly before taking them into the treatment room); 2) Situational violations, which can be perceived as a pragmatic approach to getting the job done (not doing the full morning check so as not to delay patient treatments); and 3) Thrill seeking violations, or taking a risk for the sake of it (uncommon in radiation therapy).

Knowledge-based (Mistakes). These occur when individuals tackle unfamiliar problems, like commissioning a new TBI technique or prescribing to a new (for you) tumor site.

The ability to relate failure types to human performance categories will be useful in later chapters that deal with Incident Learning Systems and Root Cause Analysis. These are also useful in the latter stages of FMEA, in which process changes and QA strategies that reduce the likelihood of error are developed.

In the above discussion, we've assumed that a failure, incident, or error has occurred, and that we are now trying to learn from it. In doing so, we've missed some of the best learning opportunities that arise in the clinic—near-failures, otherwise known as "free lessons." Here, a failure, incident, or error is detected before it affects the patient. Some part of the process has gone awry, but another has functioned as planned and prevented the failure from affecting the patient.

The IAEA and the WHO offer the following definitions for near-failures. A "near miss" is defined as: a potential significant event that could have occurred as the consequence of a sequence of actual occurrences but did not occur owing to the plant conditions prevailing at the time (IAEA 2007). A near miss is an incident that did not cause harm (also known as a close call) (WHO 2009).

As discussed in later chapters, failures are almost never the result of single events. Rather, they are a culmination of small events that combine to cause negative effects. Weick et al. discuss the following in relation to this concept. "In the absence of this understanding, people must wait until some crisis actually occurs before they can diagnose a problem, rather than be in a position to detect a potential problem before it emerges. To anticipate and forestall disasters is to understand regulations in the ways small events can combine to have disproportionally large effects" (Weick 1991).

By examining and analyzing near-failures, we can work proactively to reduce the likelihood that a failure occurs, without having to wait for staff or patients to suffer negative consequences.

3.6 Taxonomies

3.6.1 Definition of Failure Taxonomy and the Rationale for Its Importance

A taxonomy is just a classification system. While any method of classification counts as a taxonomy, in a scientific sense, the classification system should reflect an organizational structure and be hierarchical. In the context of patient quality and safety, the taxonomy also needs to be useful. The usefulness of a taxonomy comes when trying to design steps to prevent errors, either proactively by following a risk assessment while establishing a procedure, or remedially, by fixing a procedure after problems have been noted. Effective solutions to a problem most often depend on the causes behind the problem. Take, for example, a therapist failing to place a bolus on a patient for a treatment. The cause of this failure may have been that the therapist failed to look in the directions for the treatment, the radiation oncologist neglected to write an order to place the bolus, or the therapist thought that the other therapist working on the patient placed the bolus. Certainly, other possible causes exist, such as being interrupted during patient setup and forgetting that step. Actions to take to prevent this failure from recurring may depend on the reason for the failure. Making a checklist for the radiation oncologists to remind them to order a bolus when needed would have no effect if one therapist thought that the other therapist placed the bolus, and neither would placing a stop in the treatment unit computer with a reminder. Classifying the failure provides guidance for selecting tools that will have the best likelihood of preventing particular failure types. For a more complete discussion of taxonomies, see Thomadsen (2012).

3.6.2 Taxonomic Organization

The utility of the taxonomy depends on its organization—how it groups similar things together or separates things by relevant categories. Any organizational structure for a failure taxonomy will have some arbitrariness since failures often have many characteristics. Sometimes the organization of a taxonomy includes the method of its use. In this section we will consider some very different taxonomies, all of which have proven useful in health care.

3.6.2.1 Lists

The simplest classification systems use lists of possible failures. Sometimes they are grouped into larger categories (New York State Department of Health 2008) and sometimes not (Wright et al. 1991, Busse and Wright 2000, Benner et al. 2002,

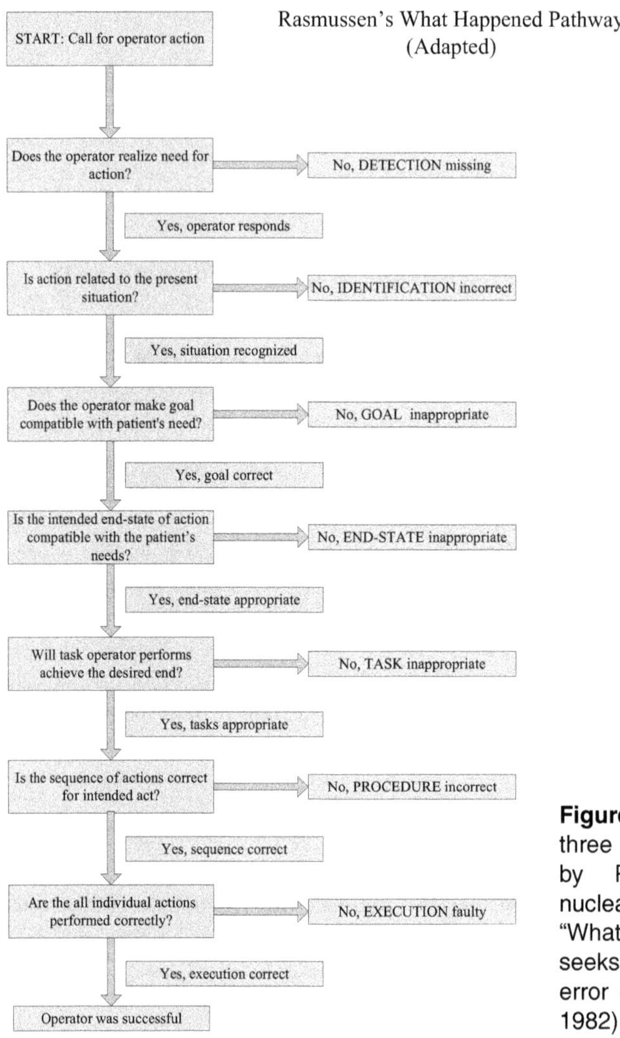

Figure 3–3a. The first of three taxonomies developed by Rasmussen for the nuclear power industry, the "What Happened" pathway seeks to determine where an error occurred. (Rasmussen 1982)

Busse and Holland 2002). Each of the lists referenced have proven useful in identifying common problems or showing trends. Most show a mix of quite specific errors and, at equivalent levels, general factors.

3.6.2.2 Multidimensional Taxonomies

Many taxonomies consider the various facets of failures as separate, possibly independent, dimensions. Often, the taxonomy method of use directs its organization. Some examples help understand the relationship.

Rasmussen's Failure Taxonomy. Rasmussen developed a taxonomy for the nuclear power industry (Rasmussen 1982). This taxonomy considers three facets of

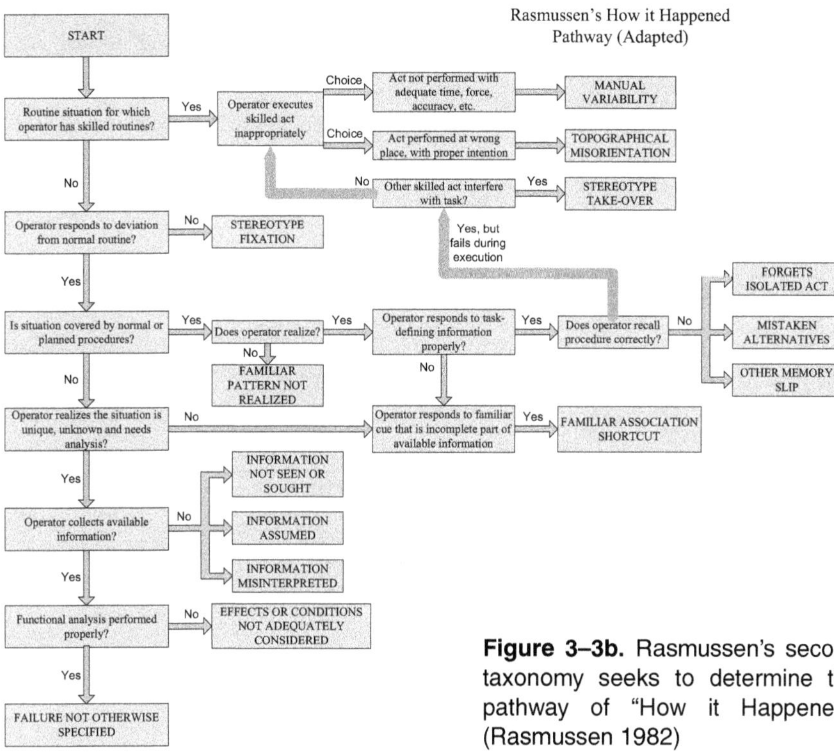

Figure 3–3b. Rasmussen's second taxonomy seeks to determine the pathway of "How it Happened." (Rasmussen 1982)

the human failure related to an event: where in the steps of performing an action the failure occurred (related to Figure 3–3a, referred to as "what happened"), what was the nature of the failure ("how it happened"), and what led the human to the failure ("why it happened"). Figure 3–3a shows the process to classify an event according to the first of the tree facets or dimensions, beginning at "Start," and following the arrows through the decisions. At each step, the analyst is asked a question about the performance of the person involved with the failure, stepping through, in order, the required steps to accomplish an action. If the answer indicates that the person succeeded at that step, the analysis proceeds to the next step, otherwise, the failure at that step classifies the nature of the failure by which step in the process the person failed to perform correctly. The figure shows the classifications in capitals.

Determining the second dimension becomes a little more involved. Figure 3–3b shows the pathway. Again, the analyst is led by a series of questions with branchings for the answers. This dimension addresses the psychology behind the failure. Some of the classifications will be obvious for a medical physicist, such as any of the classifications dealing with a lack of information or forgetting parts of a procedure. Other classifications use terms that may not be familiar:

Figure 3–3c. Rasmussen's "Why It Happened" pathway. (Rasmussen 1982).

Manual variability. Simply the failure to perform a skill adequately, such as not tightening the locking screw sufficiently that holds a Farmer chamber in place in a water scanner.

Topographic misorientation. Performing the intended act in the wrong location, for example, turning the wrong screw when adjusting the output of a linear accelerator.

Stereotype take-over. An incredibly common failure mode where the intended actions get preempted by more common actions that begin the same way. The classic example is intending to stop at a store to buy milk on the way home from work. On getting into the car, pulling out of the parking lot at work, and starting on the road toward home, the next conscious thought is being at home without the milk. Driving home often becomes so automatic that, once begun, can take over the less common intended action.

Stereotype fixation. Another very common failure mode, this is when an operator notices that something is unusual, but interprets the situation in a way that seems normal. An example of such a failure from the external-beam event in Section 3.2

where the MLC failed to shape the IMRT fields was when the therapists noticed that the on-screen cartoon did not show MLC movement, but assumed that that was due to the new software. In creating that model, everything seemed normal.

Familiar pattern not recognized. This occurs when an operator does not recognize that the situation is one covered during training. An example would be not taking the appropriate actions if an HDR source breaks free from the cable during a patient's treatment.

Familiar association shortcut. After receiving only partial information, a person erroneously jumps to a conclusion about the situation because the partial information mirrors a better-known situation.

The final dimension is why the person failed, determined using the pathway in Figure 3–3c. This set of classifications seems less well developed than the other two, and reflects again that the setting for which this was determined assumed a human reacting to a machine failure. As such, there are no classifications for autonomous, self-generated failures, something that occurs frequently in medicine. In this classification pathway, spontaneous human variability is similar to manual variability in the last pathway.

This taxonomy reflects the bias mentioned above—that the model for action assumes that something happens to which the operator must respond. Also, as noted before, while this is common in medicine as well as in a nuclear power plants, in some situations the operator begins the action. This taxonomy also assumes that equipment failure is not the real cause for an event. Equipment will fail, but in environments such as a nuclear power plant, any equipment failure will have back-up equipment that takes over or procedures for the operator to address a failure. As such, any system failure is a human failure. Medicine is just beginning to recognize that all critical equipment needs backups or fail-safe design so that when equipment fails, it is human actions that make the difference between an event and a near event.

The SMART taxonomy

Van der Schaaf (1992) reflected the importance of human intervention in a taxonomy originally developed for the petrochemical industry, but which was later modified and applied to medical settings.

Figure 3–4 shows the overview of his taxonomy, where the origin of a developing event may be from a technical, human, or organization failure, or often a combination. As the potential event unfolds, if the system has adequate quality defenses, the situation may resolve itself. If the defenses are weak, the event may progress until the incident becomes a near event. If the human recovery fails, the incident may result in an actual event.

The pathway for the medical version of van der Schaaf's taxonomy, SMART, is shown in Figure 3–5 (van der Schaaf 1992, Anderson 1997, van Vuuren et al. 1997, van der Schaaf 2005). Unlike Rasmussen's taxonomy, this taxonomy is divided into systemic units rather than performance characteristics.

To determine the classification for a failure, again, begin at "Start." The first choice is whether there was a technical failure. If there was, take the path to the right to determine the classification; if not, move down to the next level. If it is not clear, go along the technical failure pathway, since if no technical classification if found, the pathway automatically moves to the next level anyway. The first possible classification is "External," for example, if the facility lost power from the utility. That is followed by "Design," "Construction," and "Materials." Each of these is a way that the technology could fail. If one applies, that would be a classification.

The next sub-pathway is for organizational failures. Chapter 2 discussed radiotherapy as a system, and it should be clear that while almost all events have a human failure component, all events are organizational failures because the organization should have provided protection from all potential failures. This sub-pathway tries to classify the nature of the organizational failure. The possibilities are "External," Knowledge transfer" (which would be failing to provide initial training or failure to keep training up to date), "Protocols" (or rather, not having appropriate procedures in place), "Management priorities" (which usually involves failure to provide adequate resources), and "Culture" (also as discussed in Chapter 2).

Human failures constitute the next sub-pathway. After the "External" classification come failures in the knowledge-based domain, which would correspond mostly to mistakes. In this domain, the only classification can be a knowledge failure. Moving to the rule-based subpathway, one might guess from the number of classifications that many events result from failures in the rule-based domain, and that would be correct. Not a lot of work in radiotherapy occurs in the knowledge

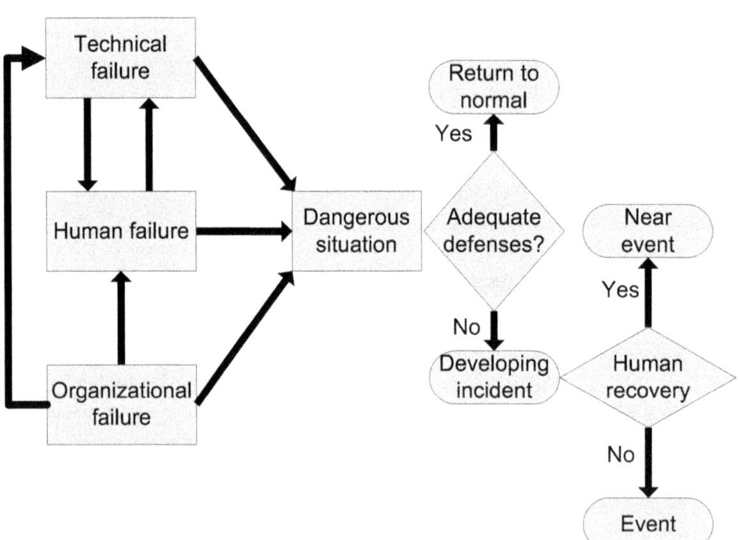

Figure 3–4. The SMART Incident Taxonomy was originally developed for the petrochemical industry and later applied to health care. (van der Schaaf 1992)

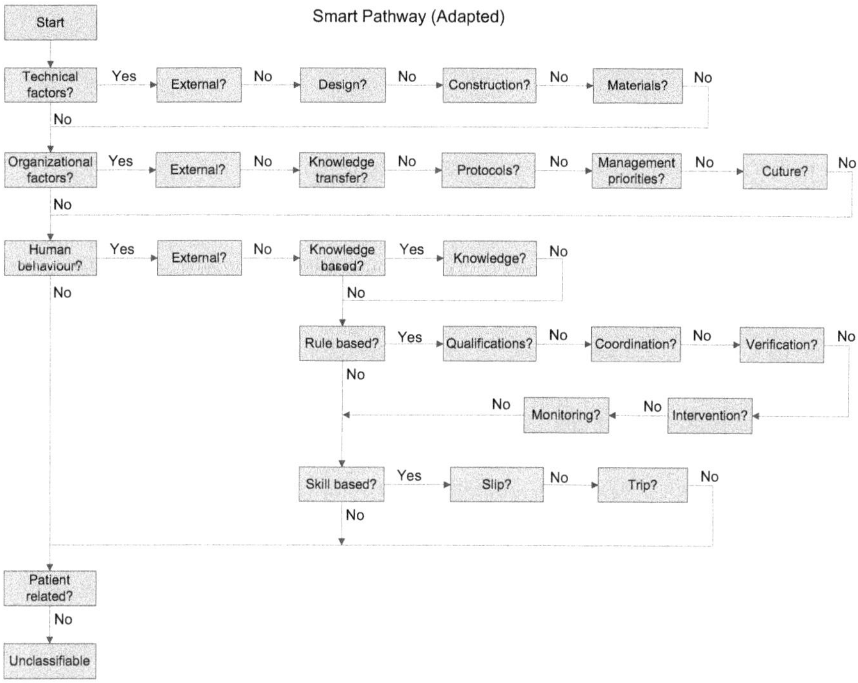

Figure 3–5. The SMART taxonomy pathway as adapted to health care. (van der Schaaf 1992)

domain, and that which does can usually be performed slowly and carefully. On the other hand, actions in the skill-based domain, as noted in Section 3.4, tend toward reliability. Urgent activities often fall into the rule-based domain, with little time for reference and without the assuredness of skill.

Some of the terminology used in the classifications in the rule-based sub-pathway requires explanation:

- Qualifications—that the person involved has had appropriate training and experience for the task
- Coordination—actions between the principals are organized and communications effective
- Verification—checking to assure quality, as typical in radiotherapy as conventional quality assurance and quality control
- Intervention—the execution of the activity
- Monitoring—following the progress of the activity

The last sub-pathway, for skill-based failures, has only errors that occur due to "Slips," mostly omissions or mistaken fine-motor actions, and "Trips," which would be large-motor failures.

The taxonomy includes "Patient related" as a classification. This is a controversial classification, since many in the field would assert that the system should anticipate inappropriate patient actions and prepare for that in the system.

The use of the taxonomy dictates its organization. The levels are set with the technical failures first because they would be the easiest to solve. Equipment can be fixed or redesigned, and even if there are organizational or human failures involved, removing the technical factors would prevent the potential failures. If there are no technical failures, organizational factors come next, because addressing an organizational issue could correct a situation that might lead to multiple events in the future. Only after failing to find technical or organizational issues to address does the analysis look toward human failures, because correcting the human failure would likely produce very narrow effects.

The organization facilitates finding the *one* classification for the potential failure mode that would lead to the most efficacious solution. As conceived, each event or potential failure would have just one classification. An alternative would be to go through the pathway and note any classification that would bear upon the failure. This approach follows the concept that intercepting any of the causes for failure is likely to prevent the event. Both approaches have their strengths, and recording all the relevant classifications—but accentuating the first encountered—can maximize the benefits.

Norman's Failure Taxonomy

Norman (1981) studied reasons for verbal slips and developed a list of reasons behind the slips, shown in Table 3.3 along with their hierarchical organization. The table shown updates some of the terminology from the original. The taxonomy applies as well to actions as to verbal slips, and the terminology has been incorporated into many subsequent taxonomies.

The Alberta Radiotherapy Taxonomy

- Cooke et al. (2006) produced a taxonomy specifically targeted at classifying events in radiotherapy. They based the taxonomy around the process for radiotherapy, particularly patient assessment, prescription, preparation (including treatment planning and verification), treatment, and follow up. The process circles back on itself with potential reassessment for further treatment. Causes for events fall into major categories of:
- Job factors—with lists of possible causes for failures dealing with standards, procedures or practices; materials, tools or equipment; and design
- Systemic or management factors—including possibilities grouped by planning, communication, and knowledge or skill

Table 3–3. Norman's failure taxonomy (Norman 1981)

Failures in the formation of intention
Mistakes: failures in determining goals and other aspects of intention.
Mode errors: erroneous classification of the situation.
Description errors: ambiguous or incomplete specification of intention.
Slips resulting from faulty activation of actions
Unintended activation: when a sequence of actions become activated for extraneous reasons
Capture errors: when a sequence being performed is similar to another, more frequent or better-learned sequence and the latter takes control
Data-driven activation: external events activate a sequence of actions
Associative activation: current sequence of actions activates others, associated sequences
Loss of activation: activated sequences of actions lose activation
Forgetting intention
Misordering the components of a sequence
Skipping steps
Repeating steps
Slips resulting from faulty triggering of a sequence of actions
False triggering: sequence triggered at the wrong time
Spoonerisms: reversal of components
Blends: combinations of components from two competing sequences
Thoughts leading to actions: triggering mean only to be thought
Premature triggering
Failure to trigger: when a sequence never starts due to:
Preemption by competing sequence
Insufficient activation
Failure of triggering condition, being poorly specified, or match with conditions insufficiently close.

- Personal factors—with subcategories covering various possible causes under capabilities or judgment
- Natural factors—such as fires and other disasters

3.6.2.3 Active and Latent Errors

Reason (1990) divides errors into two major classes. The first class is *active error*, which corresponds to some person failing to perform some function adequately. This would include failures in the Human Failure pathway in the SMART taxonomy. Reason's other class is *latent error*. These errors are characterized by conditions in the work environment that lead an individual to make an active error or which permit an active error to propagate into an event. Latent errors are systemic failures, and they would fall along the Organizational Failure pathway of SMART. They are often hard to find and even more difficult to eliminate, but one latent error can lead to many active errors.

3.6.3 Performance-shaping Factors

As discussed in Section 3.4.3, the environment in which an activity takes place influences the outcome of the activity. Some environmental features, such as lighting or background noise, can interfere with communication or a person's ability to simply monitor what is happening. Other factors, such as how long someone has been awake and working, relate to alertness and the ability to make difficult decisions. These factors do not cause failures, but they can increase or decrease the likelihood of success. Such contributing factors are called performance-shaping factors (PSFs). Those PSFs that strongly degrade performance and increase the probability for failure are called enabling factors. A partial list of the most common PSFs include:

- Tangible factors
    - background noise
    - discussions
    - lighting
    - smells
    - cleanliness and clutter
    - isolation
    - milieu
- Intangible factors
    - day of the week
    - time of day
    - holidays
    - personal problems
    - health problems
    - interactions with supervisor

The organization of the facility can also strongly affect individual performance. Some of the organizational enabling factors include the following, based on Thomadsen (2012):

- Knowledge of leader
- Management priorities
    - providing staffing
    - providing equipment
    - providing maintenance
    - providing supplies
    - safety culture (with further divisions)
- Communication system
- Knowledge transfer (training inadequate)
- Matching abilities to job
    - tasks exceed physical ability

- tasks exceed mental ability
- task exceeds experience

- Competition for attention
  - background
    - lack of staffing or time
    - other, competitive goals
  - immediate
    - other, competitive duties
    - lack of staffing
    - other inputs into the system
  - previous
    - mental fatigue due to complex work concern over previous, unaddressed problems

3.6.4 The AAPM's Consensus Recommendations for Incident Learning Database Structure

Given the number and diversity of taxonomies and event databases for radiotherapy reporting (see Chapter 6), the AAPM's Work Group for Error Prevention—with input from several other professional organizations, government agencies, non-governmental advisory agencies, and consultants—generated a taxonomy to provide uniformity and guide the formation of a nationwide radiotherapy event database (Ford et al. 2012). The data elements that a responder would enter are listed in Chapter 6, Appendix E. Behind the information entered by the respondent lies a taxonomy for coding the failures that occurred. The coding combines much from several other taxonomies, notably that of Alberta (Cooke et al. 2006) and Madison (Thomadsen and Lin 2005), the latter having incorporated concepts of Rasmussen (1982), van der Schaaf (1992), Norman (1981), and Reason (1990). The group further built upon the experience of the ROSIS radiotherapy event database. The Alberta system particularly gave guidance by classifying the location of the failure in the radiotherapy process. Table 3.4 gives the causality classifications from the AAPM taxonomy. While quite extensive, this part of the database would be completed by the analyst, not the reporter from the faculty with the event.

3.6.5 Taxonometric Guidance to Interventions

Rasmussen, van der Schaaf, and many others have suggested that the failure classifications can provide guidance for designing interventions. Certainly there are interventions that will not help prevent certain types of failure modes. For example, if someone had been trained in going from the setup notes for an accelerator treatment to placing a bolus on a field, additional training will not prevent omission of the bolus because the therapist did not look at the setup note. While Van der Schaaf provides a coarse grid relating classification to possibly effective interventions, Thomadsen and Lin (2005) and Thomadsen (2012) crafted a much more thorough taxonomy and intervention grid (see Figure 3–6).

Table 3–4. Causality classifications from AAPM Consensus Recommendations (Ford et al. 2012)

Causality classifications
1. Organizational management
a. Planning for program maintenance or expansion
i. Inadequate human resources
1. Staffing not consistent with professional clinical recommendations
2. Staffing not consistent with vendor specs
3. Staffing not consistent with regulations
4. No provision for incremental increases in activities (reports, additional duties, committee participation)
ii. Inadequate capital resources
1. Inadequate budget for equipment
2. Inadequate support/service contracts
3. Inadequate training support
4. Insufficient educational budgets
5. Insufficient IT infrastructure
6. Inappropriate or inadequate equipment
iii. Admin/contractor negotiations for support or staff
b. Policies, procedures, regulations
i. Relevant policy nonexistent
ii. Policy not implemented
iii. Policy inadequate
iv. Policy not followed
v. External regulation not followed
vi. Conflicting policies
c. Training: acquiring and transmitting knowledge and skills
i. Appropriate skills not acquired from facility training
ii. Appropriate skills not acquired from vendor provided training
iii. Inadequate periodic assessment of staff competencies
iv. Lack of continuing education
d. Communication
i. Poor, incomplete, unclear or missing documentation
ii. Inadequate communication patterns designed
iii. Inappropriate or misdirected communication
iv. Failure to request needed information
v. Outside medical records old/incorrect/incomplete/absent
vi. Lack of timeliness
vii. External factors
viii. Verbal instructions not supported by written documentation
e. Physical environment
i. Physical environment inadequate (poor lighting, excessive sound, etc.)
ii. Distracting environment
iii. Interruptions

Table 3–4, cont. Causality classifications

f. Leadership and external issues
i. Inadequate safety culture
ii. Failure to remedy past known shortcomings
iii. Environment not conducive to safety
iv. Hostile work environment
v. Inadequate supervision
vi. Lack of peer review
vii. Leaders not fluent in the discipline
viii. Outdated practices
2. Technical
a. Proper acceptance testing and commissioning of new equipment
i. Not following explicit referral to best-practice documents (AAPM TG reports, ASTRO, ACR, IPEM, COMP, etc.)
ii. Lack of independent review
iii. Lack of review of preexisting reports
iv. Lack of effective documentation (vendor or self)
b. Equipment design and construction issues
i. Inadequate policies and procedures for quality assurance and quality control
ii. Poor human factors engineering
iii. Interoperability problems
iv. Networking problems (IT)
v. Software operation failure
vi. Hardware failure
vii. Poor construction (physical)
c. Equipment maintenance issues
i. Failure to report problems to vendor
ii. Failure to follow vendor notices (field change orders)
iii. Failure to provide adequate preventive maintenance
iv. Failure on the vendor's part to share failure/safety issues in a timely manner
v. Unavailability of local and field support as needed
d. Environment (within the facility)
i. Ergonomics (room layout and equipment setup)
ii. Machine collision issues (room specific)
iii. Environment (water, HVAC, electrical, gas)
iv. IT infrastructure and networking issues (including compliance to expected security and capacity standards)
v. Delay in corrective actions for facility problems
3. Human behavior involving staff
a. Acting outside one's scope of practice
b. Slip causing physical error (failure in performance of highly developed skills as intended or maintained)
c. Poor judgment (e.g., failure to carry out quality control on a patient due to time limitation)
d. Language and comprehension issues

Table 3–4, cont. Causality classifications

e. Intentional rules violations (sabotage/criminal acts, criminal intent, intentional violation)
f. Negligence (risky behavior, poor judgment in failure to address issues or extreme demands, lack of vigilance; recklessness)
4. Patient-related circumstances
a. Misleading representation
b. Cognitive performance issues
c. Non-compliance
d. Language issues and comprehension
e. Patient medical condition (inability to be positioned or remain still)
5. External factors (beyond facility control)
a. Natural environment
b. Hazards
6. Procedural issues
a. Failure to detect a developing problem
i. Environmental masking
ii. Distraction
iii. Loss of attention
iv. Lack of information
b. Failure to interpret the nature of the developing problem
i. Inadequate search
ii. Missing information
iii. Incorrect information
iv. Expectation bias
c. Failure to select the correct rule to address problem
i. Incomplete or faulty rule
ii. Old or invalid rule
iii. Misapplication of a rule
1. Similarity bias/stereotype fixation
2. Familiar pattern not recognized
3. Familiar association short-cut
d. Failure to develop an effective plan
i. Information not seen or sought
ii. Inappropriate assumptions
iii. Information misinterpreted
iv. Side effects not adequately considered
v. Mistaken options
e. Failure to execute the planned action
i. Stereotype take-over/faulty triggering
ii. Plan forgotten in progress
iii. Plan misinterpreted
iv. Plan too complicated (bounded reality)
7. Other

	Physical Tools					Information Tools			Measurement Tools									Computerization			Knowledge Tools		
	Interlocks	Bar Codes	Alarms	Barriers	Communication Devices	Reduce Similarity	Labels	Signs	Check Off Forms	Operational Checks	Comparison with Standards	Increase Monitoring	Automate Monitoring	Add Status Check	Redundant Measurement	Independent Review	Acceptance Test	Computerized Verification	Computerized Order (Data) Entry	Computerized Order Entry with Feedba	Training	Experience	Instruction
Topographic Misorientation (SB)	X																			X			
Stereotype Fixation (SB)	X	X	X	X	X	X	X	X	X	X	X	X	X	X	X	X		X					
Stereotype Takeover (SB)	X		X	X		X	X	X	X		X	X	X	X	?	X		X					
Familiar Association Shortcut (RB)	X		X	X		X			X				X	X	?			?					X
Familiar Pattern not Recognized (RB)		X			X	X	X	X	XX		X		X		X			X				X	X
Mistakes Alternatives (RB)									X							X		?		X	X	X	?
Mistake Consequence (RB)									?							X		?		X	X	X	?
Forget Isolated Act (RB)	X	X	X	X	X		?	?	X		?		?	?	?	X		X				?	
Condition or Side Effect not Considered (KB)	X		X	X							X	X		X		X		X			?	?	
Information not Sought - Assumed, Negligent Omission, Not Reealized It Was Needed	X	X	X	X	X	?	?	?	X		X			X		X		X	X				
Lack of Vigilance (Arousal, Commitment, Complacency)	X	X	X	X					?			X	X			X		X		X			
Organizational - Knowledge of Leader																X					X	X	
Organizational - Management Priority																X							
Organizational - Communication System					X																		
Organizational - Knowledge Transfer																					X		X
Organizational - Ability																	X				X	X	
Competition for Attention (Background) - Lack of Staff or Time	X		X	?					?				X			?		X		X			
Competition for Attention (Background) - Other Goals	X		X	?					?				X			?		X		X			
Competition for Attention (Immediate) - Other Duties	?		?	?					?				X			X		X		?			
Competition for Attention (Immediate) - Lack of Staff	X		X	?					X				X					X					
Competition for Attention (Immediate) - Too Many Inputs to System	X		X	X		?			X		?		X			X		X		X			
Competition for Attention (Previous) - Fatigue due to Complex Work	X	X	X	X		X	X	X	X	X	X		X	X	X	X		X		X			

Figure 3–6. A small portion of the Thomadsen and Lin matrix. Failure classifications are in the first column, and possible interventions are along the top row. An "x" in the box at the intersection of a classification and an intervention indicates that that intervention could be useful in addressing a failure with the given classification. A blank in the box means there would be little utility in using the indicated intervention. A question mark implies that there is insufficient data to indicate the likely effectiveness of the intervention with the particular failure classification.

3.7 Summary

- While the rate remains low, there continue to be medical events in radiotherapy (Section 3.2).
- Several organizations have published guidance documents to assist facilities in the pursuit of quality and safety (Section 3.3).
- As discussed in the rest of this book, preventing failures from propagating to untoward effects for the patient is best addressed through risk analysis at a given facility.
- The Safety Profile Assessment under development by the AAPM's Work Group on Prevention of Error can help facilities evaluate their safety program (Section 3.3.3).
- Understanding how humans perform activities helps develop strategies to facilitate success. Some models of human actions (Section 3.4) use the following divisions:
    - The arenas of human actions are observation, identification, interpretation, definition, and execution.
    - The levels of human action are skill based, rule based, and knowledge based.
- Consistent and clear terminology helps both in thinking about quality and safety issues and in communications. In this volume:
    - An *incident* is an unwanted or unexpected change from a normal system behavior that causes, or has the potential to cause, an adverse effect to persons or equipment.
    - An *event* is an incident that is intercepted or interrupted but which affects the patient in some undesirable way.
    - A *good catch, close call,* or *near miss* is an incident in which the circumstance or actions were intercepted or interrupted before affecting a patient.
    - A *failure* is not achieving the desired outcome.
    - An *error* is a failure to execute correctly and completely a planned action or activity.
    - A *mistake* is the implementation of a plan unlikely to achieve its intended outcome, even if executed correctly.
    - *Sporadic* errors occur once or in an unpredictable fashion.
    - *Systematic* errors occur predictably under similar circumstances.
    - *Active errors* are errors by humans during actions or activities.
    - *Latent errors* are conditions or policies (or lack thereof) that may lead to active errors, or at least not prevent them.
- Taxonomies help us understand the nature of failures that could happen or have taken place, minimize ambiguity in communications, and provide guidance in selecting tools to prevent similar failures (Section 3.6).

References

Almond, P.R., P.J. Biggs, B.M. Coursey, W.F. Hanson, M.S. Huq, R. Nath, D.W. Rogers. (1999). AAPM's TG–51 protocol for clinical reference dosimetry of high-energy photon and electron beams. *Med Phys* 26:1847–70.

Anderson, T. "A case study and conceptual framework for medical event reporting systems." Master's thesis. University of Wisconsin, 1997.

Baker, G.R., P.G. Norton, V. Flintoft, R. Blais, A. Brown, J. Cox, E. Etchells, W.A. Ghali, P. Hebert, S.R. Majumdar, M. O'Beirne, L. Palacios-Derflingher, R.J. Reid, S. Sheps, and R. Tamblyn. (2004). "The Canadian adverse events study: The incidence of adverse events among hospital patients in Canada." *CMAJ* 170:1678–86.

Benner, P., V. Sheets, P. Uris, K. Malloch, K. Schwed, and D. Jamison. (2002). "Individual, practice , and system causes of errors in nursing: A taxonomy." *J Nursing Admin* 32:45–48.

Bogdanich, W. "Case Studies: When Medical Radiation Goes Awry." *The New York Times*, January 27, 2010.

Bogdanich, W. "Radiation Offers New Cures, and Ways to Do Harm." *The New York Times*, January 23, 2010.

Busse, D., and B. Holland. (2002). "Implementation of Critical Incident Reporting in a Neonatal Intensive Care Unit." *Cognition, Technology & Work* 4:101–106.

Busse, D.K., and D.J. Wright. (2000). "Classification and analysis of incidents in complex medical environments." *Top Health Inf Manage* 20:1–11.

Clark, B.G., R.J. Brown, J.L. Ploquin, A.L. Kind, and L. Grimard. (2010). "The management of radiation treatment error through incident learning." *Radiother Oncol* 95:344–349.

Cooke, D., M. Dubertz, R. Heshmati, S. Iftody, E. McKimmon, J. Powers, R. Lee, and P.B. Dunscombe. *A Reference Guide for Learning from Incidents in Radiation Treatment.* Health Technology Assessment Unit #22, Edmonton, Alberta: Alberta Heritage Foundation for Medical Research, 2006.

Dunscombe, P. (2012). "Recommendations for safer radiotherapy: what's the message?" *Front Oncol* 2:129.

Ford, E.C., L. Fong de Los Santos, T. Pawlicki, S. Sutlief, and P. Dunscombe. (2012). "Consensus recommendations for incident learning database structures in radiation oncology." *Med Phys* 39:7272–90.

Hendee, W.R., and M.G. Herman. (2011). "Improving patient safety in radiation oncology." *Prac Radiat Oncol* 1: 16–21.

Institute of Medicine, Committee on the Quality of Health Care in America. *To Err is Human: Building a Safer Health System.* Washington, DC: National Academy Press, 1999.

Institute of Medicine, Committee on the Quality of Health Care in America. *Crossing the Quality Chasm: A New Health System for the 21st Century.* Washington, DC: National Academy Press, 2001.

International Atomic Energy Agency. *IAEA safety glossary: Terminology used in nuclear safety and radiation protection.* 2007 Ed. Vienna: IAEA. Available at: http://www-pub.iaea.org/MTCD/Publications/PDF/Pub1290_web.pdf.

International Atomic Energy Agency. *Lessons learned from accidental exposures in radiotherapy.* Safety Reports Series 17. Vienna, Austria: IAEA, 2000.

New York State Department of Health. "New York patient occurrence reporting and tracking system" (NYPORTS), Annual reports 2005–2008. Available at http://www.health.ny.gov/facilities/hospital/nyports/.

Norman, D. (1981). "Categorization of action slips." *Psychol Rev* 88: 1–15.

Ostrom, L.T., P. Rathbun, R. Cumberlin, J. Horton, R. Gastorf, and T.J. Leahy. (1996). "Lessons Learned from Investigations of Therapy Misadministration Events." *Int J Radiat Oncol Biol Phys* 34:227–34.

Peiffert, D., J.M. Simon, and F. Eschwege. (2007). "Epinal radiotherapy accident: passed, present, future." *Cancer Radiother* 11:309–312.

Phillips, E.H. (1994). "United's AQP training to stress human factors." *Aviation Week & Space Technol* 140:50.

Rassumsen, J. (1982). "Human errors: A taxonomy for describing human malfunction in industrial installations." *J. Occupational Accidents* 4:311–335.

Reason J. *Human Error.* Cambridge: Cambridge Univesity Press, 1990.

Reason, J. and A. Hobbs. *Managing Maintenance Error: A Practical Guide*. Farnham, Surrey, United Kingdom: Ashgate Publishing, 2003.

ROSIS: Radiation Oncology Safety Information System. Available at http://www.rosis.info. Accessed January 4, 2013.

Scottish Ministers for the Ionizing Radiation. (2006). "Unintended overexposure of patient Lisa Norris during radiotherapy treatment at the Beatson Oncology Centre, Glasgow in January 2006."Available at www.scotland.gov.uk/Resource/Doc/153082/0041158.pdf.

Thomadsen, B. "Medical Failure Taxonomies." In *Handbook of Human Factors and Ergonomics in Health Care and Patient Safety*, 2nd ed., P. Carayon, ed. Boca Raton, FL: Taylor and Francis, 2012.

Thomadsen, B., S.-W. Lin, P. Laemmrich, T. Waller, A. Cheng, B. Caldwell, R. Rankin, and J. Stitt. (2003). "Analysis of treatment delivery errors in brachytherapy using formal risk analysis techniques." *Int J Radiat Oncol Biol Phys* 57:1492–508.

Thomadsen, B., S.-W. Lin. "Taxonomic Guidance for Remedial Actions." In *Advances in Patient Safety: From Research to Implementation*, vol 2., K. Henriksen, J.B. Battles, E. Marks, and D.I. Lewin, eds. Rockville, MD: Agency for Healthcare Research and Quality, 2005.

U.S. Nuclear Regulatory Commission (USNRC). http://www.nrc.gov/reading-rm/doc-collections/event-status/event/. Accessed January 9, 2013.

van der Schaaf, T.W. "Near miss reporting in the chemical process industry." Doctoral thesis. Eindhoven, Netherlands: Eindhoven University of Technology, 1992.

van der Schaaf. T.W. (2005). PRISMA Medical: a Brief Description. (Unpublished).

van Vuuren, W., C.E. Shea, and T.W. van der Schaaf. *The Development of an Incident Analysis Tool for the Medical Field.* Eindhoven, Netherlands: Eindhoven University of Technology, 1997.

Weick, K. "The Vulnerable System: An Analysis of the Tenerife Air Disaster." In *Reframing Organizational Culture*, P.J. Frost, L.F. Moore, M.R. Louis, C.C. Lundberg, and J. Martin, eds. London: Sage Publications, 1991.

Wood, R.H. (1991). "Investigators focus on eliminating errors, not preventing accidents." Aviation Week & Space Tech 135:83–84.

World Health Organization (WHO). International Classification for Patient Safety (ICPS) version 1.1. 2009. Accessed January 4, 2013. Available at http://www.who.int/patientsafety/implementation/taxonomy/icps_technical_report_en.pdf.

Wright, D., S.J. Mackenzie, I. Buchan, C.S. Cairns, and L.E. Price. (1991). "Critical incidents in the intensive therapy unit." *Lancet* 338: 676–678.

Zietman, A.L., J.R. Palta, and M.L. Steinberg. *Safety Is No Accident.* Fairfax, VA: American Society for Radiation Oncology, 2012.

Chapter 4

Risk Assessment Using the TG–100 Methodology

Bruce Thomadsen, Ph.D.[1], Derek Brown, Ph.D.[2], Eric Ford, Ph.D.[3], M. Saiful Huq, Ph.D.[4], and Frank Rath, MSIE[5]

[1]Professor, Departments of Medical Physics, Industrial and Systems Engineering, Engineering Physics, and Biomedical Engineering, University of Wisconsin, Madison, Wisconsin

[2]Adjunct Associate Professor, Departments of Oncology and Physics and Astronomy, University of Calgary, Calgary, Alberta

[3]Associate Professor, Department of Radiation Oncology, University of Washington Medical Center, Seattle, Washington

[4]Professor and Director, Department of Radiation Oncology, Medical Physics Division, University of Pittsburgh Cancer Institute Pittsburgh, Pennsylvania

[5]Director, Department of Engineering Professional Development, University of Wisconsin Madison, Wisconsin

4.1 Introduction

Risk assessment is the process of analyzing the hazards involved in a process. The purpose of risk assessment is to minimize the effects of failures in radiotherapy by being conscious of the risks associated with the system in which it operates.

The term risk is frequently defined as the set of answers to three related questions (American Society of Mechanical Engineers 2002/2003, U.S. Nuclear Regulatory Commission 1998):

1. What can go wrong?
2. How likely is it to go wrong?
3. What are the consequences if it goes wrong?

The first question is addressed by identifying all (or as many as possible) of the potential failure modes—or ways in which a process could fail—for a given process. The second question is answered by developing models that estimate the probability or frequency of occurrence of the failure modes resulting from various causes, such as human factors or errors. The third question is addressed by assessing the consequences of each failure mode on the outcome of the process, assuming that the causes for each failure mode are not detected and corrected. The goal of a quality management program is then to focus on risk assessment for the entire process involved in the treatment of a radiation therapy patient and put control measures in place which will inhibit failures and prevent degradation of treatment quality if there is a failure.

Risk assessments may be *a priori,* before much is known about the process and its potential failures, or *a posteriori,* considering data collected on events that have happened. An *a priori* analysis tends to weight the probability of all failures fairly equally, without reasons to differentiate. "Expert opinions" serve to give some guidance to the probabilities. An *a posteriori* analysis can be much more precise, with probabilities based on gathered data. For radiation therapy processes, the development of probabilistic risk assessment models is a challenging task because the equipment and therapy procedures are highly reliable, and there is a lack of measured data on occurrence and detection probabilities for undesired events. Most radiotherapy risk analyses studies are thus based on expert consensus opinion to subjectively estimate the required probability data.

New approaches are being developed to address the quality management (QM) needs of both existing and emerging technologies in radiation therapy. These are based on risk assessment that consists of analysis of the hazards involved in the entire radiation therapy process (critical or less critical) and accounts not only for the devices used for radiation treatment and the individual steps employed in each process (and their consequences, should they fail), but also for interactions between them. They reflect jointly the probability of a failure occurring and the severity should it occur. Thus, they allow a radiotherapy department to develop a systematic QM program that acknowledges its current level of technology and clinical practice. Task Group 100 (TG 100) of the Therapy Physics Committee of the American Association of Physicists in Medicine (AAPM) has adopted such an approach and developed a risk assessment-based framework for QM activities in radiation therapy [Huq et al. 2008, Huq et al. 2013). This chapter will briefly present the methodology of this approach.

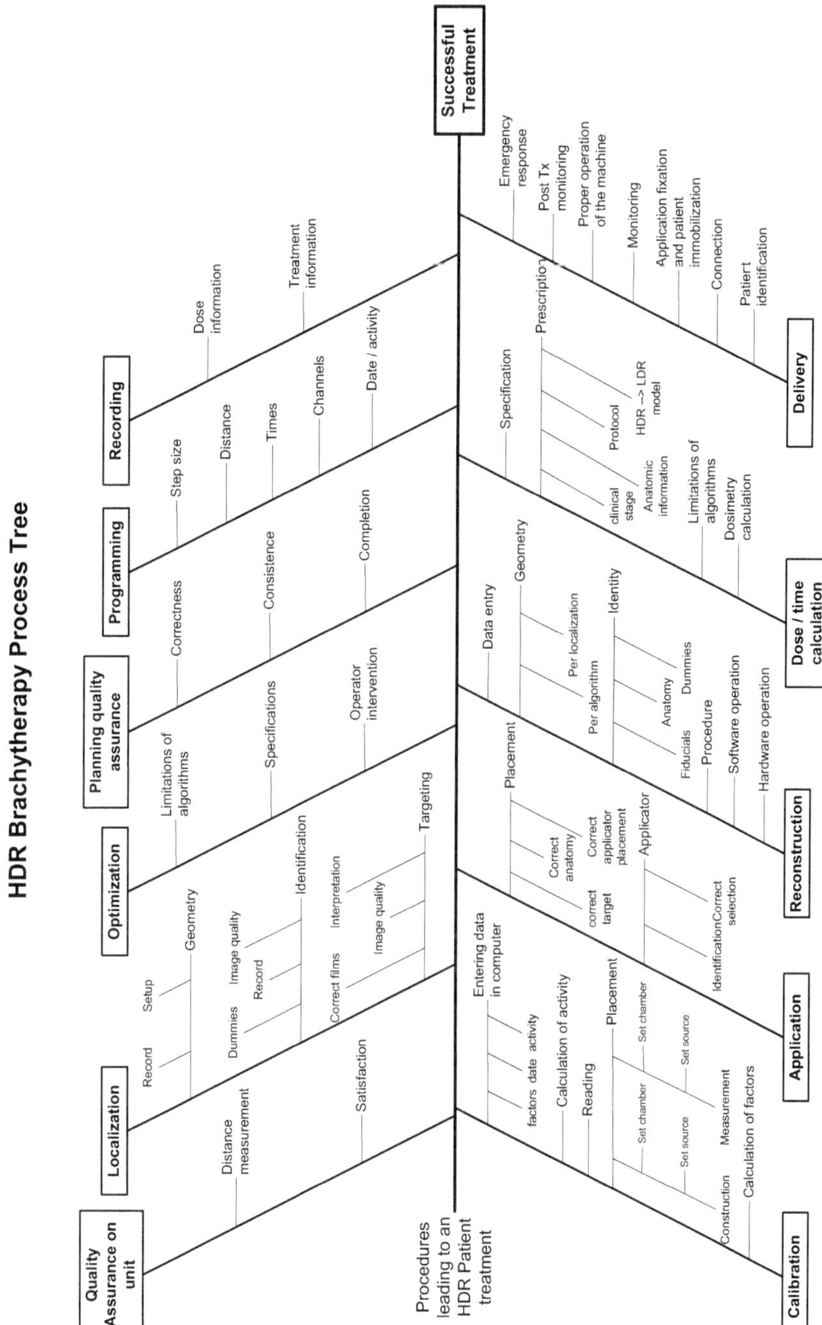

Figure 4–1. Example of one form of a process tree. Figure courtesy of Bruce Thomadsen.

4.2 Risk Assessment and the TG–100 Approach

Although many risk assessment and analysis techniques exist in industry, the framework for quality management in radiation therapy developed by TG 100 is based on the use of three industrial engineering-based tools. These are 1) process mapping, 2) failure modes and effects analysis (FMEA), and 3) fault tree analysis to create a quality management program that will mitigate the most important risks that are identified in the previous analyses. Successful implementation of this methodology will require the participation of a clinic's entire radiation therapy delivery team—consisting of radiation oncologists, medical physicists, dosimetrists, therapists, nurses, engineers, and IT personnel as appropriate—to develop an FMEA-based risk-aware QM program for all clinical processes by contributing to the analysis of process steps and failure modes that involve their work.

The following sub-sections briefly introduce the concepts of process mapping, FMEA, and fault-tree analysis.

4.3 Process mapping

4.3.1 Introduction

The purpose of this section is to 1) define "process map," 2) discuss the reasons for creating and using process maps, 3) present an example map and case scenario, and 4) suggest techniques and tips for creating process maps.

We take the term "process map" (sometimes called a "process flow chart") to mean any visual representation of the steps involved in a process. In doing so, we lump together terms like "process tree," "fishbone diagram," etc.—all different ways of presenting the same information, each with their own benefits and drawbacks. Technically, graphical approaches like fishbone diagrams or affinity diagrams are not really process maps since they are not temporally ordered and do not depict the connections between various steps in the workflow. However, we believe in a practical approach and leave it to the reader to use his or her preferred style of visual representation. An excellent example of a process tree is shown in Figure 4–1. The trunk of the tree describes the process, while the branches depict subprocesses that feed into the main process. There are also branches off of branches. This is a great way to handle detail in a process flow chart, as it is amenable to both high-level and detailed process descriptions.

Process maps offer an excellent starting point for those considering quality and safety improvement initiatives (The Victorian Quality Council 2007). They can be used to shape current clinical practice, as well as facilitate future quality and safety initiatives. Often, the simple act of sitting together to develop the process maps can produce immediate benefits in terms of communication and understanding between professional groups. Process maps also form the basis of other quality and safety tools, including Failure Modes and Effects Analysis (Fore et al. 2009) and Fault Tree Analysis (Ekaette et al. 2007), as well as Incident Learning Systems and Root

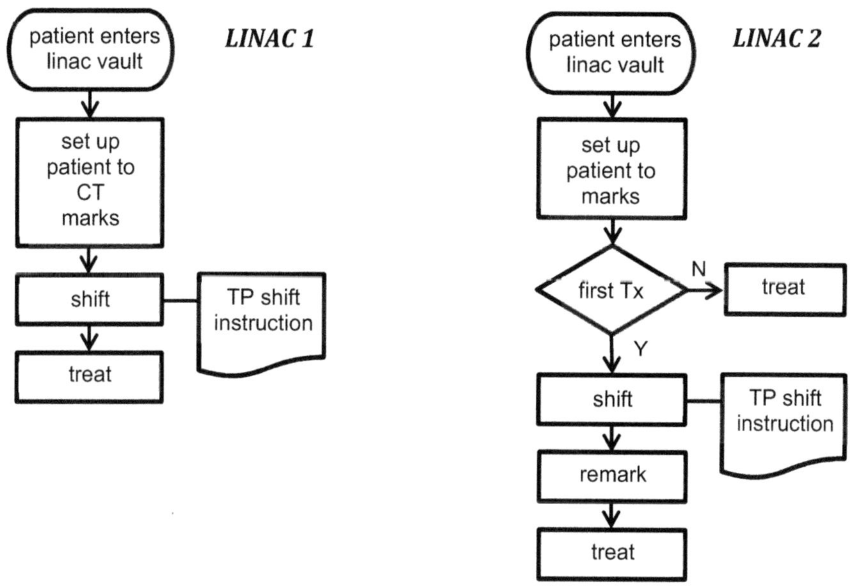

Figure 4–2. General process describing steps taken from the time the patient enters the treatment room until the time the beam is turned on. Different workflow processes are employed on linac #1 (left) vs. linac #2 (right). Figure 4–3 gives explanations for the symbols used in this type of process map.

Cause Analysis (Cooke et al. 2006). All of these often use process descriptions as a starting point.

4.3.2 Example Process Map Scenario

The many benefits of creating and using process maps can be illustrated using a simple, realistic clinical example. Consider the process of patient setup and treatment in an external beam radiation therapy clinic. The workflow is illustrated in Figure 4–2 in a simplified process map. Such a map might be created, for example, as part of a root-cause analysis in the wake of a patient mistreatment incident. The example in Figure 4–2 shows a clinic that has two treatment units (linac #1, left, and linac #2, right) each employing slightly different workflow processes.

To make the example more concrete, we consider two radiation therapists working in a medium-sized, 4-linac clinic: Greg (lead therapist on linac #1) and Marsha (lead therapist on linac #2). Greg on linac #2 follows the process on the right in Figure 4–2: he re-marks patients on the linac table after the first treatment setup. On subsequent treatment setups, he simply sets the patient up to the new marks. On linac #1, Marsha sets up the patient to CT marks and then shifts the patient according the treatment planning instructions for every fraction. The existing documentation on this process at the clinic is poorly written and confusing.

It's a busy day, and Marsha's teammate has called in sick, so Greg is asked to join Marsha on linac #2. Greg and Marsha are in the process of setting the patient up when Marsha is called away to deal with urgent questions from the spouse of another patient. Greg finishes the patient setup using the workflow that he is accustomed to (right panel of Figure 4–2). He sets the patient up to her marks and does not make a further shift. When Marsha returns, Greg has completed the setup and is at the treatment console. Marsha knows that Greg has significant experience, and she assumes he has made the correct patient shifts, but as a courtesy, asks him if everything is ready to go. Greg, always happy to help out, asserts that everything is ready to go. They beam on and treat, not realizing that Greg has not made the shifts that were required.

There is a clear divergence in this process. Two employees working in the same clinic, treating the same types of patients, are following two different processes. A root-cause analysis would reveal the multifactorial causal factors behind this incident, including communication, distraction, and deficiencies in policy and procedure. However, even before an incident occurs, creating process maps like those in Figure 4–2 would demonstrate a hazardous situation: the clinic is operating with diverging workflows.

In summary, process maps are valuable for several reasons:

- Improving communication—different professional groups, and different individuals within those groups, now have the same understanding of the process and how it fits together (providing a clear picture of where hand-offs occur).
- Harmonizing clinical practice and ensuring that everyone operates with a shared model.
- Improving efficiency. Workflow inefficiencies can become obvious when mapped out visually.
- Process maps also allow one to codify where along the workflow incidents are occurring, which is potentially powerful information for incident learning.

The section below suggests a recipe for process mapping and highlights some of the potential pitfalls that occur when creating a process map. In the workshop,

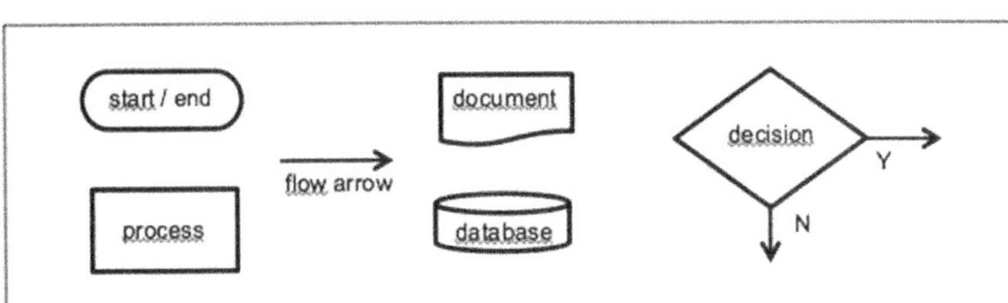

Figure 4–3. A dictionary of process map symbols.

the IMRT treatment planning process will be mapped as an exercise, selected because studies have shown that many errors originate in the treatment planning process (Clark et al. 2010). Here, the various steps involved are discussed.

4.3.3 Techniques and Tips for Creating Process Maps

Step 1: Decide what process to map. The scale of the process is an important concern here. Mapping the entire external beam radiation oncology process, for example, is a large project that could take many weeks. It is beneficial to focus on smaller sub-sections. Often the needs of the analysis dictate the scope of the mapping project.

Step 2: Form a group and identify a team leader. It is vital that all professional groups are represented in this process. This may include administrators and managers, as well as clinical staff.

Step 3: Create an initial process map. It is often useful to make a first draft that does not attempt to capture the entire process in detail, but rather the workflow at a more general level. The standard symbol dictionary outline in Figure 4–3 may be useful. Programs such as Microsoft Office® (Microsoft Corp, Redmond, Washington) or OpenOffice® (The Apache Software Foundation, Forest Hill, Maryland) have these tools built-in. Dedicated software platforms for process mapping are also available, such as Microsoft Visio® and Lucidchart® (Lucid Software, Chenaai, Tamil Nadu, India). However, hand drawing is often easier and sufficient.

Step 4: Iterative mapping. The process map is refined with the input of all staff involved.

Step 5: Check with external resources to make sure that no steps have been missed. Examples here include the incident reporting white paper from the AAPM Work Group on Prevention of Errors, which has a general list of process steps for external beam radiotherapy as well as brachytherapy (Ford et al. 2012). Process mapping is further discussed in TG 100 when it becomes available (Huq et al. 2013).

Step 6: Use the process map. Examples of this use can be found later in this chapter on FMEA analysis.

4.3.4 Tips for Creating Useful Process Maps

1. It is often useful to look at processes from the patient's perspective.
2. For clinical processes, a *multi-professional team* is necessary for the development of a valid map.
3. The number of subprocesses identified should be the *smallest number needed* to meet the objective.
4. The users of the map should have the *same understanding* of the meaning of the subprocesses.
5. Choose the right level of detail. A map that is too general loses its utility, while one that is too detailed becomes unmanageable and staff can lose the big picture.

6. Don't get hung up on fancy graphics. There is value in the *process of creating the map.*

4.3.5 Process-mapping Conclusions

It is important to note that process mapping is less about coming up with "the right" map and more about the process of creating the map—getting different groups to communicate and producing a map that accurately reflects the nuances of a particular clinic. We believe that process maps may find a future utility as a means for clinics to compare themselves to each other. To our knowledge, this has not yet been attempted in any significant way, but it may be a very effective way of communicating and quantifying differences in standard practices.

4.4 Failure Modes and Effects Analysis

4.4.1 Introduction

After outlining a process, the next step is to assess the potential risks involved in that process. FMEA is a technique that is used to perform a quantitative assessment of risks. A good FMEA defines, identifies, and helps to eliminates or intercept known and potential failures (failure modes) and errors from a process before they reach a patient. FMEA assesses the likelihood of failures in each step of a process, identifies the causes and effects of each failure mode, determines the likelihood that these failure modes will not be detected, and assesses how severe a failure's consequences might be on the patient. The overall risk of each identified failure mode is then scored according to a risk priority number (RPN) so these failures can be prioritized. A good FMEA thus facilitates problem follow-up and corrective actions required to prevent failures from reaching the patient, thereby assuring the highest quality and reliability in rendering service to the patient.

For quantitative FMEA analysis, numerical values are assigned to three components for each cause of failure:

- O, for the likelihood of occurrence
- S, for the severity of the effect of failure
- D, for the likelihood that the failure considered would go undetected

TG 100 used numbers between 1 and 10 for each of these indices, with 1 assigned to no appreciable danger and 10 to the most severe. The product of these three indices form the risk priority number, or RPN. The higher the RPN, the more likely the failure mode is to occur, the less likely the failure mode will be detected, and the more severe the results of a failure mode will be.

Each of the significant hazards requires the addition of controls or a quality management step to eliminate the dangers, i.e., bring the hazard rating to an acceptably low value. Fault tree analysis is a useful tool in determining the placement of these controls.

FMEA is a proactive methodology used to systematically evaluate and identify risks or weaknesses in product design or other process and to identify process controls that address them. The origin of FMEA can be traced to the late 1940s to U.S. Military Standard MIL-P-1629 (Department of Defense 1980). During the 1960s, FMEA methodology was used in several NASA programs, including Apollo, to ensure robust designs and processes. In the 1970s, FMEA was adopted by the automotive industry (Automotive Industry Action Group) to better manage quality in product designs and throughout the supply chain. More recently, the use of FMEA in health care appears to have grown significantly, based on my own observations. For further reading on FMEA, see Stamatis and the Veterans Administration guide (Stamatis 1995, U.S. Veterans Administration).

In industry, there are four distinct types of FMEA:

1. *Design FMEA*—used to evaluate a product design process
2. *Process FMEA*—used to evaluate a manufacturing or other type of process
3. *Application FMEA*—used to evaluate the use of a product design in a customer's process
4. *Service FMEA*—used to evaluate the serviceability or maintainability of a product design

Each type of FMEA is independent, focusing only on the design, manufacture, use, or service/maintenance of a specific product. Companies designing equipment or devices should:

- Use Design FMEA to identify design risks that could result in suboptimal performance or hazardous results and take steps to eliminate those risks or mitigate the effects of those risks.
- Use Application FMEA to evaluate the use of their designs by their customers and identify potential problems their customers might experience.
- Use Service FMEA to make sure that their designs can be easily and adequately serviced or maintained.

Manufacturers of equipment or devices should:

- Use Process FMEA to evaluate their production processes to ensure that they are capable of meeting the specified requirements of an equipment or device design.
- Request that suppliers who produce parts, components, sub-systems, and systems use Process FMEA to ensure that their production processes are capable of meeting design requirements/specifications.

Healthcare organizations or hospitals should:

- Use Process FMEA to evaluate treatments, procedures, and processes and to identify specific process steps that are likely to or could produce suboptimal or hazardous outcomes.

FMEA is considered to be a living or continuous process and should be updated whenever any changes occur affecting designs or processes.

4.4.2 Overview of Process FMEA in Health Care

Process FMEA provides a "picture" of a process at the time it was performed, showing the weakest or riskiest areas of the overall process. Those areas are then addressed and more effective quality controls are developed and implemented to improve process performance and overall process quality.

4.4.3 Steps in Completing a Process FMEA in Health Care

A process FEMA should begin with a thorough and complete description of the process being evaluated in the form of a flow chart, process map, or process tree. It should be developed by a team of process experts who actually participate in the process, and it should be led by an experienced facilitator. All opinions of what actually happens during the process should be considered and valued. A visual representation or picture showing the entire process can be very useful. People involved in the analysis (and the process) can see how what they do fits into the process and gain an understanding of what is done upstream and downstream from their part of the process. Sometimes that knowledge and insight can lead to ideas on how to improve a process.

The typical or standard process FMEA form used to guide an analysis is shown in Table 4–1. The form normally has a general information header for identifying who participated in the analysis, when it was done, the revision number of the process FMEA, etc. The style and design of the general information header varies, and it is up to the healthcare organization to decide what information should be included. The analysis begins at the left of the form and progresses to the right until the risk priority number is calculated for each process step/failure mode/cause combination.

Step 1. Identify the major subprocesses. The process being analyzed is broken into major subprocesses. These major subprocesses correspond to steps in the process flow chart or map. The process can be broken into highly detailed subprocesses or larger or more macro subprocesses. Breaking the process into larger, or macro-level subprocesses, reduces the work required to perform the process FMEA and directs the team to the subprocesses that have the highest risk of sub-optimal performance or hazardous results. A more detailed process FMEA can then be performed on those areas with the highest risk.

Step 2. Identify each failure mode for each subprocess step. A failure mode is defined as the way in which a failure occurs or is observed, or the way in which a sub process step can fail to meet its intended purpose. Typical failure modes in radiotherapy are wrong dose (too high or too low) delivered, dose delivered to the incorrect location, etc. The failure mode should be defined in terms of the impact it could have on the patient.

Step 3. Identify the effects that could result from the failure mode. Effects are paired with failure modes—the failure mode occurs and produces the effect. Effects should be defined in terms of the patient. For the failure mode "dose delivered to the wrong location," the effects could be damage to tissue or organs (unintended). Effects should be listed if they could occur, even if the likelihood is

Table 4–1. Standard or typical FMEA form

Treatment or Major Subprocess	Process Step	Potential Failure Modes	Potential Effects of Failure Modes	Potential Causes of Failure Modes	Current Process Controls	Occurrence of Causes	Detection of Failure Modes	Severity of Effects From Failure Modes	Risk Priority Number	Corrective Actions	Resulting Occurrence of Cause	Resulting Detection of Failure Mode	Severity of Effects Resulting From Failure Mode	Resulting Risk Priority Number

minimal. Multiple effects can be listed for a single failure mode, but only the most severe effect will be used in the analysis.

Step 4. Identify each potential cause that could produce each failure mode. Each failure mode will have several causes. Causes of failure modes include, but are not limited to, the following:

- Inadequate training
- Lack of procedures
- Lack of work instructions
- Stressful environment
- Lack of time to perform work properly
- Improperly calibrated or maintained equipment or devices
- Malfunctioning equipment or devices

Causes are linked to failure modes, not effects. In some instances, teams confuse failure modes and effects and attempt to link causes to effects.

Step 5. Identify and list all process controls. Process controls are quality control measures put in place to control specific steps in the process. There are three categories of process controls:

- Controls that prevent the causes of failure modes from occurring.
- Controls that detect the failure mode of a process step (defined in terms of when the failure mode will be detected and how likely the failure mode will be detected).
- Steps taken to reduce the severity of effects resulting from a failure mode.

Process controls are listed for each failure mode and all causes of that failure mode. In some instances, teams confuse the detection of a failure mode with the detection of a cause of a failure mode. Detection is always associated with a failure

Table 4–2. FMEA ranking scales for occurrence, detection, and severity

Rank	Occurrence – Cause: Probability that the cause will occur and produce the failure mode	Detection – Failure Mode: Probability that the failure mode will be detected before resulting in the effect	Severity – Effect: Seriousness of the effect resulting from the failure mode
1	Remote probability	Always	No effect
2	Low probability	High likelihood	Minor effect
3			
4	Moderate probability	Moderate likelihood	Moderate effect
5			
6			
7	High probability	Low likelihood	Serious effect
8			
9	Very high probability	Very low likelihood	Injury
10	100% probable	Never	Death

mode, not the cause of a failure mode. Process controls include, but are not limited to the following:

- Formal technician or operator training
- Certification programs
- Standard work procedures, work instructions, or operating procedures
- Outcome metrics for the process or specific steps in the process
- Process audits (checks to ensure the work is being performed properly)
- In-process inspection (the least effective process control)
- Statistical-based monitoring of process parameters, such as monitoring the delivery of radiation via sensors during the radiotherapy process (statistical process control)
- Visual training aides

Step 6. Rank each combination of cause, failure mode, and effect, and develop recommended actions to improve quality control of the process. At this point in a process FMEA:

- Each potential failure mode has been identified
- The resulting effects have been defined
- The potential causes of each failure mode have been listed

The next step is judge how effective the process controls are at:

- Preventing the various causes of a failure mode
- Detecting the occurrence of a failure mode
- Minimizing the effects resulting from the occurrence of a failure mode

Three ranking scales are used to determine the effectiveness of the process controls: occurrence, detection, and severity. Each ranking scale runs from one to ten,

Table 4–3. Descriptions of the O, S, and D values used in the TG–100 FMEA

Rank	Occurrence (O)		Severity (S)		Detectability (D)
	Qualitative	Frequency	Qualitative	Categorization	Estimated probability of failure going undetected
1	Failure unlikely	0.01%	No effect		0.01%
2		0.02%	Inconvenience	Inconvenience	0.2%
3	Relatively few failures	0.05%			0.5%
4		0.1%	Minor dosimetric error	Suboptimal plan or treatment	1.0%
5		<0.2%	Limited toxicity or tumor underdose	Wrong dose, dose distribution, location or volume	2.0%
6	Occasional failures	<0.5%			5.0%
7		<1%	Potentially serious toxicity or tumor underdose		10%
8	Repeated failures	<2%			15%
9		<5%	Possible very serious toxicity or tumor underdose	Very wrong dose, dose distribution, location or volume	20%
10	Failures inevitable	>5%	Catastrophic		>20%

with one being the best case and ten the worst. There are many versions of each ranking scale in use. The ranking scales shown in Table 4–2 are typical and are, in fact, often used in industry and health care. After judging the effectiveness of the controls, there will be an occurrence ranking, a detection ranking, and a severity ranking for each failure mode, effects, and cause combination. Next a risk priority number (RPN) is calculated for each combination by multiplying the occurrence, detection, and severity ranking. RPNs range from one to one thousand. High RPN values indicate the weakest areas of the process and should be addressed first. The team should hold a brainstorming session to identify process controls that will reduce the likelihood of a cause occurring, improve the detectability of a failure mode, or reduce the severity of an effect resulting from a failure mode. After addressing the items with the highest RPN values, teams should move on to the next highest and develop recommended actions to improve those process steps, continuing until arriving at those process steps that are adequately controlled, as indicated by their low RPN values. There are no firm guidelines that are used to determine what level of RPN indicates that corrective actions need to be taken. Instead, most organizations use PFMEA to guide continuous process improvement. Teams often have heated debates during the ranking and evaluation effort, and facilitators need to develop a way to defuse these debates and keep the analysis moving forward. For example, if one faction of a team believes that an occurrence ranking for a specific cause should be five, while a second faction believes the ranking should be a six, a facilitator could choose to always use the higher ranking.

4.4.4 General Observations

How do you know if a PFMEA is a good one?

- Was it a team effort on done by one person?
- What is the level of detail—is there a lack of specific details for process definition?
- Is there a lot of uniformity—very few or a limited number of causes, etc.?
- Are there few, if any, serious effects?
- Low or very low cause occurrence or failure mode detection rankings?
- Does the PFMEA look like a "cut and paste" effort?
- Doesn't pass the "reality" check? Are the overall results of the PFMEA what you would expect?

4.4.5 Alternative Ranking Scales

Occasionally teams will use occurrence, detection, and severity ranking scales that range from one to three or from one to five. The main problem with using ranking scales that do not go up to 10 is a reduced level of overall analysis sensitivity. Table 4–3 shows the ranking scale used by TG 100.

4.4.6 Using a Trained Facilitator to Lead a PFMEA

Although performing a PFMEA is not difficult, it is best to have a facilitator experienced in PFMEA to lead the effort. It is not unusual for teams to get confused during the process mapping, failure mode definition, effects definition, cause, definition, listing process controls, and ranking phases of a PFMEA. Having a trained facilitator will definitely streamline the process and increase the likelihood of a successful PFMEA.

4.5 Fault Tree Analysis (FTA)

The next step in the overall process is to evaluate the propagation of failures using fault tree analysis. A fault tree complements a process tree and gives a visual representation of the propagation of a failure in the procedure, and it also helps identify intervention strategies to mitigate the risks that have been identified. Fault tree analysis allows one to visualize potential locations in the process for effective and efficient quality management (QM) measures. Quality management is discussed in Chapter 5. A Fault tree starts with a potential error (i.e., a failure mode) on the left and works backwards in time (moving right) to study what could possibly have caused that error. Figure 4–4A shows an extremely simplified example of a fault tree illustrating the propagation of errors that result in an error in a calculated value for patient. For simplicity, only four potential failures are considered: error in data, error in data input, error in the calculational algorithm, or error in the prescription. Because an error in *any* of these can propagate into an error in the calculated value for the patient, they all enter into the process through an OR gate (here, the gate serves as the representation for the process). Figure 4–4B shows the same fault tree with added quality management protection. Parallel to each of the boxes indicating errors in the inputs are boxes indicating failures of QC associated with the process. Each of the "failure of QC" boxes enter an AND gate with their respective error in

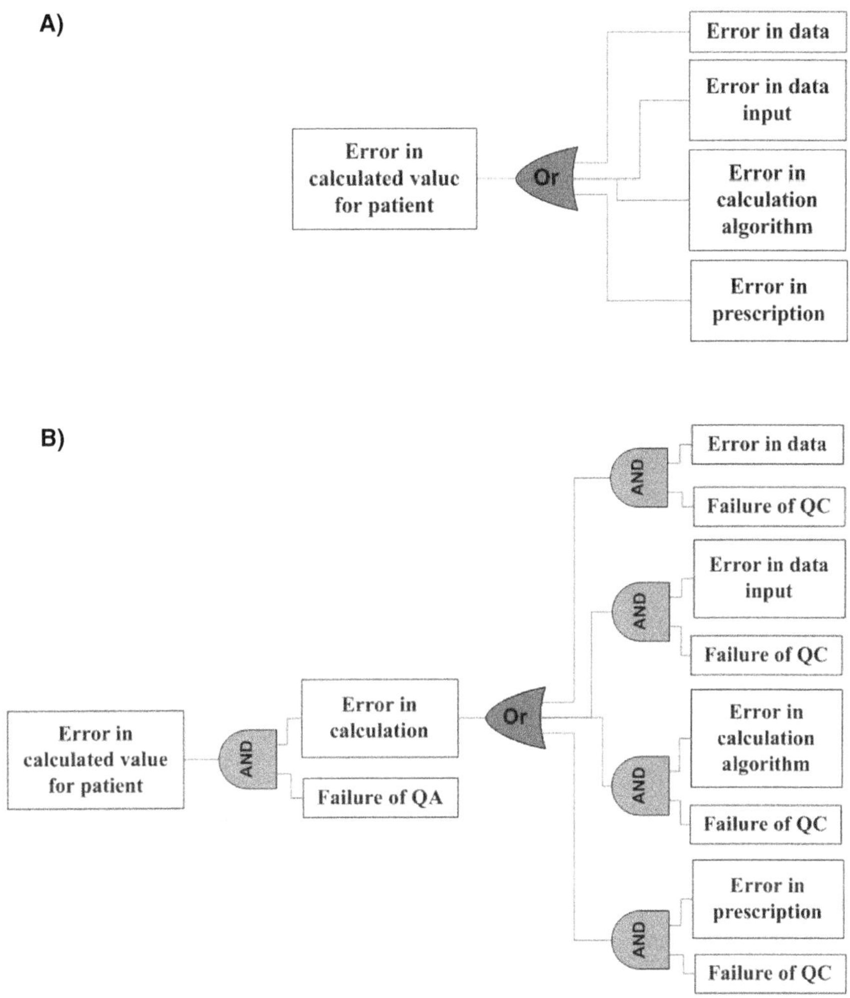

Figure 4–4. Example of a fault tree. A) A fault tree for a process with four inputs to calculate some quantity for a patient. Because an error in any of the four inputs can propagate into an error in the calculation, they all enter into the process through an OR gate. B) The same fault tree with quality management interventions. Parallel to each of the boxes indicating errors in the inputs are boxes indicating failures of QC associated with the process. Each of the "failure of QC" boxes enter an AND gate with their respective error in input box. This indicates that for the error in the input to pass into the calculation process, there must be a concomitant error in the QC that works on that input. The output of the process is interrogated with QA to provide confidence that the output of the process is correct. Again, propagation of any error requires a concomitant failure of the QA to pass the AND gate.

input box. This indicates that for the error in the input to pass into "Error in calculation," there must be a concomitant error in the QC that works on that input. The likelihood of simultaneous failures in both the processes becomes considerably smaller than an error in either (if proper guidelines for the QC are followed), adding a great deal of protection from error. If an error were to pass into, and thus through, the OR gate, it should be caught by carefully conceived QA on the outbound side. In order for the error to propagate into an actual (realized) error in the calculation, the QA must also fail. The figure shows this as the error in output from the OR gate passing into an AND gate along with the failure of QA. In general, multiple potential failures connected to an OR gate signals hazards because any of the failures can lead to the failure to the left. The more potential failures leading to an OR gate, the greater the difficulty in controlling all the inputs.

The FMEA helps lay out the fault tree, and the fault tree helps see steps that are not covered by QM. In general, it is not a good idea to rely on a single box to interrupt the flow of failures. Although it is tempting to insert a QA step as an efficiency measure to block the propagation of errors from many steps combined, failure of that one QA step would leave the procedure completely unprotected. In addition, detection of the problem from that one QA step 1) may happen after many incorrect steps and much wasted effort, and 2) it may be hard to identify which of the upstream steps actually led to the problem, although this must be known in order to correct the problem. Thus, both QM program efficacy and overall process efficiency are enhanced by incorporating multiple QM measures along the way between a possible failure mode and the final process outcome. These redundant measures reduce the possibility of an error going undetected due to a failure in a single QM measure and, as described earlier, also provide an opportunity for detection of errors early in the process, thus avoiding wasted time and effort. Again, Chapter 5 discusses the design of QM.

While making a fault tree, asking what could cause a given failure can almost always move another step farther to the right. At some point, the deeper cause may fall outside of the control of the facility. For example, a computer record may be disrupted by a power failure. In plotting the fault tree for this failure, among others causes for the disruption would be the power failure. However, if the power failure occurred because the power company had an outage, knowing what caused the outage would do the facility no good since they would not be in a position to do anything about it. The facility would need to address what to do in case of an outage, but not to prevent the outage. The point at which the facility no longer controls the situation defines the edge of that facility's universe for this analysis. In makes no sense to probe beyond the universe.

During the analysis, particular attention should be paid to far-right causes that begin more than one propagation path. Such causes only become apparent after construction of the tree and searching for identical causes. An example could be that power failure which, in addition to disrupting a treatment record, could also result in a high dose-rate source retraction failure. Addressing causes that begin multiple paths provides some efficiency in quality management.

Often, a cause that appears to start several propagation paths may actually be different manifestations in each path. Obviously, a hardware failure in a linac and hardware failure in the scheduling computer are very different causes. Notwithstanding the independence of the different pieces of hardware, seeing multiple occurrences of this potential cause highlights the importance that the facility should place on preventative maintenance.

4.6 Summary

- Risk assessment is an analysis of a process to identify its more hazardous aspects.
- Many tools exist to assist in risk assessment. This chapter presents three recommended by AAPM Task Group 100 for applications in radiotherapy.
- Process mapping presents a visual representation of a process with the goal of understanding the procedure and the interactions between the people, equipment, facilities, and organization involved.
- Failure Modes and Effects Analysis facilitates identification and ranking of potential failures. It can suggest which potential failures to address first and what aspects of the failures to try to modify.
- Fault tree analysis graphically shows how failures propagate through a procedure, and it can help identify interventional strategies to reduce hazard and improve quality.

References

American Society of Mechanical Engineers (ASME). 2002. *Standards for Probabilistic Risk Assessment for Nuclear Power Plant Applications.* RA-S-2002. New York: ASME.

American Society of Mechanical Engineers (ASME). 2003. *Addenda to Standards for Probabilistic Risk Assessment for Nuclear Power Plant Applications.* RA-S-2002, RA-S-2003. New York: ASME.

Clark, B.G., R.J. Brown, J.L. Ploquin, A.L. Kind, and L. Grimard. (2010). "The management of radiation treatment error through incident learning." *Radiother Oncol* 95:344–9.

Cooke, D.L., M. Dubetz, R. Heshmati, S. Iftody, E. McKimmon, J. Powers, R.C. Lee, and P.B. Dunscombe. *A Reference Guide for Learning from Incidents in Radiation Treatment.* Alberta Heritage Foundation for Medical Research. Health Technology Assessment Unit Initiative Series #22. 2006.

Department of Defense (DOD). *Military Standard: Procedures for Performing a Failure Mode, Effects and Criticality Analysis.* Washington, D.C.: Department of Defense, 1980.

Ekaette, E., R.C. Lee, D.L. Cooke, S. Iftody, and P. Craighead. (2007). "Probabilistic fault tree analysis of a radiation treatment system." *Risk Anal* 27:1395–410.

Ford, E.C., L. Fong de Los Santos, T. Pawlicki, S. Sutlief, and P. Dunscombe. (2012). "Consensus recommendations for incident learning database structures in radiation oncology." *Med Phys* 39:7272–90.

Ford, E.C., R. Gaudette, L. Myers, B. Vanderver, L. Engineer, R. Zellars, D.Y. Song, J. Wong, and T.L. Deweese. (2009). "Evaluation of safety in a radiation oncology setting using failure mode and effects analysis." *Int J Radiat Oncol Biol Phys* 74:852–8.

Huq, M.S., B.A. Fraass, P.B. Dunscombe, J.P. Gibbons, G.S. Ibbott, A.J. Mundt, S. Mutic, J.R. Palta, F. Rath, B. Thomadsen, J.F. Williamson, and E.D. Yorke. Application of Risk Analysis Methods to Radiation Therapy Quality Management: Report of AAPM TG–100 of the Therapy Physics Committee of the AAPM. College Park, MD: AAPM, 2013.

Huq, M.S., B.A. Fraass, P.B. Dunscombe, J.P. Gibbons, Jr., G.S. Ibbott, P.M. Medin, A. Mundt, S. Mutic, J.R. Palta, B.R. Thomadsen, J.F. Williamson, and E.D. Yorke. (2008). "A method for evaluating quality assurance needs in radiation therapy." *Int J Radiat Oncol Biol Phys* 71:S170–3.

Stamatis, D.H. *Failure Mode and Effect Analysis: FMEA from Theory to Execution.* Milwaukee, WI: ASQC Quality Press, 1995.

U.S. Nuclear Regulatory Commission (USNRC). White paper. *Risk-Informed and Performance-Based Regulation.* SECY-98-144. Washington, DC: Government Printing Office, 1998.

U.S. Veterans Administration. February 13, 2013. *The Basics of Healthcare Failure Mode and Effects Analysis (HFMEA).* Available from http://www.patientsafety.gov/SafetyTopics.html#HFMEA. (Accessed 2-27-2013.)

Victorian Quality Council. *Process Mapping: A Guide for Health Service Staff.* Melbourne, Australia: Department of Human Services, 2007.

Chapter 5

Quality Management in Radiation Oncology

Eric Ford, Ph.D.[1], Jean-Pierre Bissonnette, MCCPM[2], Luis Fong de los Santos, Ph.D.[3], Frank Rath, MSIE[4], George Sherouse, Ph.D.[5], Jatinder Palta, Ph.D.[6], Bruce Thomadsen, Ph.D.[7], and Ellen Yorke, Ph.D.[8]

[1]Associate Professor, Department of Radiation Oncology,
University of Washington Medical Center,
Seattle, Washington
[2]Department or Radiation Oncology, University of Toronto,
Department of Radiation Physics, Princess Margaret Cancer Centre
Toronto, Ontario
[3]Department of Radiation Oncology, Mayo Clinic,
Rochester, Minnesota
[4]Director, Department of Engineering Professional Development,
University of Wisconsin
Madison, Wisconsin
[5]Landauer Medical Physics
Glenwood, Illinois
[6]Professor and Chair of Medical Physics, Virginia Commonwealth University,
Chief Physicist, VHA National Radiation Oncology Program,
Richmond, Virginia
[7]Professor, Departments of Medical Physics, Industrial and Systems Engineering,
Engineering Physics, and Biomedical Engineering, University of Wisconsin,
Madison, Wisconsin
[8]Attending Physicist, Medical Physics Department,
Mem1orial Sloan Kettering Cancer Center
New York, New York

5.1 Introduction

In 2004, the Radiation Oncology Department of Cox Health in Springfield, Missouri commissioned a new stereotactic radiosurgery system. In the process, something went wrong. In measuring outputs for field sizes smaller than 1.1 cm, the physicist used a chamber that was too large. The result was a measured output that was approximately 50% too low (CoxHealth 2010). This situation went unnoticed, and patient treatments began in 2004. Over the years, 152 treatments were delivered, until September 2009 when the problem was discovered after a second physicist received training on the particular stereotactic equipment. A subsequent evaluation of treatments showed that 76 patients had received an overdose of approximately 50%, though no patient symptoms were noted (Bogdanich and Rebelo 2010). The organization responded promptly and reported the error to the patients and the public, even though, interestingly, there are no regulations in the state of Missouri with regard to linear-accelerator-based radiotherapy and reporting.

In the wake of such an accident, many questions arise, but the most important question is: how can we ensure that such an accident is not repeated? The answer is quality management (QM), which is the subject of this chapter. Quality management refers to an organizational structure to ensure consistency in practices and improvements in performance through standard control measures.

First, it must be recognized that quality management is just that: management. It is difficult to imagine effective quality improvement without a management structure. “Safety is everyone’s responsibility” practically means “safety is no one’s responsibility.” This chapter will focus on the concepts and tools of quality management. It must be remembered throughout this discussion that a successful implementation of any of these approaches requires effective management. A clear management structure is needed and should include clear personnel responsibilities, regular review of policies and procedures, and internal and external audits, all documented appropriately. The necessity of such a formal structure was recognized in AAPM Task Group 40, published in 1994 (Kutcher et al. 1994) and is a requirement for accreditation under the American College of Radiology (ACR) and the

American Society for Radiation Oncology (ASTRO) practice accreditation system (Ellerbroek et al. 2006).

Second, it must be recognized that quality management goes beyond just the routine performance of a set of prescriptive tests, such as calibration checks of linear accelerators, dosimetry verifications, and the like. Such tests are important, but they only address part of the picture. Typical monthly and annual tests of a linear accelerator, for example, would not have uncovered the output error described above unless these checks were truly independent in some way. Also, a purely technology-centered view ignores the fact that most radiotherapy misadministrations are not the result of purely technical failures, but rather involve the entire process of care and all its complexities. Quality management provides a structure for addressing safety hazards from this broad perspective. It is also directly related to the risk analysis such as failure modes and effect analysis (FMEA) or fault tree analysis (FTA) discussed in Chapter 4. Quality management interventions are the natural and direct outgrowth of such risk analysis exercises.

This chapter will describe the basic underpinnings and tools for effective quality management. Section 5.2 outlines general principles of quality management from the perspective of industrial psychology and operations. This is followed in Section 5.3 with a discussion of quality management tools and what makes for a good quality management tool. The next sections present two specific forms in which these tools are used: commissioning (Section 5.4) and audits (Section 5.5). Section 5.6 considers other specific tools for quality management that are in use in radiation oncology. Finally, Section 5.7 presents a case-example of a ^{90}Y microsphere-based liver cancer treatment service which demonstrates the development of a quality management approach based on risk assessment analysis.

5.2 Concepts in Quality Management

A widely accepted approach for quality management was defined by Shewhart in the 1930s (Shewhart and Deming 1939) and later popularized by W. Edwards Deming. He proposed a cyclical approach involving the following steps: plan, do, study the results, then act (PDSA). The PDSA concept for quality improvement is discussed in more detail in Chapter 7, but the basic objective is to aim at small, gradual improvements rather than seeking out the breakthrough that will simultaneously resolve all problems. Deming also enriched quality management by postulating a series of 14 values (Deming 2000) that include key concepts such as:

- breaking down barriers between departments and subunits within an organization,
- management taking on leadership issues,
- shifting the role of the supervisor from one of discipline and compliance to one of helping people and machines do a better job,
- improving constantly using the cyclical approach described above, and
- instituting a vigorous program of education and self-improvement.

These values and concepts are grouped to form the four fundamental principles of the Deming management theory: system thinking, knowledge of variations, theory of knowledge, and psychology (Deming 1994).

5.2.1 Appreciation of a System

All organizations operate as a system of interrelated components, units, and professions. This concept is discussed in more detail in Chapter 2. These systems have aims that are generally known across all components. To manage a system, individuals need to understand the inter-relationships within the system and nurture cooperation among the different components of the system. Without management, system components become competing entities that provide obstacles to the aim of the system. Thus, the role of the manager is to improve communication and ensure the best results for every system component so overall objectives are met. This involves balancing costs, resolving internal conflicts, and removing obstacles to collaboration.

5.2.1.1 Technical Quality Control in Radiation Oncology

Medical physicists have a professional responsibility to ensure the safe, reliable, and reproducible performance of radiation therapy equipment. A layer of mandatory regulatory tests have promoted a culture of prescriptive quality control (QC) of equipment that has been prevalent in our field since the publication of the first codes of practice in 1969. In continuation with this established culture, more recent guidelines for technical quality assurance (QA) reinforce the utilization of tests to verify that equipment complies with specifications established by a vendor or by expert opinion (Kutcher et al. 1994; Thwaites et al. 1995; Klein et al. 2009). There are several limitations to recommendations from professional societies:

- updates are relatively infrequent and not synchronized with performance improvements from equipment manufacturers,
- regulatory agencies misinterpret the intent of the reports and demand exact compliance from users despite obsolescence,
- reports may be biased since they are often drafted from a literature review from early and first-time users, and
- there are high costs to running such QC programs (Palta et al. 2008).

Experts have developed technical quality control guidelines in isolation, often lacking the statistical evidence to justify test frequency and tolerances, while ignoring the impact of other components of the radiotherapy process. Finally, tests are often not linked to identified potential hazards. It is hoped that the forthcoming AAPM Task Group 100 will help rationalize existing and future reports from professional societies (Huq et al. 2008).

5.2.1.2 Process Quality Control

The commercial introduction of new technology, such as modulated arc therapy and image-guided radiotherapy, has profoundly transformed the way we administer radiotherapy. Notably, small mistakes that occur at the scanning and contouring phase can impact the outcome of treatment. In parallel, manufacturers have worked extensively to improve the reliability of their equipment. Despite the costly QC program described in the above section, significant errors have still occurred (Peiffert et al. 2007; Williams 2007; Bogdanich 2010).

The rapid implementation of novel technologies has also revealed another issue with the current QC paradigm: years may elapse between the commercialization of new technologies and the publication of quality control guidelines. Thus, early users are left without guidance, relying on initial reports of technology performance and, perhaps more wisely, dry runs to ensure that equipment performs as intended and that staff is comfortable with the new technology and its practices. Self-audits help maintain consistent practice until guidelines are published (Brundage et al. 1999; Boxer et al. 2009).

Growing evidence demonstrates that high-quality practices impact the outcome of radiation therapy, and that such practices should not be assumed. An audit of plans and treatments delivered under the auspices of an international collaborative trial on chemoradiotherapy of head and neck cancers has demonstrated a markedly inferior outcome for plans that did not meet study objectives compared with those that were compliant (Peters et al. 2010). This underscores the value of quality improvement; the study noted that "the value of good radiotherapy is substantially greater than the incremental gains that have been achieved with new drugs and/or biological."

5.2.2 Knowledge of Variation

The second of Deming's four fundamental principles involves variation. Variation is ubiquitous. It lies in devices, processes, and people. Understanding sources of variation can lead to safer, more efficient, and higher-quality treatment by removing or reducing the effects of these sources. Using statistical methods, graphical tools have been developed to reveal sources of variations that can be acted upon.

There are two main classes of variations: random causes and special causes. Random causes are inherent to a process and cause small variations about a well-defined mean. One way to reduce a source of variation would be to replace existing equipment with new devices that have tighter specifications. Special causes, meanwhile, are not inherent to a process and cause large variations. One example would be the failure of a monitor chamber in the head of a medical linear accelerator. When only random causes are present, the process can be described as in statistical control because its mean and variance are stable with time. Once statistical control is achieved, one can compare the mean with the objective: is the process on target? When it is confirmed that the process is on target, one can then work on minimizing variations across the mean. Chapter 7 will more fully address knowledge of variations and tools to manage them.

5.2.3 Theory of Knowledge

Humans can strive on deepening knowledge with the help of models, frameworks, plans, and theories. In this context, a theory—usually derived from statistical analysis and by understanding variations—helps predict the outcome of future events within a process. A theory might be perfected, perhaps following several revisions or iterations of the PDSA cycle, as it becomes more accurate in predicting the behavior of a system. Theories remain vulnerable to new knowledge and innovations, changing contexts, "unknown unknowns," or human behaviors like entrained thinking and complacency.

5.2.4 Knowledge of Psychology

Psychology helps us understand how a person makes decisions. Cognitive psychology tells us that people learn differently (Redish 1994; Kolb and Kolb 2005) and at different speeds. People also have different intrinsic and extrinsic motivators. The need to relate to other people, to be loved and esteemed, to have dignity, and the natural inclination for humans to learn are powerful intrinsic motivators that wise managers know how to nurture. Extrinsic motivators—such as money, a merit system, and numeric goals—may work for a time, but they have been shown to have little impact in the long term and eventually lead to cynicism, loss of motivation, and low self-esteem.

Humans are also known to have limited attention spans (Johnstone and Percival 1976) which may also apply to the work context. Some studies suggest that human vigilance has a reliability rate of only 80% in the face of monotony and fatigue (Juran and Gryna 1993). In healthcare organizations, failure to call an action an error when it falls very close to specification is common in order to avoid the perceived high cost of rework or delaying treatment as noted above. Errors can go one of two ways—a good result can be rejected just as a bad result can be accepted.

Finally, adults do not learn in the same ways as children do; a straightforward transfer from teacher to learner is not sufficient, as adult learners need to process the new knowledge in terms of what they already know, integrating it into their mental framework. Recall of new knowledge is much more effective when tied to emotionally significant, well-known bits of knowledge (Redish 1994; Regehr and Norman 1996). Conversely, when integration of new knowledge is poor, learners can make seemingly bizarre errors (McLeod et al. 2003). Since human error is so prevalent, manufacturers would be well advised to design products with good ergonomics, intuitive user interfaces, and fail-safe positions in case of malfunctions (Cafazzo et al. 2009; Chan et al. 2010).

In summary, quality management must be understood from a broad perspective at the systems level and with an appreciation of the underlying phenomena. The following sections build on these more general points and consider the specifics of quality assurance and quality control as used in radiation oncology (Section 3.2) and then some very specific tools for quality improvement (Section 3.3), including process checklists, audits, and the role of information technology in quality management.

5.3 Introduction to Quality Management Tools

Quality management programs consist of actions taken to control specific process steps to improve the probability that they consistently produce optimal, desired, and safe outcomes or results. In some cases, the corrective actions are predefined, while in other cases corrective actions must be developed. Sometimes corrective actions and quality improvement are the direct result of a risk analysis, such as FMEA or FTA (see Chapter 4), and it must be appreciated that quality management interventions are the natural and direct outgrowth of such exercises.

In considering corrective actions, it is helpful to keep in mind the three fundamental categories (Grout 2006):

1. ***Actions that prevent errors from happening.*** This means designing a process or product such that errors are less likely to happen. Forcing functions and automation belong to this category, but the category may also include improving the work environment in more general ways. This approach corresponds to improving the *occurrence* score in an FMEA consideration of errors.
2. ***Actions that make errors quickly and easily detectable when they occur.*** This approach corresponds to improving the *detectability* score in an FMEA consideration of errors.
3. ***Actions that mitigate impact.*** The goal here is to develop design features or process that will temper the impact of an error when it does occur and reaches the patient. This approach corresponds to improving the severity score in an FMEA consideration of errors.

This list clearly demonstrates that there is a link between risk analysis and the tools of quality management. Performing risk analysis allows one to define weak points either prospectively (e.g., through failure modes, as described in Chapter 4), based on observed failure patterns (e.g., through incident learning, as described in Chapter 6) or through root cause analysis (as described in Chapter 7). In a quality management program, each of these hazards is addressed with corrective actions.

It is also clear from the above list that some corrective actions are more effective than others. This is explored in more detail in Chapter 7, but a useful concept here is the "hierarchy of safety interventions" developed by the Institute for Safe Medical Practices (Vaida 1999). The ISMP ranks safety interventions in order of effectiveness:

- *Forcing function.* These prevent errors, often with a physical design barrier that prevents a person from performing the wrong action.
- *Automation and computerization.* These either prevent errors or make them more visible by "lessening human fallibility by limiting reliance on memory" (Vaida 1999).
- *Protocols and standard order forms.* These are intended to prevent errors by providing a guide and aid to safe usage. They are not as strong as forcing functions or automation. One example of a protocol is a directive to

limit the dose per fraction for on-call treatments with the intent of limiting the damage done by a wrong dose calculation.

- *Independent double checks.* These are often performed by a second person. These make an error more visible.
- *Rules and policies.* While necessary, rules can sometimes be a superficial reaction to an incident with an intent to "fix people." Rules and policies should be used primarily to support other, more effective, interventions. Rules and policies might be considered to be one type of protocol (see above). As used here, however, the term 'rule' relates to personal performance and behavior and not to conventions for care delivery.
- *Education and information.* These are important, supportive interventions, but they should be coupled with other strategies for error reduction.

In addition to the interventions list above, AAPM Task Group 100 also refers to key core components, that is, "essential aspects of any program required for safe and effective operation." These include 1) training, 2) standardized procedures, and 3) communication (Huq et al. 2013). The first two overlap with the ISMP list above, while the third key component concerning the facilitation of communication is discussed below in Section 5.3.5.

One final, key core component discussed in TG 100 and elsewhere is resource allocation and staffing. The contributions of workload and stress to the safety environment in radiation oncology have been well documented (Mazur et al. 2012). One major contributor to events is a rushed environment, often due to understaffing (Thomadsen et al. 2003). Achieving quality and operating safely requires adequate resources, including staffing and equipment. Recognizing that funding is always a problem and that almost no program has all the resources it desires, embarking on a program of any sort while unable to assure adequate control on the execution and full attention of those involved can lead to disaster. It should be remembered that the cost of one event often could have paid for that event's prevention several times over.

5.3.1 QM Approach #1: Forcing Function and Error Prevention by Design

Forcing functions and automation are considered the most effective method of safety intervention. A simple clinical example is interlock systems on a linear accelerator. If an incorrect wedge or applicator is inserted into the machine, an interlock is enabled that prevents treatment. Automation and computerization (also listed above) are similar, philosophically, to forcing functions in that the goal is to prevent errors by directing action.

Such error proofing by design, however, is often the purview of the manufacturer, and not something that the clinical medical physicist has direct influence over. Designers of equipment and devices often adequately capture the main or most critical outputs, such as the amount and location of dose for radiotherapy devices, etc. However, designers may not adequately define less critical outputs that cover use of their product. Can dials be easily read? Is the design ergonomically sound? Is the equipment or device easily maintained and serviced?

Redesign for error proofing can also be considered from the workflow process point of view. This is somewhat different than the "forcing functions" mentioned above in the list from ISMP, but it is similar in spirit in that the goal is to remove the possibility of error. In some cases, the process can be redesigned so that failure modes no longer exist, and this is particularly attractive for the higher-ranking modes in an FMEA. Such a redesign would consider the system design principles discussed in Chapter 2. Examples would consist of rearranging the equipment in a procedure room, changing the order of steps, or changing the modality by which information passes between the players. If the redesign was performed because of an FMEA risk analysis, then the FMEA should be revised, not only because of the elimination of some potential failure modes, but also because any redesign can produce new potential failure modes. As with any change, redesign requires consultation with all of those who participate in the procedure to assure that changes do not make their work more difficult. Consultation also increases buy-in with the change (see Chapter 8 on managing change).

5.3.2 QM Approach #2: Protocols and Standardization

One of the greatest contributors to medical events is not having standard procedures, or performing every function as if each time it were unique. Most clinical work is not unique, even though each patient presents differently. Having standard procedures, processes, and protocols eliminates haphazard actions, clarifies expectations, and aids in fulfilling those expectations. Standard protocols cover a wide variety of disparate topics. Obviously, many patients have treatment under protocols as part of national trials, and the entry of patients into trials should be encouraged. Even if not eligible for any trial, most patients (but certainly not all) fall into general types—such as early prostate, post tylectomy breast, or palliative bone irradiation—that would be treated similarly. Having *written* protocols for these types of patients facilitates the detection of discrepancies in individual patients that may be errors or mistakes, as well as providing guidance at each step of the patient's treatment.

Many other activities in radiotherapy should be covered by standard protocols. Physicists are quite accustomed to following protocols for calibrating treatment units or brachytherapy sources or adapting professional protocols for quality assurance. Any national or international guidelines need to be adapted to each facility in a more specific set of procedures that include the details applicable to that facility's equipment and operation, e.g., specifying the ionization chamber to use for calibration.

Much of the operation in the given department should be covered by standard protocols. Protocols can address activities such as cone-beam CT image guidance with instructions that spells out how the images are made, the alignment of the CT images with the planning images, and the application of patient shifts. Problems arise when various personnel perform the same activity in different ways, because

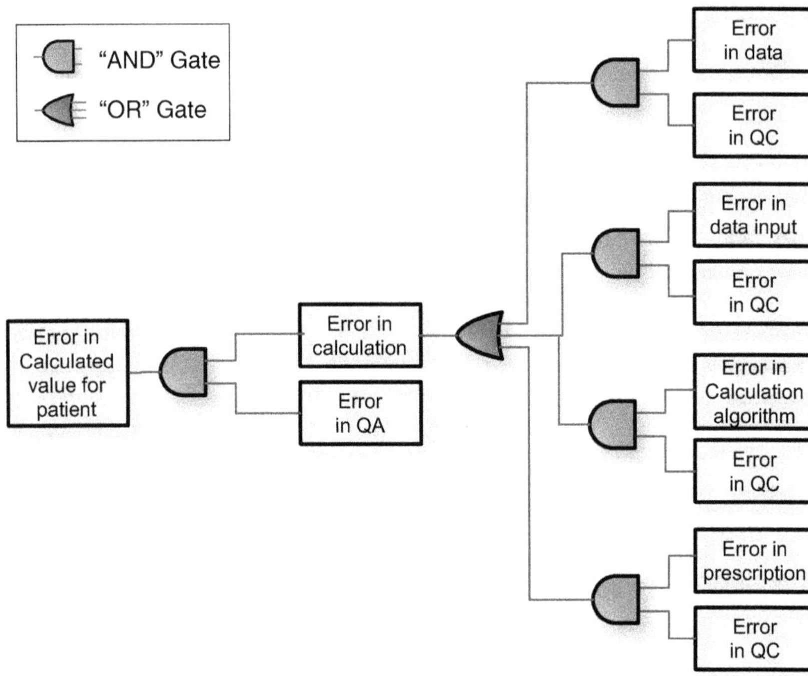

Figure 5–1. Example of a fault tree from TG 100. The figure shows a process with four inputs, each with QC to maintain the integrity of the process, and QA to provide confidence that the output of the process is correct. The symbols represent "OR" and "AND" gates. Because an error in any of the four inputs can propagate into an error in the calculation, they all enter into the process through an OR gate. Parallel to each of the boxes, indicating errors in the inputs, are boxes indicating errors in QC associated with the process. Each of the "error in QC" boxes enter an AND gate with their respective error in input box. This indicates that for the error in the input to pass into the calculation process, there must be a concomitant error in the QC that works on that input.

that opens the door for misunderstandings, inappropriate assumptions, and confusion.

5.3.3 QM Approach #3: Quality Assurance and Quality Control

Quality Control (QC) and Quality Assurance (QA) are considered in this section. In one sense, QA and QC activities can be considered to be a form of "independent double check," which is the fourth item on the above list of safety interventions from ISMP. However, QA and QC extend well beyond this, encompassing many approaches and tools. This section presents these concepts in detail.

First, we propose some definitions. Although physicists often refer to their quality management activities as "QA," they are actually involved in both Quality

Control (QC) and Quality Assurance (QA). The distinctions are often blurred between these very similar activities, and there are various definitions in play. This chapter adopts descriptions consistent with those used by Task Group 100: QC activities deal with the inputs to a process to make sure that everything that goes into a process is correct, while QA activities deal with the outputs of a process and demonstrate that the goals are adequately met. For example, in the setting of an individual's treatment plan, QC includes reviewing organ and target contours, looking at the prescription to see that it makes sense—making sure one is using the proper beam data tables (prescribed energy and linac) and making sure one is using the correct dose calculation algorithm. QA includes such things as the independent monitor unit check, patient-specific IMRT measurements, and making sure that the plan meets the prescribed target coverage and OAR protection goals. The distinctions between QA and QC can be confusing because the output from one process sometimes becomes the input for another. That is, the QA from the first step becomes the QC for the second step. The distinction between QA and QC is useful, however, because the two function differently. To take an example, consider a form used for a preliminary prescription or order for treatment planning that includes all the information needed to perform the plan. This is really a QC form, guiding the agent along in the next process step. If it were a QA form, it would be structured somewhat differently, and it would address whether the prescription and directive coming from the physician were really what was intended. QA and QC therefore function somewhat differently: either to provide guidance going into a process (QC) or provide an indication that the process worked correctly (QA).

A final term to distinguish is Quality Management (QM), which encompasses all the activities needed to maintain a desired level of quality. In health care, safety is an important (though definitely not the only) component of quality, and QM is the overarching management system under which QA and QC are conducted.

Both QA and QC are needed for safe, high-quality treatments. Having QC for each input to a process often requires more resources than QA. On the other hand, well-designed QC is very specific in detecting the failure for which it is designed—for example, a prescription that is very discrepant with the department's norm may become quite obvious in the physics plan review. Also, as seen in the Figure 5–1, single large-scale quality management efforts, such as good linac commissioning, automatically provide QC for numerous individual patient treatments. A single QA activity is efficient in that it can directly detect a failure in output. However, QA does not specifically determine the cause of that failure, and investigations (and the need to repeat all or part of the process) can be labor-intensive. In this regard, confidence that equipment is well commissioned is helpful in narrowing the list of most likely failure causes (though one must always be alert to rude surprises). Also, if the output receives only a single QA step (as in Figure 5–1), a failure in that QA step can allow an error to propagate until it affects the treatment. Typically, a treatment plan that has been approved by a physician receives the "Physics plan check," which is designed to catch major errors. But the radiation therapy QM process has evolved to make it highly likely that the plan will be reviewed independently by other eyes before the patient finishes treatment. The therapists typically review the chart before the first treatment, the plan is often peer-reviewed early in the treatment course in chart rounds, and a physicist or dosimetrist performs a weekly chart review. Anyone who has been in the field for a while has seen plan errors that were caught through such later QA steps. For example, a therapist notes a discrepancy in setup depth, which points toward monitor units that were calculated with the wrong TMR; or chart rounds peer review discloses that the fields are too tight to encompass the designated target volume; or a weekly chart check reveals that even though bolus was requested in the prescription, that request was not in the plan write-up or transcription to the treatment management system.

It must be recognized that QA checks apply not only to processes, but also to devices. Consistently achieving optimal and safe process outcomes or results requires that effective quality controls or tests are defined and deployed. Quality control of equipment or devices should begin with the definition of performance specifications and functional requirements. In manufacturing environments, the evaluation of production equipment is an important component of an effective quality management program. Equipment specifications must meet production requirements, such as line speeds or production rates, process tolerance capabilities, life cycles, etc. Specifications should also address the use of equipment in manufacturing, such as the ease of setup or changeover, service or maintainability, programming, the loading and unloading of parts, and that the equipment meets ergonomic use requirements. It is the responsibility of a manufacturing company to make sure that their equipment (new or in place) meets their performance requirements. Manufacturers typically request and review the Design Failure Modes and Effects Analysis (DFMEA) of equipment from equipment designers to ensure that:

- equipment specifications meet performance requirements,
- all potential failure modes (ways in which equipment can fail to meet performance or safety requirements) have been identified and addressed, and
- design verification plan testing of equipment (performance testing of equipment to verify that the equipment meets performance specifications) has been completed.

In this context, it is useful to note that equipment QC includes periodic preventive maintenance inspection, or PMI.

The radiation therapy community should consider adopting a similar approach to verifying that devices meet all performance and use requirements and that potential failure modes be identified and addressed as part of device commissioning, with ongoing review when the device is in full clinical use. Although not often performed, an Application Failure Mode and Effects Analysis could give insight into a device's design and it's utilization in a specific process, and it should be considered.

5.3.4 QM Approach #4: Education and Training

Initial training for persons involved in a procedure is essential for correct performance. This training should have validation, such as an examination, to assess whether the trainee has mastered the concepts and skills required. While training is necessary, it is not sufficient to prevent failures. In most events, the people who failed had training and knew what they should have done, but for various reasons, they did not perform correctly. Any interventions to prevent a potential failure mode must realize that training does not prevent human failures. Periodic retraining provides dubious benefits, and it should not be a cornerstone for a quality program. Forced retraining of people involved in an incident almost never reaps the expected results—first, because the people often already know the information, and second, such retraining is perceived simply as punishment. However, if a deficiency in training is identified, training in that particular topic is certainly warranted.

5.3.5 QM Approach #5: Facilitate Communication

During the design of a procedure, great attention must be paid to lines of communication and how communications will flow. Who passes what information to whom must be clearly defined, and anyone involved in information transfer must understand what information is involved and in what format. Forms can serve well to facilitate information transfer, reminding the person entering data what information is needed and minimizing misunderstanding. As needed, a form can pass from step to step in the procedure. The form should have blanks for any information needed. If, during use, marginal comments frequently occur on a form, new blanks to accommodate the information being entered marginally should be added. At the same time, blanks that are never used should be deleted. Commissioning of a procedure (see Chapter 6) allows each participant to ensure that the information they will need will be available. Communication also entails that all members of the team make it clear what they will do and what they expect others to do.

5.4 Quality Management and Commissioning

The incident outlined at the beginning of this chapter illustrates the possible consequence of errors in commissioning. There, 76 patients receiving stereotactic radiosurgery at a clinic in Missouri were overdosed by approximately 50% due to the use of an oversized chamber for output factor measurements (CoxHealth 2010). This incident is not unique, however. On April 2, 2005, the H. Lee Moffitt Cancer Center and Research Institute announced that 77 patients treated with stereotactic radiosurgery had received overdoses of ≥50% because a depth dose factor was not used in calculating the dose per monitor unit when employing the AAPM TG–51 protocol (Bogdanich 2010). This was discovered when the department undertook credentialing by the Radiological Physics Center (RPC) for participation in a cooperative group trial. A further example of an error in commissioning is the well-known miscalibration of ^{60}Co output at Riverside Methodist Hospital in Columbus, OH in 1972 which resulted in at least 10 deaths and numerous serious injuries (Rubin 1978). The accident is described in more detail in Chapter 1, but the error occurred when a new physicist was hired and began calculating ^{60}Co decay from linear graphs instead of semi-log plots. These examples are some of the most well-known cases of errors in commissioning, but it is likely that the problem is more extensive than these few reports indicate. Recent RPC data indicates that approximately 11% of institutions have had at least one beam fall outside the RPC's 5% dose agreement criteria (Ibbott et al. 2008), and such an audit is only sensitive to a very particular type of discrepancy: the output of a 10 x 10 cm field under standard conditions. Errors made in the process of commissioning are especially pernicious because they can affect many or all patients treated and can go undiscovered for years.

The clinical practice of radiotherapy is becoming increasingly complex, and with this complexity comes a wide range of conditions treated, technologies used, and patient and physician preferences. There are currently more than seven IMRT delivery methods (Moran et al. 2011). Clinical workflow of IMRT treatments depends on disease site, institutional experience, technology available, and individual physician preference. Along the same lines, several image-guided radiotherapy techniques are currently in clinical use to improve the precision and accuracy of conformal radiotherapy (Potters et al. 2010; Bissonnette et al. 2012). The complexity of advanced radiotherapy techniques is further compounded by the multiple steps involved in the clinical process, by intra- and inter-departmental structure, by the large amount of published data with disparate conclusions, by the continual changes brought about by clinical outcomes research, and by rapidly introduced new technologies. It is a daunting task to develop generic acceptance testing, clinical commissioning, and QA protocols for the ever- widening range of possible treatment techniques and delivery equipment. Furthermore, the state-of-the-art radiotherapy planning and delivery systems are, by themselves, very complex. They typically process thousands of input parameters for every radiation treatment delivered to a patient. These systems require an elaborate infrastructure to manage complex treatment planning and delivery parameters safely and accurately. The consequence is that the acquisition, acceptance testing, and clinical commissioning of these systems and associated infrastructure are much more involved. The scope of work depends on clinical requirements and utilization, functional performance characteristics of a system, and how that system integrates with other systems, which include imaging, treatment simulation, treatment planning, treatment delivery, treatment management, electronic medical records, and treatment verification.

It is fair to conclude that acceptance testing and clinical commissioning of radiotherapy systems is very much dependent on the clinical process. The generic acceptance testing and clinical commissioning guidance for these systems is not possible. Historically, professional and scientific societies have provided guidance and recommendations for acceptance testing, commissioning, and quality assurance of radiotherapy equipment, and this guidance has at times been turned into prescriptive requirements by the regulatory agencies. A major deficiency of traditional approaches of providing guidance on acceptance testing and commissioning is its emphasis on device-specific procedures at the expense of other sources of errors. Many drivers of error exist that are not specific to devices, including inadequacies in process design, information flow, training, and documentation. Although many serious radiation therapy errors involve the incorrect or inappropriate use of devices due to miscommunication or misunderstanding, traditional physics QA often overlooks these factors. While it is important that each device used in planning and delivering radiotherapy treatment should perform according to specifications and expectations, a better understanding of the interactions between the clinical processes, human users, individual devices, and the impact of various 'failures' on treatment outcome will help distribute resources more efficiently and effectively. Furthermore, such strategies are not optimal for the new generation of radiotherapy equipment because the amount of effort to test everything is likely to be daunting. It is worth noting that the recommendations from AAPM on QA requirements for linacs increase the number of daily, monthly, and annual checks by over 60% each, mostly to account for technologies such as IMRT and on-board imaging that were not in clinical use at the time of the writing of earlier reports (Klein et al. 2009, Kutcher et al. 1994).

This section focuses on the use of quality management structure to manage commissioning of new devices and procedures, with the goal of prevented errors like those highlighted above. This section does not lead the reader through step-by-step instructions for commissioning. For more detailed background on commissioning, the reader should consult the relevant AAPM Task Group reports and other literature which include:

- Commissioning and ongoing QA of treatment machines (Das et al. 2008) and simulators (Mutic et al. 2003)
- Commissioning brachytherapy equipment (Kubo et al. 1998; Pfeiffer et al. 2008)
- Commissioning of helical tomotherapy (Langen et al. 2010; Bissonnette et al. 2012; Willoughby et al. 2012), image-guidance CT systems (Bissonnette et al. 2012), and non-radiographic localization systems (Willoughby et al. 2012)
- Commissioning of treatment planning systems (Fraass et al. 1998), including validation measurements in typical clinical scenarios (Miller et al. 1995; Ezzell et al. 2009)
- Commissioning of treatment management systems (record and verify). This is partially the subject of an upcoming AAPM Task Group 201 Report, "Quality Assurance of External Beam Treatment Data Transfer."

This section does not attempt to reproduce these reports in any detail, but rather highlights broader strategies for quality management, which, although often mentioned in Task Group reports and other documents, can be viewed as something of an afterthought.

5.4.1 Device Commissioning

A vast and growing literature exists to help the medical physicist and other staff members in the process of commissioning. Many of the important issues and pitfalls are outlined somewhere in a task group report. To take the example of output measurements for small field sizes, AAPM Task Group 155, formed in 2007, is in the process of formulating recommendations, but the issue is already discussed extensively in the report of TG 42 from 1995 on stereotactic radiosurgery, which

Table 5–1. Sample checklist for quality management in device commissioning

✓	Quality Management Item
☐	Systematic review of background information including AAPM Task Group reports, other reports, and training materials
☐	Develop a plan for acceptance testing and commissioning
☐	Develop a plan of training for medical physicist and other staff
☐	Evaluate staff time commitment and manage appropriately
☐	Perform phantom-based tests, including end-to-end tests
☐	Test the effects of not following the correct procedures, and also establish the operational limits on the equipment, outside of which the equipment's reliability diminishes
☐	Produce a commissioning report, which is dated and signed
☐	Audit the commissioning data and report by an independent, qualified medical physicist
☐	Check the beam outputs of radiation-producing units by an independent agency
☐	Check phantom dosimetry by an independent agency, if appropriate
☐	Develop a plan for roll-out to staff, including dry-run test cases where appropriate

advocated the use of a detector smaller than 1 to 2 mm, or even film, for field sizes less than 1 cm. The issue is also taken up in more recent reports, such as TG 106 on beam data in linear accelerator commissioning (Das et al. 2008), TG 101 on stereotactic body radiosurgery (Benedict et al. 2010), and other reports, such as tools for IMRT (Low et al. 2011). The point being that an extensive literature is available with a wealth of practical advice for the technical issues encountered in device commissioning.

In addition to these reports, the medical physicist also has many training opportunities available. The issue of training is considered in more detail in Chapter 8, but some of the training resources available for commissioning include vendor-sponsored courses, educational programs sponsored by AAPM and ASTRO, and site visits to other institutions that have experience with the device or software being commissioned.

The training and preparation outlined above requires a significant investment of time, which highlights the issue of resources and institutional commitment. In commissioning a linear accelerator, for example, AAPM Task Group 106 (Das et al. 2008) estimates that approximately six weeks are required, depending on the number of energies and other capabilities, and this does not consider the complex imaging sub-components now available on many clinical linacs. It is important, also, for administrators and department leadership to recognize the importance of training and sound workflow procedures in their frequent rush to "go live" with revenue-generating new equipment (Zeitman et al. 2012).

The other task for management is to develop a plan to manage commissioning. This is a crucial feature of quality management. Table 5–1 provides a checklist, drawn from AAPM reports and other sources, for some of the key features in the management of commissioning. Several items listed in Table 5–1 relate to reports and independent checks of the commissioning process. Independent measurements include end-to-end tests or validation measurement, such as verifying the planning system output in typical clinical scenarios when commissioning a treatment planning system (Miller et al. 1995; Ezzell et al. 2009). Peer review is also listed in Table 5–1, and it is a crucial feature of quality improvement at all levels (Marks et al. 2013), commissioning being no exception. In some cases, an outside audit by an independent agency, such as the RPC or a colleague, may be appropriate, as discussed further in Section 5.5.

In summary, quality management in commissioning employs the foundations discussed above in Section 5.3. Efforts are made to prevent errors from happening in the first place through training, preparation, and resource commitment. Recognizing that, in spite of this, errors will still occur, quality management includes steps that uncover errors. This includes phantom tests, independent checks, and outside audits. In establishing commissioning procedures for new equipment, particularly before guidance documents become available, the techniques discussed in Chapter 4 of process mapping, FMEA, and fault tree analysis can help find particular aspects to concentrate on during commissioning.

5.4.1.1 Radiotherapy equipment purchase, acceptance testing, and clinical commissioning process

The process of purchase, acceptance testing, and commissioning of any advanced radiotherapy equipment is a major undertaking that can require a considerable amount of time, effort, and expense. Therefore, it is crucial that a great deal of thought and care go into the initial planning. The primary objective should be that the equipment specifications meet the clearly defined needs of the facility over the projected lifetime of the equipment, which can be up to 10 years. The process of selection, acceptance testing, and commissioning of any radiotherapy equipment should involve:

- Systematic review of current and projected clinical needs and types of patients who will be treated with the equipment
- Assessment of personnel needs and their qualifications
- Assessment of training and credentialing requirements for safe operation of the equipment
- A careful review of deliverables, functionality, technical and physical specifications, and cost of all commercially available equipment
- Evaluation of non-obsolescence, warranties, and maintenance contracts offered by the vendors over the lifetime of the equipment

- Evaluation of the quality of the vendor's service and technical support
- Design and construction or renovation of the facility to house the new equipment
- Installation of the equipment, safety checks, and initial tests
- Acceptance testing of the equipment
- Commissioning of the equipment for clinical use
- End-to-end testing and clinical validation of the equipment
- Development of standard operating procedures and checklists
- Training of the staff in the safe and efficacious use of the equipment
- Risk-based assessment and establishment of the baseline quality assurance parameters and schedule
- Final report and documentation

It is critical that the selection process for the equipment includes input from radiation oncologists, physicists, therapists, and facility engineers. These tasks are best accomplished by the formation of an ad hoc committee in the department that includes all the stakeholder groups. The criteria for selecting any radiotherapy equipment can become quite controversial, complex, and time consuming. Sales representatives from the vendors at times do not clearly distinguish between what is currently deliverable on the equipment and what may be planned for it in the future. It is the responsibility of the equipment selection team to contact personnel at facilities that have and use similar machines clinically. It is good to contact only those facilities that have technical resources and patient distributions comparable to yours. Site visits to the factory or to a vendor's designated facility are rarely useful unless a special and new modality or option of treatment delivery is under consideration.

Equipment purchase

A generic decision tree for the purchase of any radiotherapy equipment is shown in Figure 5–2. Such a strategy is very useful in establishing the criteria for selecting and evaluating various options. It can take considerable time and effort to go through some of the steps described in this figure. It is important, however, not to skip any of the steps. The decision tree ensures that all clinical requirements are considered and that the physical infrastructure is adequately evaluated for the planned equipment purchase. A critical review of needs and the physical infrastructure during the planning stage can save aggravation, time, and money later on.

Equipment specifications and purchase agreement

The recommendations of the equipment selection committee are followed by the development of comprehensive equipment specifications and a binding purchase agreement. If the equipment is purchased through a bidding process, then the equipment specifications are developed before the decision is made to select a vendor. Otherwise, the final specifications may be developed in close collaboration with the vendor's representative. All vendors have developed product specifications for the functional performance of their equipment in response to requirements from poten-

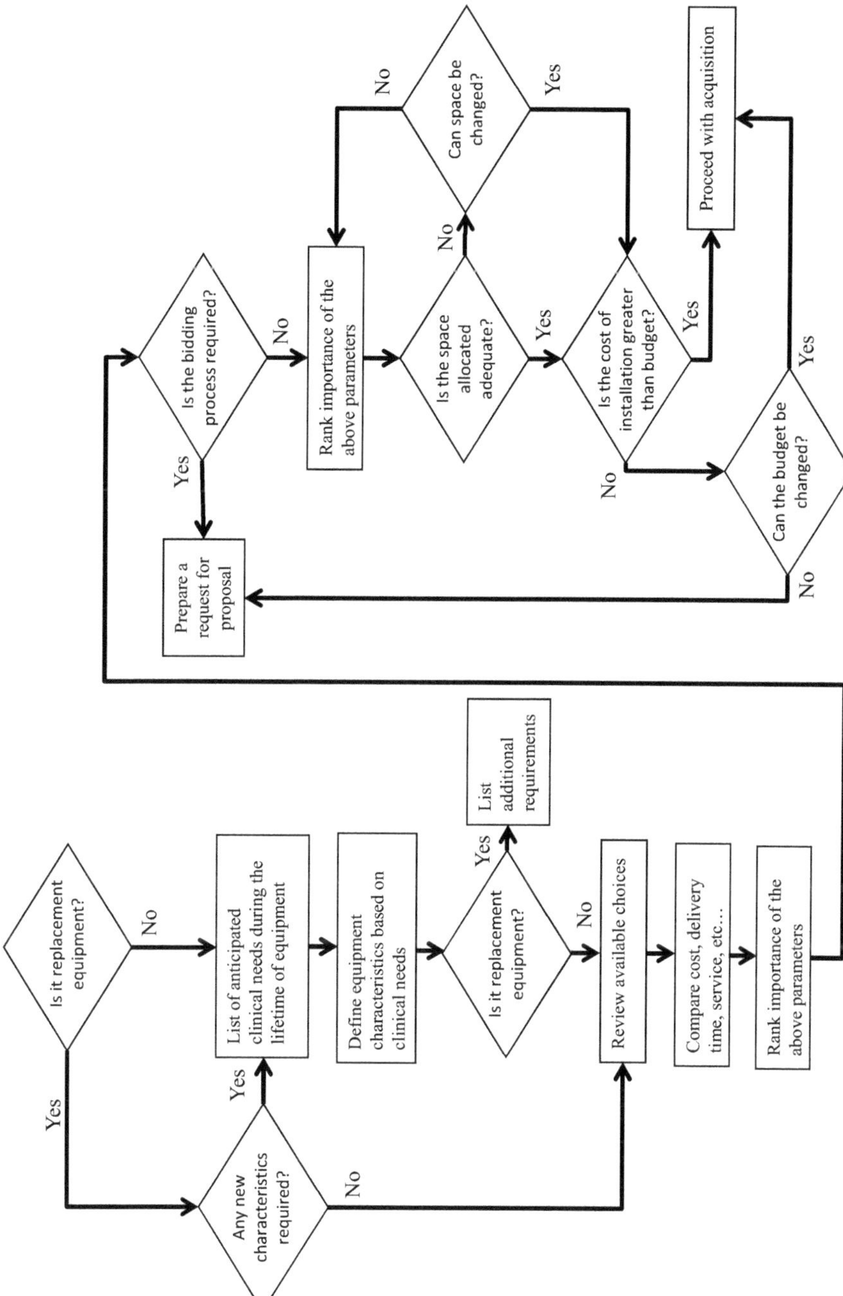

Figure 5–2. Process tree for the purchase of radiotherapy equipment.

tial users and in commercial competition with other manufacturers. These are available in the form of product data and specification sheets, which can serve as a good starting point for the purchase agreement. Any special requirement can then be added as an addendum to the purchase agreement. This saves time and effort from being expended on repetitive work.

It is imperative that the facility's physicist develops a comprehensive acceptance-testing document with a detailed test procedure to verify each term of the agreement and machine specifications. This document should be shared with the manufacturer's representative before the installation begins so that all ambiguities are clarified in advance. It is not prudent to depend on the manufacturer-supplied acceptance test procedure exclusively. However, the manufacturer's procedure should be reviewed thoroughly before acceptance testing. It is essential that the physicist review the facility layout with the planning and installation department of the equipment vendor. They can provide very useful information on workflow, equipment layout, and special requirements. A joint meeting of the equipment planning coordinator, architect, contractor, and physicist in the earlier stages of construction is very helpful and productive. This meeting can resolve all potential problems regarding electrical power supply, electrical conduit layout, air conditioning, and other requirements for the equipment. In the case of the installation of radiation-producing equipment, the shielding design and its final approval are solely the responsibility of the physicist, even if generic room designs and shielding barrier thicknesses are available from other sources.

Equipment Installation

The physicist and facility engineer (if available) should work closely with the installation engineer. A close collaboration during the installation can reduce the acceptance testing time considerably. It is important that the facility's personnel do not interfere in the work of the installation engineer, but observe the progress in the background. In the case of the radiation-producing equipment, as soon as the equipment is capable of producing a radiation beam, a series of tests should be conducted to assure the safety of all concerned. These include:

- Testing of door interlocks
- Testing of proper operation of the emergency off switches
- A preliminary calibration of the equipment output in all modes
- A radiation survey in both controlled and uncontrolled areas around the treatment vault at the highest available radiation output and under worst irradiation conditions (without phantom)

For radiation-producing machines, a full radiation survey after a full calibration will still have to be completed to comply with regulatory requirements. The preliminary survey is done to assure the safety of individuals during the acceptance testing and commissioning.

Acceptance Testing

The installation is followed by acceptance testing by the physicist to ensure that the equipment meets the product specifications and the purchase agreement. These tests are conducted according to the acceptance testing procedure agreed on between the vendor's representative and the facility's physicist. Each facility should have the equipment necessary for acceptance testing. It is important to know that all major radiotherapy equipment, such as a linac, arrives with the functional performance test values performed in test cells in the factory. These are helpful for comparison during acceptance testing. Sometimes the functional performance values of some parameters are required to be more stringent than the vendor-declared values for special application of the equipment. For example, it is not unusual to require a radiation isocenter tolerance within 1 mm diameter of the linac scheduled to be used for high-precision radiation therapy and radiosurgery.

Equipment with special modes must be tested separately for each modality. For example, on a linac, test total-body irradiation mode for maximum monitor-unit and dose-rate; test high-dose-rate total skin electron therapy mode for maximum MU, dose rate, and field size interlocks; test electron arc mode for MU/degree and dose rate, etc.

Other important aspects of acceptance testing are to assure the safety of the patients and machine operators and to provide critical baseline data for future quality assurance tests. As mentioned before, there is very little risk-based guidance from professional organizations at the present time. Therefore, it is essential that each physicist who embarks on acceptance testing and commissioning radiotherapy equipment carefully review and implement methodologies presented in this summer school proceeding.

Commissioning

Satisfactory acceptance testing simply assures that the accelerator satisfies all agreed-upon specifications and pertinent safety requirements. The process of commissioning any radiotherapy equipment for clinical use includes comprehensive measurements of input parameters and clinical validation, preferably employing end-to-end testing procedure. For example, the commissioning of a linac entails the measurement of dosimetric characteristics of all clinical beams and measurements of parameters necessary to model these clinical beams in the treatment planning system. Linac-specific beam data for commissioning is highly dependent on the dose calculation algorithms used in the treatment planning systems. The model-based dose calculation algorithms, such as convolution/superposition, require much less measured data than correction-based algorithms. Irrespective of the dose calculation algorithm, it is essential to have a minimum dataset that includes percentage depth dose, isodose distribution, and output characterization for a series of field sizes. It is very important that the measured dosimetric characteristics of the commissioned linac are compared with published data on the same make and model, if available. Radiological Physics Center (RPC) in Houston is an excellent resource for such data. RPC keeps measured data on practically all types of clinical machines in its database.

Commissioning also includes the development of standard operating procedures and checklists for all radiotherapy techniques. The training of all designated staff in the safe operation of the equipment is also an essential component of the commissioning process.

End-to-end testing verifies that a piece of equipment performs as designed under normal conditions. Many events occur when operators use equipment outside of the normal procedures. It is extremely important to know where the limits of operation fall. Thus, during commissioning, equipment should be challenged by deliberately pushing wrong buttons, ignoring warnings, and overriding interlocks, if possible. One cannot be sure of the reliability of a unit until its failure point has been established.

Physicists must resist the pressure to initiate clinical treatments as soon as the acceptance testing is finished. Rushing into clinical implementation without completing proper commissioning can potentially cause harm to patients. Therefore, an appropriate amount of time must be set aside for this activity based on the projected use of the equipment. It is imperative that the physicist have proper instrumentation to collect all necessary data. The report of AAPM Task Group 106 provides guidance on this subject (Das et al. 2008). The commissioning of special treatment techniques such as IMRT, IGRT, and SBRT require special considerations. The reports of AAPM Task Groups 119, 120, 179, 148, and 101 discuss the nuances of these treatment techniques (Ezzell et al. 2009; Benedict et al. 2010; Langen et al. 2010; Low et al. 2011; Bissonnette et al. 2012).

5.4.2 Procedures Commissioning

It is increasingly appreciated that there is an important "human use" component of commissioning. This includes user training and the development and documentation of clinical workflows that support safe and effective use of the new equipment. These concepts have gained wide acceptance as described, for example, in the ASTRO document *Safety is No Accident* (Zeitman et al. 2012) and the references therein. These concepts are not new; the report of AAPM TG 40 (Kutcher et al. 1994) describes the importance of a team approach, resource requirements, professional responsibility, and education. Sometimes these operations work on the input (QC) and sometimes on the output (QA) of a sub-process, but there is no doubt that they are QM. Several known serious errors have been partly attributed to users misunderstanding their tools, including situations involving the inverse square law. The increasing complexity of treatment machines—along with the growing interest in image guidance and adaptive re-planning that requires users to make more rapid decisions—makes human use commissioning particularly important.

Commissioning a procedure entails many of the same steps as commissioning a piece of equipment:

- Develop the process map with the team performing the process so all steps are understood and all players agree with the expected flow
- Perform the FMEA to look for hazards in the process and possibly redesign the process to eliminate those hazards (being sure to revise the FMEA looking for new hazards resulting from the redesign)

- Make a fault tree to understand the propagation of the failures

Much of the commissioning of a procedure performs an end-to-end walk-through, or dry run, of the procedure. During this, pay attention to the following:

- Make sure that the process map matches the procedures
- Assure that all players know what they should do, and have the required training and *understanding* of what they do, to adapt if the process departs from the expected path
- Establish the lines of communication so everyone knows what information they need to provide to whom and how that information is passed (forms may prove useful to ensure that no information is neglected)
- Check that all equipment, instrumentation, and devices are present and ready, and that the staff know where to find replacements if necessary
- Verify the presence and operation of any interlocks, safeguards, barriers, or warnings, and ensure that all present know what to do if a warning occurs
- Plan for recovery and mitigation strategies for potential failures
- Compile all information that might be necessary in a convenient place, including phone and pager numbers, and alert all principals that the information is in that location
- Check that all steps can be performed in the time allotted for them, and perform dry runs often enough so much of the operation moves from rule-based to skill-based execution.

After the dry run, the team should meet to discuss their thoughts about the execution and possible modification of the procedure to improve reliability and efficiency. Backup staff should also participate in the dry run. The commissioning of procedures addresses many potential failure modes, and it helps maintain the quality in most processes.

5.5 Quality Management Tools: Audits

5.5.1 Impact and Importance of Audits

The first type of QM tool considered here is an audit, which may be considered a special form of the independent checks presented in Section 5.3 above. Auditing, or the systematic review of practices and records by a qualified and empowered party, is a critical element of any quality management program. Audits provide an essential feedback mechanism to every aspect of the quality program, including, but not limited to:

- Evaluation of the *design* of the program, relating to its initial and ongoing appropriateness and adequacy
- Evaluation of the *implementation* of the program, relating to how well it meets the design goals in the face of practical and conceptual limitations

- Evaluation of *compliance* with the program's details and spirit
- Evaluation of the overall *effectiveness* of the program in controlling quality and safety of the care delivered

These elements are, of course, also interrelated, and it can be very difficult to readily categorize and tease them apart in practice.

Audits can take many forms, as discussed in detail below, but perhaps the most visible are the audits conducted by the Radiological Physics Center (RPC) (Ibbott et al. 2008). The RPC, established in 1968 in an agreement between the AAPM and NCI, now monitors 1532 radiation therapy facilities. Among the auditing activities of the RPC is a program to remotely monitor beam outputs in external beam radiotherapy using TLDs or OSLDs sent by mail (Kirby et al. 1986). Audit data from the RPC in 2006 showed that approximately 11% of institutions had at least one beam that fell outside the RPC's 5% dose agreement criteria (Ibbott et al. 2008). This was an improvement over earlier data, but clearly highlights the need for such independent audits. It must be appreciated that the remote RPC audit only measures beam outputs under standard conditions, e.g., 10 x 10 cm fields at a standard SSD, and does not assess the many other possible sources of error in the process of radiotherapy delivery.

Another form of audit is the credentialing that is required for participation in some cooperative group clinical trials (Ibbott et al. 2008). This credentialing process is only required of an institution that participates in cooperative trials, e.g., RTOG. When it is performed, however, it provides valuable feedback. Credentialing is performed by the RPC working in conjunction with the NCI-sponsored cooperative trial group. The process consists of a questionnaire, the submission of example plans, and a test run on an anthropomorphic phantom supplied by the RPC. The phantoms are site-specific, and they include films and TLDs or OSLDs. They therefore provide objective feedback on the quality of planning and treatment delivery. The RPC credentialing results are sobering. In 2008, Ibbott et al. reported that approximately 30% of institutions participating in a head-and-neck study failed to deliver a dose distribution that agreed with their own planning system within a 7%/4 mm criterion (Ibbott et al. 2008). These results underscore the need for improvement and the role that an external audit can serve.

There are also data from cooperative group trials that support the value of independent audits. In the cooperative group trials that required outside credentialing, the rates of protocol deviations were approximately three times lower than in trials that did not require such credentialing (Ibbott et al. 2008). A recent analysis of European data also suggests that institutions that perform phantom tests (or "dummy runs") may provide better-quality radiotherapy (Fairchild et al. 2012). In that study, institutions that previously participated were three times more likely to pass the dummy run on their first attempt ($p<0.001$).

Another form of quality audit is practice accreditation. By this we do not mean the certification by the Joint Commission—which rarely probes the services of a radiation oncology department in any depth—but rather accreditation that is specific to radiation oncology and aimed at improving the quality of care. A voluntary practice accreditation system for radiation oncology administered by the American College of Radiology (ACR) has been in existence since 1986 (Ellerbroek et al. 2006). Until recently this system was administered in collaboration with ASTRO, though this has now changed, and it remains to be seen what the future form of this accreditation will be. In any case, roughly 10% of radiation oncology clinics participate in the program, but participation is expected to grow as the other professional societies determine how they will contribute and administer such programs. Practice accreditation provides the opportunity to examine one's own practice from the total quality-of-care perspective with independent, expert input, including an on-site audit.

5.5.2 Issues and Considerations in Audits

It's a cold reality that an all-encompassing, multi-factorial audit—one that would provide the central nervous system to a perfectly engineered quality program—is too expensive, too time-consuming, and requires way too much expertise to be implemented and supported by real-world health care operations, or by pretty much any real-world operation. Instead, and particularly in health care, audits occur in a piecemeal fashion, often in a relatively strongly politicized context, and often with an emphasis on practice elements that are irrelevant to the QM program. It is almost always the case that the audit process has to be perceived to "pay for itself" in some plausibly demonstrable way, or audits are tolerated only because of a regulatory requirement. Where there is an economic or regulatory imperative, there is always an accompanying political mandate that may well be at odds with the "pure" engineering-based process that QM theorists (including many of us) spend much energy advocating.

This is the reality in which we operate and. as with any political context, the hallmark of effective management is to balance the dues that must be paid against the values that can be derived. In some ways, effective management of available audit resources amounts to optimizing the use of the wrong tools to solve the right problems. Given that, it is important to understand the dynamics, costs, and values of audit resources, particularly those provided by or demanded by extra-institutional organizations. Serious failures can readily arise in the cracks if the value and scope of audit programs are overestimated.

There are many axes along which the characteristics of audit programs could be decomposed. Let us discuss a few somewhat arbitrary ones (it's far from clear what might constitute eigenvectors).

5.5.2.1 Internal vs. External

An internal audit is us checking on us. As such, an internal audit should be the most finely tuned tool in our armamentarium for evaluating how well our QM program is performing and for providing the essential feedback to shape and prioritize changes. It is truly impossible to implement and maintain a sophisticated QM program without the engagement of a management team that approaches auditing as a central part of the program. We are accustomed to doing this in limited ways with, for instance, mortality and morbidity conferences and regular safety committee meetings. Expanding that role to evaluate in a unified, systematic way the performance of the entire QM program is a significant undertaking. There are challenges in identifying the right people, allocating the time, defining roles and rules, maintaining enthusiasm, and engraining habits.

There is also a constant pressure for an internal audit process to become entwined with the HR performance evaluation and discipline process. As we all know, a well-engineered QM program will anticipate and accommodate so-called "human error" and, as such, will tend to desensitize the care delivery process to the specific quality of performance of a particular individual. But in the reality of limited resources, the performance of individuals does matter and is a legitimate concern to QM management. In addition, we recognize that a "blame-free environment" is useful in encouraging the reporting of incidents, and such incident reports are the lifeblood of the audit program. But incident reporting and review inevitably also become the fodder of HR decisions. The balance is extremely hard to find, and it requires the diligent nurturing of an appropriate philosophy and culture, a balance perhaps rarely achieved in practice, but very important to constantly strive toward.

By contrast, external audits are almost by necessity ignorant of the details of the local QM program and instead, they bring their own set of rulers. At the same time they do not have to concern themselves with day-to-day operation details like those that play into budgeting and HR. This gives an external audit the chance to bring a truly useful broader perspective to the table for use by local management. But it is important for auditors and local management alike to remain clearly cognizant that all QM is local, and what may make sense as a philosophical recommendation—or as an observation that "institution X does that this way"—is not and should not be seen as implementable per se. It is material for consideration by the internal audit and management team at best. There is no plug-and-play in QM.

External audits have the advantage of being automatically perceived as more authoritative than internal audits. (Nobody is considered an expert at home.) Again, this may be problematic if the external audit is conducted with a political agenda behind it ("we need to clean house"), with an inadequate respect for local circumstances, or, frankly, just done poorly. On the other hand, a close collaboration of the local audit committee with external reviewers can sometimes result in useful and empowering recommendations in a final external audit report.

5.5.2.2 Integral vs. Stochastic

An audit program that is integral to the QM program should be part of the design of the QM program. Building in feedback ("administrative servos") is just good engineering practice. Careful thought as to how we will know that what we've designed and implemented is actually working needs to be part of the design process, not an afterthought. The audit component of the QM program needs to be designed in tandem with the rest of the QM program.

Audits that are not designed as part of the program are going to be, in some sense, random. They will sample an essentially random set of observations and test them against essentially random benchmarks. This is not without value. Random sampling of water quality at a treatment plant tells us something that may be actionable about the operation of the plant at that moment with the inputs of the moment. But negative findings could easily be false, and positive findings tell us essentially nothing about the problem. Without knowledge of the system, observations of system failure are interesting, but not instructive.

5.5.2.3 Engaged vs. Perfunctory

The traditional "talking cure" approach to psychotherapy is perhaps the ultimate example of engaged audit. A skilled therapist will spend a great deal of time evaluating one's psyche exactly "where it lies" before attempting to suggest modifications that might result in a psyche that more reliably produces a quality product. This type of analysis involves making some standardized assessments, but mostly it is a slow and essentially organic process of probing and observing with a great deal of meta-analysis, noting what is said but also seeking the true meaning of what is said. It takes forever. One can imagine a similar approach to quality audits involving a Zen master QM expert listening and hearing deeply before rendering opinions that are keenly focused on the system "where it lies." Wouldn't that be great?

Many real-world audits amount to a perfunctory race through a prepared generic checklist that has nothing specific to do with the local QM operations. There are flights to catch. Again, there is value in these comparisons against generic benchmarks, but we must be vigilant in resisting the urge to over-interpret findings and turn them into specific, implementable recommendations. We must also be extremely wary of the false-negative trap of feeling all has been proven to be well if a checklist does not result in an observation of a deficiency. The series of approximately 100 consecutive misadministrations of prostate brachytherapy at the VA Medical Center in Philadelphia stands as a poster child—in the aftermath, hospital administrators cited accreditation by the Joint Commission as evidence that they had been duly diligent (NRC 2012). The details of a radiation oncology QM program are well beyond the scope of a Joint Commission accreditation.

5.5.2.4 "Friendly" vs. "Hostile"

Audits always come with a point of view that, among other aspects, may be primarily collaborative and focused on improvement, or may be primarily vested in finding and reporting deficiencies to enhance the perception that the audit was valuable. To get a grip on whether an audit is essentially "friendly" or "hostile" (that is, on our side or on their own side) it is helpful to follow the money. Particularly with external audits, someone is paying someone for something. Whoever is paying has an expectation, and whoever is taking the money will feel some pressure to satisfy the vested interest.

The healthcare facility may be paying for a certificate that carries with it competitive advantage or, in some cases, the ability to bill for certain services. In that case, the auditing entity has a financial pressure to grant the certificate. This is not unlike the much-discussed pressures of "grade inflation" in higher education. Parents are very clear with administrators that because they are paying for a degree, they expect that Johnny and Jane will get passing grades. Granting of the grade, or certificate, is not automatic, and the process must certainly maintain some degree of credible objectivity, but the pressures are real to deliver what has been paid for, and inflation is inevitable.

It may be the federal government in the form of National Cancer Institute (NCI) that is paying for the audit. The RPC, which is a tremendously valuable resource to the radiotherapy community, is paid by the NCI to assess quality of the physics programs at institutions that participate in NCI-funded collaborative trials. While the local institution pays a tax that goes toward support of the audit program, the ultimate customer for RPC services is NCI. RPC's practices and postures have to, at the end of the day, provide adequate value to the NCI to support continued funding of their government contract. Again, the integrity of the program is not in question, but it is useful to remember that when RPC audits your work remotely or on-site, they are performing a mandatory inspection in the official capacity of government-funded police and judging your work against their own internally derived benchmarks.

Perhaps you can persuade a colleague to audit your practice for free, or in exchange for in-kind services, or perhaps as a formal service for a fee. This sort of arrangement aligns the money more favorably with the interests of the organization. If the colleague has some deep knowledge of QM program design and implementation, and especially if the audit is of both broad and deep scope, this sort of arrangement could be extremely valuable (Halvorsen et al. 2005) But there are also traps. A truly friendly audit might be more friendly than honest, especially if you are about to go audit the auditor's program in return. And unfortunately, not all colleagues are qualified auditors. A subcommittee of the AAPM spent several years trying to design a clearinghouse that would vet and match qualified auditors to medical physicists desiring TG–103-style program audit (of just the physics program, not the whole QM program). The subcommittee finally disbanded in frustration, having failed to find a viable structure for such a program that would satisfy all of the competing concerns. Let this serve as a tale of caution.

It can be a little harder to follow the money that underwrites internal audits. But it is a useful consciousness-raising exercise to nonetheless try to understand where the money comes from and what expectations are attached.

5.5.2.5 Evaluative vs. Authoritarian

As already discussed, in a perfect world, audits would be deep and focused on a skilled and knowledgeable evaluation of a program's design and implementation. That's hard to come by, but it is still important to be aware as to whether an audit is designed more to learn what you do or rather to tell you what to do. The latter is common and not all that useful. One must sometimes go through the motions of satisfying an imposed requirement in the context of a mandatory audit, but that should not be confused for care and feeding of the QM program.

With these and other considerations in hand, it seems clear that there are three categories of audits as triaged by their value in support of a QM program:

1. Internal audit designed as part of the program and fully supported financially and culturally by the organization.
2. Third party audits, both internal and external, both voluntary and mandatory, that are not well integrated with the program, but which provide independent assay.
3. Third party audits that provide no value to the QM program.

Category 1 audits are really not optional. Category 3 audits are a fact of life, but must not be confused for QM at the risk of self-inflicted failure. Category 2 audits are where the work of opportunity lies. Much like the problem of adapting the functionality of an information management system like current generation EMRs to support a well-engineered QM program, there may be elements of third party audits that can be used in support of the audit program of category 1. To identify and leverage those resources becomes, once again, an engineering problem best addressed in the context of the design of the QM program.

5.6 Other Quality Management Tools

This section builds on the underlying principles outlined above and considers specific tools used in quality management in radiation oncology. This list of tools is not meant to be exhaustive, but rather to encompass the most commonly used tools. This section will consider the following tools:

- Interlocks
- QA and QC tests in radiation oncology
- Safety barriers and stops
- Checklists

There are other specific QM tools that are in use, but which are not discussed in this chapter. These include:

- Alerts and warning signs
- Computer checks (discussed in Chapter 8)
- Training and refresher training (also discussed in Chapter 8)

5.6.1 Interlocks

Interlocks, or "forcing functions" (Norman 1988), are considered to be one of the most effective safety interventions (Vaida 1999; Grout 2006). An interlock forces one to perform an action a certain way. Examples include mechanical interlocks on linear accelerators such as mechanical touch guards that prevent physical injury to patients, or the hardware interlocks on some stereotactic radiotherapy devices that force the use of the intended collimator size.

These are hardware interlocks that prevent a very specific type of error, but there are other examples of more interlocks or automation that operate at a more general level. An example is a patient registration system that employs bar codes on registration cards. The patient presents the card, the bar code is scanned, and the patient's treatment record is brought up. This is intended to prevent the unintended treatment of a patient with the wrong treatment plan (Ford et al. 2009).

Software interlocks exist as well. An example is the functionality available in some radiation oncology treatment management systems that enforce document review. In at least one system, the software settings can be configured to require that certain documents have been signed prior to treatment. For example, a treatment plan document must be approved by a medical physicist and a physician before treatment can be initiated. This program stop enforces the review of the intended plan prior to treatment. A patient's bar code and plan approval interlock should be viewed as being in the second tier of error prevention, since there are ways to subvert them. It is possible in some systems, for example, to initiate a treatment without using the bar code system.

Quality improvement must evolve toward the selection of quality management tools such as interlocks that make certain errors impossible. It is only when such tools are *not* available that double checks and other forms of corrective action should be considered.

5.6.2 QA and QC Tests in Radiation Oncology

The basic theory behind QA and QC tests is presented above in Section 5.3.2. This section discusses specific QA and QC tests in radiation oncology and how they are used. These tests fall into two general categories: general and patient-specific.

5.6.2.1 General QA

General QA includes much of the work of a clinical physicist. Many AAPM and other organizational reports—as well as individual group-refereed publications—provide guidance on these tasks. General QA can be thought of as QA during the commissioning process (discussed in Section 5.4 above) and "Periodic QA." Periodic QA of all systems (e.g., the TG–142 report for linac QA) is one of the most recent and influential of the references, but simulators, planning systems, and auxiliary devices also need periodic verification of their continued, safe functionality (Klein et al. 2009)). AAPM TG reports (e.g., TG 142) typically set forth generic test tolerances and frequencies for relevant parameters, but room for customization remains, and TG 142 was to serve until the techniques of TG 100 were developed at a facility (Huq et al. 2008). For example, if there has been no major work on a linac, *and* commissioning QA has been checked by some independent means, *and* the beam quality indicators (e.g. TMR20/10) remain at baseline, what is the QA value of checking photon or electron output factors annually? What is the QA value of checking transmission factors annually, such as for physical wedges, once one has verified that their positions have not changed? On the other hand, given the increasing frequency of single fraction, high-dose treatments, especially in the curative setting, should some beam dosimetry and geometry tests be made more frequently?

5.6.2.2 Patient-specific QA

Patient-specific QA encompasses much of a clinical physicist's daily work. Dosimetrists, therapists, and physicians are also deeply involved, as has long been acknowledged (Kutcher et al. 1994). QM steps are necessary at each step of the radiation therapy process, and there are many steps. Within a similar framework, Ford et al. identified 269 steps or "nodes" (Ford et al. 2009), and TG 100 identified 215 steps for IMRT alone. QA of the physician's up-front work of prescription, site specification, and target and risk organ delineation is best done through peer review; similarly, peer review of individual treatment plans selected by an individual physician is important QA, as is review by a planning physicist or senior dosimetrist. Physician peers and physics and dosimetry personnel are likely to spot different shortcomings in plan quality. The physician's required monitoring of the patient's progress and his or her review of treatment unit images are also QA on the patient's progress through treatment. Physicists often refer only to pretreatment dosimetric measurements as "patient-specific physics QA," but, in fact, the overall quality and correctness of a treatment plan also calls for physics or dosimetrist QA (the "physics plan check"). As well, there are risks in palliative treatments. Although these often use simple beam arrangements and do not require sophisticated dose calculations or graphic plans, they may need careful consideration of previous treatments and, if set up on the fly in an emergent setting, may risk blunders in monitor unit calculation (dose doubling being the worst).

The therapist has an important role in patient-specific QA through the therapist's pretreatment plan review. These reviews ensure practical, comfortable, and correct patient setup (at simulation and during treatment), call anomalies in machine behavior to the attention of the physicist, and forward patient complaints to the attention of the nurse or physician. It is important that department culture encourages and takes therapist input seriously. The reader should refer to Chapter 2 where safety culture is discussed.

5.6.2.3 Desirable characteristics of QA/QC checks

QA and QC checks are a large part of the clinical physicist's responsibility and, as treatment complexity increases, they will become a larger part. To perform these tasks effectively and accurately, it is desirable to design tests that seek to identify several different errors or failures in a single operation. Several companies offer array devices for daily checking of beam characteristics (accompanied by automatic report generation) in a single beam delivery. There are other, more sophisticated, array devices for efficient pretreatment measurement and gamma analysis of IMRT, VMAT, and tomotherapy. A recent study by some of us (Ford et al. 2012) quantified the effectiveness of pretreatment QA checks and other checks through an analysis of an incident reporting database from two institutions. That study concluded that QA tests of IMRT treatments could detect less than 2% of all high-severity incidents in the database. This underscores the need to perform such QA tests in an efficient and cost-effective manner. In the future, software checks may reduce or eliminate the need for such measurements (Litzenberg et al. 2002; Hunt 2012).

Several groups have reported on EPID-based transit dosimetry for on-treatment QA. This technique improves on current patient-specific QA because it identifies treatment problems that occur anywhere during treatment. These treatment problems include both machine delivery problems and anatomical factors, such as weight changes, tumor growth, or tumor shrinkage, as discussed in the references (Mans et al. 2010; Mans et al. 2010; Berry et al. 2012).

The implementation of semi-automatic segmentation tools in planning systems may provide QC at least for normal organ delineation (providing these tools are themselves subject to QA before going into general use). And methods have been proposed to partially automate routine aspects of the physics plan-check process. If the planning system and treatment management system databases are separate, such software can reduce the physicist's time spent checking items such as field size or wedges or, depending on preferences, serve as an extra set of QA "eyes" in addition to those of the physicist (Azmandian et al. 2007; Furhang et al. 2009; Siochi et al. 2009).

5.6.3 Safety Barriers and Stops

Safety barriers and stops are actions in the workflow that are intended to prevent an error from propagating. They can be considered as somewhat redundant to the quality assurance measures presented in Section 4.2, but since they are not typically discussed as part of QA, they are considered separately here. One list of various safety barriers in the workflow process can be found in the white paper on incident reported recently developed by the AAPM Work Group on Prevention of Errors (Ford et al. 2012). A somewhat modified version of this list, appropriate for external beam radiotherapy, appears in Table 5–2. In some cases, these barriers are explicitly advocated in a report from a professional society, such as an AAPM task group or the like. These cases are listed in the reference column. Of note is the fact that Task Group 40, published in 1994, already included mention of many of these barriers (Kutcher et al. 1994).

Little data exists on the operations of safety barriers or stops. Critical questions remain as to which barriers are the most effective or how the barriers should be used in combination. A recent study by one of us (Ford et al. 2012) attempted to address this question by examining an extensive incident reporting database from two institutions. The most effective safety barrier in that study was pretreatment chart check by a physicist, though even that barrier could only detect approximately 60% of errors in a best case scenario. The study did suggest, however, that when safety barriers are used in combination, error detection rates can be very high and approach 100%. Further work is needed to understand which safety barriers are the most effective and how their use can be optimized.

Table 5–2. Safety barriers in external beam radiotherapy (from AAPM Task Group 40 and other references)

Number	Safety Barrier	Reference
1	Verification of patient identification	
2	Verification of pretreatment conditions (e.g., previous Rt, chemotherapy, pregnancy, etc.)	
3	Verification of image data set for treatment planning	
4	Verification of reference points	
5	Physician peer review	3, 53–55
6	Review of treatment plan by physician	3, 53, 54
7	Review of treatment plan by physicist	3, 53, 54
8	Review of treatment plan by radiation therapist	
9	Independent confirmation of dose (handcalc or patient-specific QA)	3, 54, 56, 57
10	Use of R&V system	
11	Verification of parameters in R&V system	
12	Time out by radiation therapists	
13	Verification of SSD	
14	Verification of treatment accessories (e.g., use of correct wedge or bolus	
15	Verification with port films	3, 58–60
16	Verification with on-line CT	36, 59, 61
17	*In vivo* dosimetry	3, 62–64
18	Monitoring patient during treatment	
19	Regular independent chart check	3, 65
20	Regular clinical patient assessment	
21	Use of checklists	
22	Independent review of commissioning	
23	Regular external audit	
24	Regular equipment performance verification	

5.6.4 Checklists and Forms

Because of the complexity of modern radiation oncology and the limitations of human performance as noted above, cognitive aids are a useful tool for reducing human error. One such tool is the checklist. This tool has been used in areas outside medicine, such as aviation (Degani and Wiener 1993) and product manufacturing.

Checklists have been extensively validated in non-medical fields for many years, and they have proven to be an effective tool in error management, a key instrument in reducing the risk of costly mistakes, and an aid in improving overall outcomes. Recently, checklists have been advocated in medical disciplines such as surgery (Lingard et al. 2008), anesthesiology (Hart and Owen 2005), and diagnostics (Ely et al. 2011) as one of the main error management tools, and their use has shown measurable improvements in patient safety and error reduction.

One of the early successes in medicine was a five-point checklist pioneered by Peter Pronovost. Pronovost's checklist significantly reduced blood stream infections acquired in the ICU as a result of central lines (Pronovost et al. 2006). The median rate of infection decreased from 2.7 per 1000 catheter-days at baseline to 0 within the first three months after the implementation of the intervention. The benefit from the intervention was sustained, and there was a reduction in the rate of catheter-related bloodstream infection of 66% at 16 to 18 months after implementation. Pronovost's group concluded that broad use of this intervention could significantly reduce morbidity and the costs of care associated with catheter-related bloodstream infections.

Building upon Pronovost's success, the World Health Organization's (WHO) Patient Safety Programme "Safe Surgery Saves Lives" developed a Surgical Safety Checklist as a means of improving the safety of surgical care around the world. The WHO surgical safety checklist included specific steps during induction of anesthesia, surgical timeout, and transfer of the patient out of the operating room. In a multinational study involving eight hospitals from diverse economic settings, the use of the WHO surgical safety checklist improved compliance with standards of care by 65% and reduced the death rate following surgery by nearly 50% (Haynes et al. 2009).

Recognizing the relevance of checklists as safety tools, AAPM members recently formed the multidisciplinary Task Group 230. Based on current evidence, this task group will provide recommendations for designing and implementing clear and effective safety checklists for radiation oncology.

5.6.4.1 The Role of Checklists in Error Management

As discussed in Chapter 3, human errors can be categorized, one of which is a slip or omission associated with distractions, exhaustion, burnout, or lapses of concentration. Checklists provide a method to reduce the risk of errors originated by slips. The airline industry is a prime example of the successful use of checklists (Boorman 2001). They have learned that when pilots and air traffic controllers are provided and trained on the effective use of evidenced-based checklists in an environment, that motivates them to follow such checklists every single time, and the likelihood of errors and accidents is drastically improved.

Checklists provide a basic memory guide and backup for those tasks that are easy to forget. In other words, checklists ensure that the "basic stuff" is not missed (e.g., wrong patient, bolus, electron block, site), thereby allowing the team to concentrate on the difficult and complex tasks that require full attention (Gawande 2009). Additionally, checklists provide a communication and workflow pause to ensure that the team or individuals are working together. Properly structured checklists facilitate systematic and consistent care delivery, thus reducing variability and improving performance. Checklists must have the right balance of information and structure in order to support clinical practice without compromising or impeding professional judgment (Sidney 2011). In review, checklists function as a supporting interface between individuals and their environment, helping and guiding a particular workflow or procedure (ARHQ 2012).

5.6.4.2 The Checklist Development and Implementation Process

Based on current literature and best practices from the airline and medical industry, the development and implementation process can be categorized on the following steps:

Clinical Need and Evidence-based Best Practices

The first step for developing checklists is to find those areas or processes where the most clinical impact is possible and that have the lowest barriers for implementation and utilization (Bosk et al. 2009). A literature review of best practices, risk assessment (as with FMEA, as discussed in Chapter 4), empirical evidence, and regulatory, local, and community input can help in the selection process. Poor selection or ambiguity on the checklist's goals, role, or tasks will most likely lead to failure on the checklist intervention (Gurses et al. 2008). Therefore, each checklist intervention needs to be associated with an explicit, concise, and unambiguous behavior. When selecting processes or procedures that will potentially benefit from checklists, it is important to consider that an excessive use of them could potentially be detrimental to the practice, leading users and teams to experience "checklist fatigue" (Hales et al. 2008). With this in mind, the selection process should concentrate on those steps that are critical, but which are often missed or overlooked, and which can potentially put the patient at the highest risk of harm if not done or missed (Weiser et al. 2010). Selection of the right areas and processes that will benefit from checklists, as well as wisely selecting the checklist elements, are fundamental steps leading to the success of the tool.

The Multidisciplinary Team

The practice of medicine is populated by different perspectives, points of view, and terminology, sometimes referring to the same item in multiple ways. Radiation oncology is no exception, since safe and efficient delivery of care is achieved when the team (i.e., physicians, therapists, dosimetrists, physicists, nurses, and supporting staff) shares a unique perspective and goal. In this context, checklists provide a path to democratizing knowledge, as well as a way to standardize and improve the communication channels and information among the care team. The checklist development process will have higher impact and a greater likelihood of success when it is done in the context of a multidisciplinary team. The pool of knowledge is filled when each team member has the opportunity to share his or her own perspective, thus facilitating the development of a workflow and terminology that is clear to everyone using the checklist. Additionally, each team member who effectively participates during the development process acquires a sense of ownership, which will have a positive impact during implementation and acceptance of the checklist into the practice.

Drafting: Content and Format Definition

The content of the checklist should be organized so it facilitates efficient workflow and provides clear guidance on what, when, how, and who should performed each item. The language and sentences used for the checklist items should be simple, direct, and unambiguous, but maintain the specialized language of the field. Checklist design should incorporate the user or team context during the design of the checklist, complementing the workflow and avoiding interference with the safety and efficiency of care delivery. Processes that call for checklists with large numbers of items are better handled by dividing them into sub-steps and then creating smaller checklists for each sub-step. Miller found that the average human memory can accurately retrieve and use no more than seven (±2) pieces of information at a time (Miller 1956). Moreover, during periods of mental fatigue, when under high stress, or when doing highly complex procedures, the tasks that an average person can handle and commit to memory decrease further (Lorist et al. 2005). When borrowing checklists from other practices, the content and format of a checklist should not be considered absolute, and it would need to be modified and evaluated to fit each practice environment and workflow.

One important element when designing a checklist is to consider the type of conducting method or combination of methods to be used. There are four main methods for formulating checklists (Winters et al. 2009):

1. *Static parallel or call-do.* Using this method, the checklist items are performed and completed as a series of read-do tasks. The checklist leads the process, and directs the team or individual through the process step-by-step. In other words, the checklist uses the "cook book" approach. An example in radiation oncology might be a procedures checklist for loading a CT planning scan into the treatment planning system.
2. *Static sequential with verification.* This method requires a challenge and response mechanism by at least two individuals. One person will use his or her experience and memory to perform a group of tasks from start to finish. Then, a second team member will call each item from the checklist, so both individuals can verify that the critical items have been correctly accomplished. This method uses the combination of initial configuration and mutual redundancies as a safety barrier mechanism. An example in radiation oncology is the verification of a patient's identity by two radiation therapists at the time of treatment.
3. *Static sequential with verification and confirmation.* This method follows the same workflow as the previous method (static sequential with verification), but it is more suited for a team-based setting. Various tasks and items are done by different members of the team. Upon task completion or during a reasonable procedural pause, a designated team member calls the items from the checklist and each responsible group verifies the completion and accuracy of their corresponding tasks. An example in radiation oncology might be a checklist for the management of patients with implanted cardiac devices. Each member of the team completes a different portion of the checklist.
4. *Dynamic.* This method frequently uses flowcharts as guiding mechanisms. It is suited for complex decision-making situations where the team is confronted with multiple options and needs to decide the optimal course of action. This method is also employed by team leaders when developing strategies that required in-depth discussions with other team members. An

example in radiation oncology might be a checklist used to guide case review in peer review chart rounds.

Research from human factors engineering provides excellent guidance for checklist development that can improve individuals' perception and memory expectancy of each list item (McLaughlin 2010). Some examples of these suggestions are to use more readable fonts (sans serif), avoid using all capital letters, minimize the use of color and images, fit the checklist onto just one page, and increase the contrast between the text and its background.

Validation and Pilot Phase
In most situations, the validation of the checklist is a continuous, iterative process requiring several revisions by the developing team. During the validation process, the development team works on reaching consensus on the usability, timing, potential risks, team interaction, format, and content of the checklist. After initial validation, the checklist must go through a thorough pilot testing process, if at all possible in a simulated clinical setup conducted by a group representing the target individuals or team. Standard quality control methods like Plan-Do-Study-Act—as well as evaluation by an expert group or subjective evaluation by potential users using interviews, focus groups, and surveys—are tools that can be used during the pilot phase to collect data and improve the format and implementation of the checklist (Winters et al. 2009). The validation and pilot phases are essential for the success of the checklist because they help the development team detect and identify problems, risks, and issues before clinical deployment, thus avoiding complications that could potentially lead to resistance and the creation of unnecessary barriers.

Training and Clinical Implementation
Optimal broad practice implementation should be accompanied by an effective training on the use of the checklist. Target users and teams must have a complete understanding of the purpose and methodology of the checklist, as well as the goal of each single item on the list. There should be consistent training to avoid misinterpretation of the checklist items and erroneous answers. During the initial time following clinical implementation, the development team should follow and monitor checklist users and teams in real clinical situations to provide guidance and gather data to further enhance the tool.

Outcomes and Performance Evaluation
Measuring performance and specific outcomes is the only way to demonstrate that the intervention—in this case the checklist—is working. It is advisable to collect baseline measurements during the pre-implementation period to be able to compare it with post-implementation data to quantify the success or failure of the checklist. Incident reporting systems provide the infrastructure to collect this information (more on this topic can be found in Chapter 6). Demonstrating with concrete evidence the success of a checklist will reinforce the utility of the tool to the group and, in some instances, can help motive skeptical individuals to use the checklist (Pronovost and Vohr 2010).

Maintenance and Refinement

Checklists should evolve with the practice and reflect the most current evidence-based data, published guidelines, end-user feedback, organizational changes, and internal institutional policies and procedures. Incident learning systems provide a quality control metric for checklists, and they can potentially flag when checklists require further development. Checklists should be considered a constantly evolving document requiring monitoring, and modification, in order to adapt to the changing practices.

To read a "Checklist for Checklists" and to find examples of checklists used in medicine and the airline industry, the reader is referred to Atul Gawande's website (Gawande 2013).

5.6.4.3 The Organizational Component of Checklists

Despite evidence supporting the successful use of checklists as safety management tools, checklists do encounter resistance in medicine. Some of the main factors of this resistance are not technical, but rather organizational, social, and cultural. Care team members can wrongly believe that checklists undermine their expertise, are infantilizing, and, in some instances, take the human thinking out of effective care delivery (Pronovost Vohr 2010). In reality, effective checklists support human thinking and creativity, allow constructive team member interactions, bring together the integral experience of the whole team, and facilitate systematic care delivery by reducing process variability. Additionally, proper checklist implementation provides an open communication environment by empowering members of the care team to speak up and stop procedures when guidelines are not followed.

In the Pronovost group's article "Reality check for checklist" (Bosk et al. 2009), the authors state, "The mistake of the 'simple checklist' story is in the assumption that a technical solution (checklist) can solve an adaptive (sociocultural) problem." Further analysis of the Michigan Keystone ICU program shows that the implementation of checklists was only one of several key elements, which combined, led to the success of the program (Dixon-Woods et al. 2011). Other elements included:

- Summarizing, simplifying, and standardizing the process
- Creating internal social networks with a shared sense of mission and mutual reinforcement mechanisms
- Gathering, measuring, and providing feedback on clearly defined outcomes
- Developing and supporting a Just and Safe Culture (more on this topic can be found in Chapter 2 on Work Systems and Culture)

Checklists are, in essence, a human intervention, thus requiring a strong organizational and social infrastructure to support them. Incorporating checklists into practices requires strong mechanisms to promote teamwork, support communication, and reinforce training and shared knowledge. Checklists alone cannot do much; checklists in the appropriate organizational environment can definitely be an exceptional safety management tool.

5.7 Example: Development of a Quality Management Program from Risk Assessment in a ^{90}Y Microsphere Liver Cancer Service

This section presents a case study that illustrates the principles and tools of quality management discussed in this chapter. The example also illustrates how a risk analysis such a FMEA (Chapter 4) can be used as the basis for quality improvement interventions.

5.7.1 Background on the Procedure and FMEA

For this discussion, we will consider a specific procedure that may be of concern for a department: liver cancer treatment using ^{90}Y labeled microspheres. Specifically, for this example we will refer to TheraSheres™ from MDS Nordion, Ottawa, ON, Canada. The discussion below focuses on an FMEA-derived risk analysis from which quality improvement interventions are derived. As such, this provides a specific example of an activity under the broader umbrella of quality management.

For those not familiar with the above procedure, it involves the injection of ^{90}Y-labeled microspheres intra-arterially to treat hepatocellular carcinoma. The normal procedure has the steps spelled out in fair detail, and it uses a very detailed checklist to walk through setting up the delivery device and preparing for the injection. A radiation oncologist, a nuclear medicine physician, or an interventional radiologist can write the prescription. The activity for the treatment, based on the volume of the target segment and the desired dose, comes from a spreadsheet provided by the vendor. The activity has to be tied to the date and time of the procedure because of the decay of the radionuclide, and all this information also comes from the spreadsheet.

The example closely follows the procedure as practiced at one facility where the physicist performs all the functions that may be executed by either a medical physicist or a radiation safety officer at other facilities. In this example, we address mainly those parts of the procedure in which either the physicist plays a major part or strongly affects the physicist's role, narrowing the focus to fit in the chapter. It would be assumed that the facility would widen the investigation in a real analysis. Further, the analysis has only been performed by one of the authors to serve as an illustration of technique, and should not be used as a template for a clinical department.

Figure 5–3 shows the process map for this department's procedure, Table 5–3 shows the FMEA, and Figure 5–4 shows the fault tree. Assembling all these documents, as described in Chapter 4, is fairly straightforward, although time consuming and possibly contentious.

While making the FMEA, particular attention was paid to failures that resulted in events reported to the U.S. Nuclear Regulatory Commission or agreement states. In particular, failures have included:

- Clumping of microspheres at connections

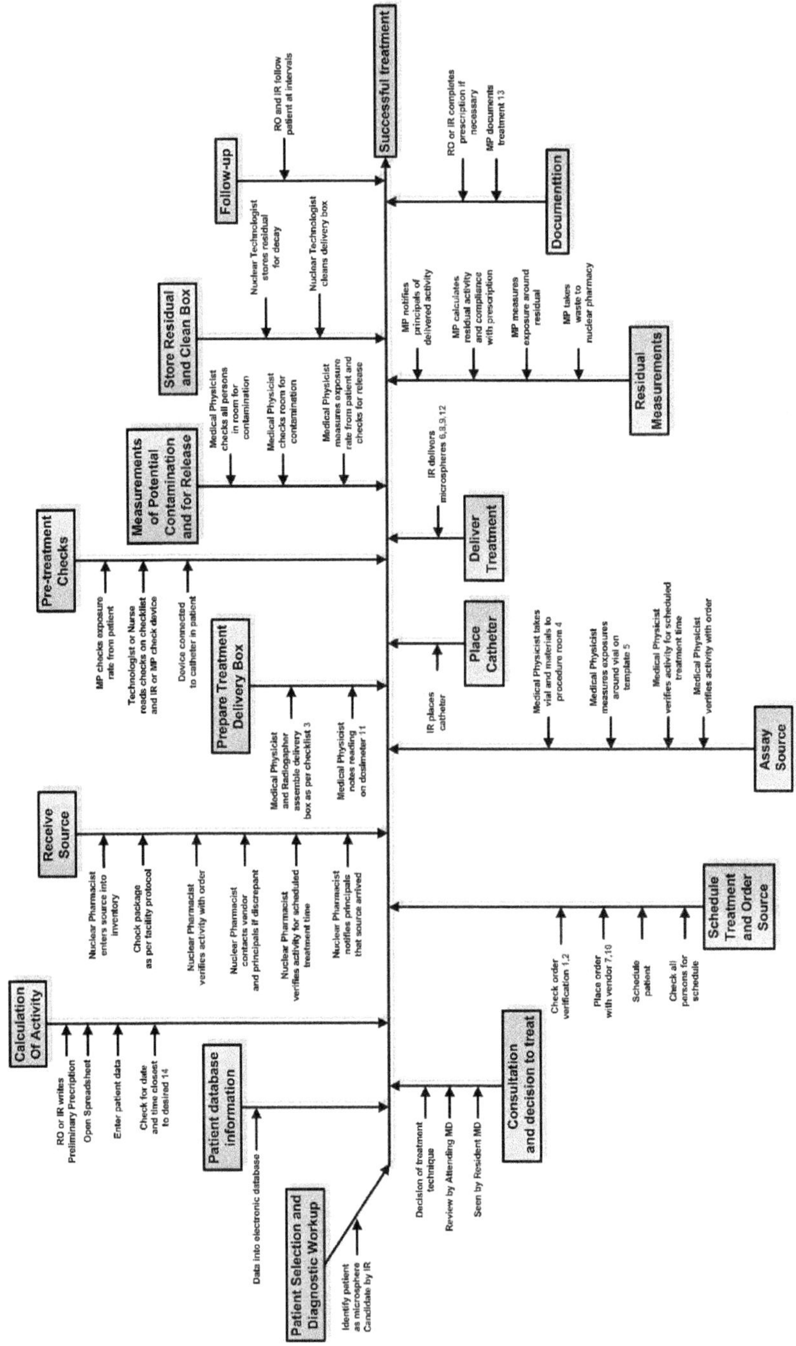

Figure 5–3. The process map for treatment of hepatocellular carcinoma using TheraSpheres™ as performed in a given clinic.

Item Number	Major Processes	Step	Potential Failure Modes	Potential Causes of Failure	Comments and descriptive scenario	Potential Effects of Failure	O	S	D	RPN	RPN Rank	S Rank
1	Decision to Treat	IR or RO decides to treat	Inappropriate decision	Incorrect patient's data accessed	Could be inappropriate presentation or history, wrong extent of disease or wrong segment	Wrong/very wrong dose or dose distribution	2	**9**	3	**54**	46	23
2	Decision to Treat	IR or RO decides to treat	Inappropriate decision	upstream physician error or clinical misjudgment in interpreting imaging studies, path reports, etc. or key data missing from EMR.	Could be inappropriate presentation or history, wrong extent of disease or wrong segment	Wrong/very wrong dose or dose distribution	3	**9**	4	**108**	38	18
3	Decision to Treat	IR or RO decides to treat	Inappropriate decision	Wrong dose chosen for presentation	Previous history, such as prior radiation treatments, not noted	Wrong/very wrong dose	3	**9**	4	**108**	39	19
4	Calculation of Activity	RO or IR writes prescribed dose	Inappropriate prescription	Incorrect patient's data accessed	Could be inappropriate presentation or history, wrong extent of disease or wrong segment	Wrong/very wrong dose or dose distribution	2	**9**	3	**54**	47	24
5	Calculation of Activity	RO or IR writes prescribed dose	Inappropriate prescription	Upstream physician error or clinical misjudgment in interpreting imaging studies, path reports, etc. or key data missing from EMR.	Could be inappropriate presentation or history, wrong extent of disease or wrong segment	Wrong/very wrong dose or dose distribution	3	**9**	4	**108**	40	20
6	Calculation of Activity	RO or IR writes prescribed dose	Inappropriate prescription	Wrong dose chosen for presentation	Previous history, such as prior radiation treatments, not noted	Wrong/very wrong dose	4	**9**	4	**144**	37	17
7	Calculation of Activity	RO or IR writes prescribed dose	Inappropriate prescription	Wrong dose chosen for presentation	Mistaken dose written	Wrong/very wrong dose	4	**9**	5	**180**	29	15
8	Calculation of Activity	MP Opesn spreadsheet	Selected old spreadsheet	Not label or sort files adequately or not receive correct spreadsheet		Wrong dose	3	**6**	2	**36**	50	45
9	Calculation of Activity	MP Opesn spreadsheet	Corrupted spreadsheet	Saved incorrect entries into spreadsheet.		Wrong/very wrong dose	2	**9**	2	**36**	51	26
10	Calculation of Activity	Enter patient data	Incorrect patient ID data	Incorrect patient's data accessed	Could be inappropriate presentation or history, wrong extent of disease or wrong segment	Wrong/very wrong dose	2	**8**	2	**32**	52	41
11	Calculation of Activity	Enter patient data	Correct patient ID data but clinical findings/images in EMR incorrect	Upstream physician error or clinical misjudgment in interpreting imaging studies, path reports, etc. or key data missing from EMR.	Could be inappropriate presentation or history, wrong extent of disease or wrong segment	Wrong/very wrong dose	2	**8**	2	**32**	53	42
12	Calculation of Activity	Enter patient data	Incorrect data entered	Typographic error		Wrong/very wrong dose	6	**8**	5	**240**	25	36
13	Calculation of Activity	Check for closest date and time for desired dose	Wrong date or time selected	Error in reading spreadsheet	Possibly detected when activity is checked	Wrong/very wrong dose	6	**8**	5	**240**	26	37
14	Calculation of Activity	Check for closest date and time for desired dose	Wrong date or time selected	Poor understanding of the spreadsheet	Inadequate training	Wrong/very wrong dose	6	**8**	6	**288**	15	32
15	Schedule Treatment and Order Source	Check all principals for schedule	Not all persons or resources available at the time of procedure	Not all persons receive notified	Examples: 1. e-mail sent with wrong address, 2 No message sent	Inconvenience if rescheduled	4	**2**	1	**8**	60	58
16	Schedule Treatment and Order Source	Check all principals for schedule	Not all persons or resources available at the time of procedure-Incorrect time sent	Error in transferring date or time from spreadsheet to notification	Transcription error	Wrong dose	4	**8**	8	**256**	20	33

Table 5–3. Failure modes and effects analysis for the procedure in Fig. 5–3

Item Number	Major Processes	Step	Potential Failure Modes	Potential Causes of Failure	Comments and descriptive scenario	Potential Effects of Failure	O	S	D	RPN	RPN Rank	S Rank
17	Schedule Treatment and Order Source	Schedule patient	Not all persons or resources available at the time of procedure- Pt scheduled at the wrong time	Transcription failure from notification to patient communication	Transcription error	Inconvenience if rescheduled	5	**2**	1	**10**	59	57
18	Schedule Treatment and Order Source	Schedule patient	Not all persons or resources available at the time of procedure- Pt scheduled at the wrong time	Transcription failure from notification to patient communication	Transcription error	Wrong dose if case procedures at the incorrect time	4	**8**	8	**256**	21	34
19	Schedule Treatment and Order Source	Place order with vendor	Wrong activity ordered	Misread spreadsheet	Reading error, detection low unless second person also calculates activity	Wrong/very wrong dose	5	**9**	9	**405**	7	8
20	Schedule Treatment and Order Source	Place order with vendor	Wrong activity ordered	Transcription failure from spreadsheet	Transcription error, possibly detected when checking the verification of the order if the person ordering remembers the calculated activity	Wrong/very wrong dose	5	**9**	8	**360**	10	9
21	Schedule Treatment and Order Source	Check order verification	Wrong activity ordered [Note, this is already a QA step. An event from this step also would need to have had a failure in the activity ordered.]	Verification not checked against spreadsheet data	Check not performed. Values assume there was an error in the activity ordered, otherwise O=0	Wrong/very wrong dose	10	**9**	9	**810**	1	5
22	Schedule Treatment and Order Source	Check order verification	Wrong activity ordered [Note, this is already a QA step. An event from this step also would need to have had a failure in the activity ordered.]	Verification not checked against spread sheet data	Difference not noted	Wrong/very wrong dose	8	**9**	9	**648**	2	6
23	Receive Source	NP enters source in inventory	Source not entered	Failure to enter		Regulatory issue	4	**1**	5	**20**	57	60
24	Receive Source	Check package as per facility protocol	Leaking source contaminates facility	Damaged box not checked		Safety issue	2	**6**	5	**60**	45	44
25	Receive Source	NP verifies activity with order	Incorrect order received and kept	Received activity is not check against that ordered: 1. Order not checked by neglect, 2. Busy and not get to check, 3 Checked incorrectly	Assumes concomitant error in shipped activity	Wrong/very wrong dose	4	**9**	2	**72**	43	21
26	Receive Source	NP contacts vendor and principals if activity is discrepant	Discrepant activity potentially used	1. Order not checked, 2. Busy and not get to calling	If source used. Likely detected by activity check later in procedure	Wrong/very wrong dose	3	**9**	2	**54**	48	25
27	Receive Source	NP verifies activity for scheduled treatment time	Discrepant activity potentially used	Order not checked for treatment time	If source used. Likely detected by activity check later in procedure	Wrong/very wrong dose	4	**9**	2	**72**	44	22

Table 5–3, cont. Failure modes and effects analysis for the procedure in Fig. 5–3

Item Number	Major Processes	Step	Potential Failure Modes	Potential Causes of Failure	Comments and descriptive scenario	Potential Effects of Failure	O	S	D	RPN	RPN Rank	S Rank
28	Receive Source	NP notifies principals that source arrived	No failure mode				0	**0**	0	**0**	61	61
29	Assay Source	MP verifies activity with order	Wrong activity taken to procedure	1. Failure to measure activity; 2. Failure to measure activity correctly; 3. Failure to check measurement activity with ordered activity	Assumes that there was a concomitant error in the activity ordered or shipped, and the source used in the procedure	Wrong/very wrong dose	4	**9**	7	**252**	23	13
30	Assay Source	MP verifies activity for scheduled treatment time	Wrong activity taken to procedure	Failure to check measured activity with that from the spreadsheet	Assumes that there was a concomitant error in the activity ordered or shipped, and the source used in the procedure	Wrong/very wrong dose	4	**9**	7	**252**	24	14
31	Assay Source	MP measures exposures around vial on template	Eventual incorrect residual measured	Failure to make measurements		Wrong compliance	3	**1**	10	**30**	55	59
32	Assay Source	MP measures exposures around vial on template	Eventual incorrect residual measured	Failure to make measurements correctly	1. Poor training; 2. Equipment failure; 3. Unnoticed high background; 4. Scale error; 5. Reading error; 6. Incorrect placement on template	Wrong Treatment if unnecessary addition treatment is given	5	**10**	10	**500**	5	1
33	Assay Source	MP takes vial and materials to procedure room	Microspheres stick to septum	MP inverts vial	1. Poor training; 2. Inadvertent motions	Wrong/very wrong dose	7	**8**	10	**560**	4	27
34	Prepare Treatment Delivery Box	MP and Radiographer assemble delivery box as per checklist	Suboptimal delivery	1. Clamps too tight or loose; 2. Needles not seated well; 3. Other assembly issues	1. Poor training; 2. Poor execution	Wrong/very wrong dose	7	**9**	9	**567**	3	7
35	Prepare Treatment Delivery Box	MP note reading on dosimeter	Premature termination if meter indicates activity delivered while some remains	1. Equipment failure; 2. Reading failure	1. Batteries failure; 2. incorrect reading; 3. Incorrect placement; 4. Wrong scale	Wrong dose (underdose)	6	**7**	8	**336**	11	43
36	Place Catheter	Identify patient	Wrong patient brought to treatment	Patient not identified correctly	1. Of multiple patients, patients brought in in the wrong order (more likely); 2. With single patient, a patient for a different procedure brought in for treatment (less likely).	Wrong dose distribution	4	**10**	7	**280**	16	2
37	Place Catheter	IR places catheter	Exposure to normal tissue	Inadequate check of flow outside of target		Wrong dose distribution	4	**10**	7	**280**	17	3
38	Place Catheter	IR places catheter	Wrong segment treated	1. Wrong artery selected (not checking plan); 2. Wrong artery entered		Wrong dose distribution	4	**10**	7	**280**	18	4
39	Pre-treatment Checks	MP checks exposure from the patient at 1 m	No failure mode				0	**0**	0	**0**	62	62
40	Pre-treatment Checks	Technologist or nurse reads checklist and IR or MP check device	QC step to possibly detect delivery device problem; no direct failure mode				0	**0**	0	**0**	63	63
41	Pre-treatment Checks	Device connected to catheter	Poor delivery	Catheter flow extremely low from poor connection		Wrong/very wrong dose - always low dose	4	**8**	8	**256**	22	35

Table 5–3, cont. Failure modes and effects analysis for the procedure in Fig. 5–3

Item Number	Major Processes	Step	Potential Failure Modes	Potential Causes of Failure	Comments and descriptive scenario	Potential Effects of Failure	O	S	D	RPN	RPN Rank	S Rank
42	Pre-treatment Checks	Device connected to catheter	Interrupted delivery due to leak	Catheter connection not tight		Wrong/very wrong dose - always low dose	3	**8**	8	**192**	28	39
43	Pre-treatment Checks	Device connected to catheter	Poor delivery	Catheter connection below level of patient, possibly table too high and device too low		Wrong/very wrong dose - always low dose	2	**8**	6	**96**	42	40
44	Deliver Treatment	IR delivers microspheres	Poor delivery	Clamp or clamp impression slows flow	Microspheres' settle and fail to flow into catheter	Wrong/very wrong dose - always low dose	6	**8**	10	**480**	6	28
45	Deliver Treatment	IR delivers microspheres	Poor delivery	syringe friction slows flow	Microspheres' settle and fail to flow into catheter	Wrong/very wrong dose - always low dose	5	**8**	10	**400**	8	29
46	Deliver Treatment	IR delivers microspheres	Poor delivery	Needle assembly problem impedes flow	Microspheres' settle and fail to flow into catheter	Wrong/very wrong dose - always low dose	3	**8**	10	**240**	27	38
47	Deliver Treatment	IR delivers microspheres	Delivery terminated prematurely	Clamped priming line leaks		Wrong/very wrong dose - always low dose	4	**8**	10	**320**	12	31
48	Deliver Treatment	IR delivers microspheres	Stasis reached	Extremely slow delivery	Attempt to adjust for patient condition	Wrong/very wrong dose - always low dose	5	**8**	10	**400**	9	30
49	Measurements of Potential Contamination and for Release	MP checks all persons in room for contamination	Contaminated persons go undetected	MP neglects check	MP busy with other activities; O value assumed there is contamination to be detected, which is very unlikely	Unnecessary exposure to personnel, coworkers and family	4	**5**	9	**180**	30	46
50	Measurements of Potential Contamination and for Release	MP checks all persons in room for contamination	Contaminated persons go undetected	MP performs check incorrectly	MP not understand instrument or procedure, or not make measurements adequately; O value assumed there is contamination to be detected, which is very unlikely	Unnecessary exposure to personnel, coworkers and family	4	**5**	9	**180**	31	47
51	Measurements of Potential Contamination and for Release	MP checks all persons in room for contamination	Contaminated persons go undetected	Detector failure	Detector malfunction; O value assumed there is contamination to be detected, which is very unlikely	Unnecessary exposure to personnel, coworkers and family	4	**5**	9	**180**	32	48
52	Measurements of Potential Contamination and for Release	MP checks room for contamination	Contaminated items go undetected	MP neglects check	MP busy with other activities; O value assumed there is contamination to be detected, which is very unlikely	Contamination carried throughout facility	4	**5**	9	**180**	33	49
53	Measurements of Potential Contamination and for Release	MP checks room for contamination	Contaminated items go undetected	MP performs check incorrectly	MP not understand instrument or procedure, or not make measurements adequately; O value assumed there is contamination to be detected, which is very unlikely	Contamination carried throughout facility	4	**5**	9	**180**	34	50
54	Measurements of Potential Contamination and for Release	MP checks room for contamination	Contaminated items go undetected	Detector failure	Detector malfunction; O value assumed there is contamination to be detected, which is very unlikely	Contamination carried throughout facility	4	**5**	9	**180**	35	51
55	Measurements of Potential Contamination and for Release	MP measures exposure rate from patient and checks for release	Patient with high exposures is released	Failure to perform check correctly or instrumentation failure	Not a likely problem	Regulatory issue and exposure to family	2	**2**	10	**40**	49	55
56	Residual Measurements	MP takes waste to nuclear pharmacy	Waste remains in occupied area with potential exposures	MP neglects to take waste	Detection is likely because residual measurement would be missing	Potential exposure to personnel and contamination	4	**4**	2	**32**	54	52

Table 5–3, cont. Failure modes and effects analysis for the procedure in Fig. 5–3

Item Number	Major Processes	Step	Potential Failure Modes	Potential Causes of Failure	Comments and descriptive scenario	Potential Effects of Failure	O	S	D	RPN	RPN Rank	S Rank
57	Residual Measurements	MP measures exposure around vial	Underdose goes undetected	MP neglects residual measurements	May not be noticed if procedure does not entail contacting principals with the results; S value assumed problem with complete delivery	Wrong/very wrong dose - always low dose	4	**9**	8	**288**	14	11
58	Residual Measurements	MP measures exposure around vial	Underdose goes undetected	Failure to make measurements correctly	1. Poor training; 2. Equipment failure; 3. Unnoticed high background; 4. Scale error; 5. Reading error; 6. Incorrect placement on template	Wrong/very wrong dose - always low dose	3	**9**	10	**270**	19	12
59	Residual Measurements	MP calculates residual activity and compliance with prescription	Underdose goes undetected	Instrumentation failure	Detector malfunction	Wrong/very wrong dose - always low dose	2	**9**	10	**180**	36	16
60	Residual Measurements	MP notifies principals of delivered activity	No failure mode		Principals likely would contact MP for delivered activity		0	**0**	0	**0**	64	64
61	Clean Delivery Box and Store Residual	Nuclear Technologist stores residual for decay	Exposure from residual and contamination	Residual left out	Exposures not likely high and would be found on survey of nuclear pharmacy	Potential exposure to personnel and contamination	3	**4**	2	**24**	56	53
62	Clean Delivery Box and Store Residual	Nuclear Technologist cleans delivery box	Box not cleaned	Nuclear Technologist busy with other duties	Box would be cleared just before next case if ready	Box not ready for next case	6	**2**	1	**12**	58	56
63	Documentation	MP documents treatment	Incorrect subsequent treatment	Other duties preempt MP		Wrong/very wrong dose	5	**9**	7	**315**	13	10
64	Documentation	RO or IR completes prescription if necessary	Incorrect prescription	MD neglects completion of prescription		Regulatory issue	9	**2**	6	**108.00**	41	54

Table 5–3, cont. Failure modes and effects analysis for the procedure in Fig. 5–3

- Clumping due to reduced flow during injections because of clamp impressions in tubing, slow injections, interrupted injections, or problems with the syringe
- Microspheres adhering to the vial septum
- Problems with needle insertion into the vial
- Leaks in the outlet line, connector to the catheter, or the priming line
- Unexpected stasis
- Transcription errors in the order
- Treating the wrong lobe or segment
- Treating the wrong patient

Most of these failures occurred more than once.

On the fault tree, the highest-ranking 20% of the potential failures based on the PRN value have been marked with their ranking value. While all potential failure modes will be covered eventually (except as noted), these pathways deserve special attention and care. Usually, the top-ranked 20% in severity would also be identified for possible intervention, but in this case, 2/3 of the potential failures have S rankings equal to or greater than eight, and of the reminder, only two result in wrong dose (and those are covered by the interventions suggested by this analysis).

5.7.2 Approaches to Addressing Failure Modes

All of the techniques discussed in this chapter are brought to bear on safety interventions to reduce the risks identified in FMEA. The first is commissioning. After any redesign, and assuring that the key core components are in place, the next step is to look at the failure modes and determine if any can be controlled through comprehensive commissioning. Often, many of the failure modes fit into this category. Using the fault tree for this step, rather than the FMEA, can simplify the identification since the hazard ranking plays little role here. Consideration here includes commissioning of both equipment and procedures.

After applying the tools above to the potential failure modes, it comes time to address each potential failure mode with QC or QA. Figure 5–4 shows the fault tree with steps noted to address each potential failure mode. There are many ways to address the failures identified through the FMEA. This example uses an approach that provides a simple way to go through the hazards in an organized manner, checking to see that all the modes of error propagation are covered by some intervention. The approach uses the fault tree as a guide.

Before beginning, some discussion is in order about the FMEA, the fault tree, and, indeed, the process of determining the interventions themselves. In each instance, there is no one, true manifestation. Any group or analyst looking at the situation will develop a different FMEA or fault tree and recommend steps to address what is found. No set of these will be exhaustive outside of an academic investigation. In practice, as noted when discussing risk assessment, most of the advantages come from the early part of the work. Before exhausting the analysis and the analytical team, actions should be implemented, thus changing the landscape and forcing the analyst to go back to the beginning. The point of the exercise is to provide a safe environment for a high-quality procedure, so practicality reigns. It would be expected that the reader will disagree with parts of this analysis, and that is welcome. This fault tree is only for those potential failure modes that result in either a wrong or very wrong dose.

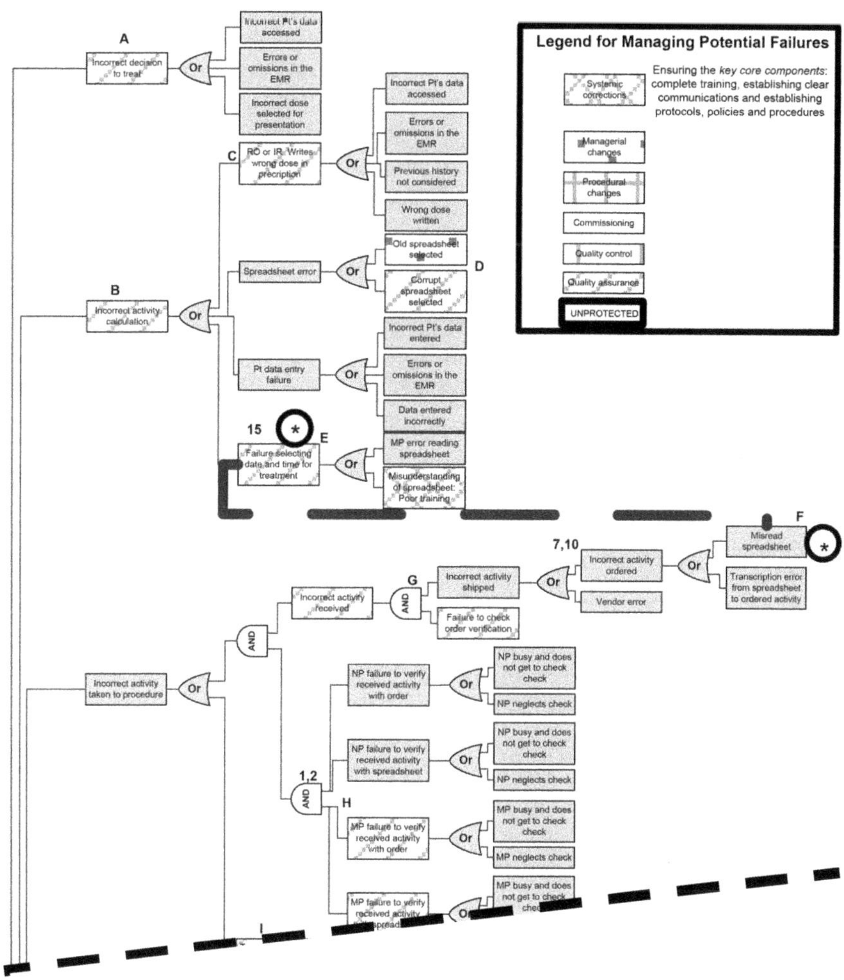

Figure 5–4. The fault tree developed for the procedure in Figure 5–3 with types of interventions indicated to address potential failures identified in the FMEA. The numbers near boxes indicate the top 20% of the potential failure modes by RPN ranking. The large letters relate to portions of the discussion in the text. (NOTE: This is only the top-most portion of this fault tree, which has been cut off at the heavy dashed line. See the book's CD for the complete figure.)

5.7.3 Safety Improvement Interventions

For the discussion that follows, the capital letters refer to the same letters in the fault tree. Let's begin at the top of the fault tree (A). The first potential failure mode involves an *incorrect decision to treat* based either on erroneous data or a bad decision on the part of the physician. Each of these are failures is very hard to prevent once the physician falls into the failure mode, and avoiding that mode is very tough. There are means, but few physicians will go to the effort of using tools such as checklists to avoid problems they do not believe will ever affect them (Pronovost and Vohr 2010). What is part of the fiber of medical practice is the holding of rounds and peer review, which should be instituted as quality assurance measure (Marks et al. 2013). The progenitor causes for these failure modes tend to be pernicious. The *errors or omissions in the electronic medical record (EMR)* are outside of the control of the medical physicist or radiotherapy. Accessing the *incorrect patient or choosing the wrong dose for the patient's presentation* are human errors, many of which appear in the fault tree. There is no way to avoid human errors. Instead, they should be expected to happen and some safety net deployed for *when* those errors occur. The safety net takes the form of some quality assurance, and the quality assurance must be in place without the need to identify that something unusual is going on. For the physician-based failures, one QA step is peer review. When routine, this step functions automatically. Similarly, the first cause for the next failure mode, *incorrect activity calculated* (B), also involves a physician's action (C), which, in itself, would suggest a QA check of the prescription is needed. One method of this would be to have both the radiation oncologist (or the nuclear physician) and the interventional radiologist determine the dose desired for the patient, providing a check for each other.

One of the other potential causes for *incorrect activity calculation* is *using an outdated spreadsheet* or one that has been *corrupted* (for example, from having saved some information from a past patient in one of the cells) (D). Management removing outdated spreadsheets upon release of revisions could address the first failure mode. Preventing the use of corrupt spreadsheets requires a QA check of the spreadsheet before each use, such as entering a test set of data and checking that the activity determined is the same as it should be, or checking using the equation embedded in the calculation. The latter usually would be the most efficient. While the check on the spreadsheet forms a good practice, the next two causes, *patient data entry error and failure selecting date* and *time for treatment* (E) are dominated by potential human errors, for which, again, little could be done to prevent them. Thus, some QA check on the calculation is in order. The simplest method is a redundant, independent calculation by a second person. A second computer program could be used, but if the same person enters the data, there is a good chance that the same input errors will find their way into the second calculation. While it does not show on the fault tree, all programs and spreadsheets most be commissioned before use.

The last failure in that grouping, *failure selecting date and time for treatment* (E), may cause the first error in the next set for incorrect activity taken to procedure, *misreading the activity from the spreadsheet* (F). This, and the human transcription error, propagates to combine with a *vendor's error*, over which the facility will have no control. Thus, some QA at the end of this branch is needed. A normal procedure is for the vendor to send a verification of the order after receiving it, and for the facility to check the verification against the intended order. This QA step enters the fault tree with an AND gate, such that there would need to be simultaneous errors in the order as processed by the vendor and in the verification by the facility for an erroneous order to be shipped (G). Reflecting back to the FMEA, the QA step reduces the (D) value since an erroneous order is likely to be detected at this point—as long as the order verification is actually checked.

The large block situated around the (H) are annotated to indicate that these steps carry the greatest and second greatest hazard according to the FMEA. This block deals with both the nuclear pharmacist and the medical physicist verifying the received activity of source by measurements made in a dose calibrator. These are QA for the activity of the source received. Of course, related to the measurements are fault-tree branches for potential failures of the dose calibrator and of the persons making those measurements. However, failure of the instrumentation or of the personnel performing the measurements will not result in erroneous treatments because a failure in the measurement most likely indicates a discrepancy between the expected activity and that of the source, which should stop the procedure until the cause of the discrepancy is determined (fail safe). The only way that an incorrect source would move forward to patient treatment is if the calibration failure exactly matched the error in the activity, but in the opposite direction. The likelihood of this is vanishingly small, and that part of the fault tree is not shown for simplicity The real problem becomes not performing the checks at all, which would permit the use of a source with the wrong activity, and that is why this failure ranks so significantly. The check has two parts, *verifying that the activity matches the order* and that the *activity matches that according to the activity determined by the spreadsheet*, when projected to the time of the procedure. While only the latter would actually affect the patient's treatment, checking both gives a measure of how well the vendor supplies the order, and it gives two chances to detect aberrant activities. Just as in the discussion in the previous paragraph, everything connected to the right of the AND gate forms QA for the received activity, so a failure in this process itself will not affect the patient's treatment unless there is a simultaneous error in the activity of the received source. The suggested QA focuses on the medical physicist because this discussion addresses mostly what the medical physicist can do to ensure quality, although the nuclear pharmacist also should check the activity.

The problem to solve at this point is not putting in place QA measurements for the source; it is assumed that that is part of the normal procedure. Rather, the problem is how to make sure the medical physicist performs the checks and performs them correctly. Two QA tools can help. The first is the simple form or checklist (see Section 5.6.4). While the vendor supplies a checklist to guide the preparation of the delivery box and a pretreatment checklist to ensure that all is in readiness for the injection, a facility-generated checklist for all the steps required to execute the case successfully helps prevent omissions (although does not guarantee omissions will not happen through some human errors). Part of the checklist would include leading the nuclear physicist and medical physicist through the assay of the source and, later, the measurement of the exposure readings around the vial and, near the end, measurements around the residual. The second tool would be technology. The University of Wisconsin migrated the checklist to a tablet device. Built into the checklist is a spreadsheet into which the medical physicist enters the assayed activity, time, and date. The program compares the activity with that ordered and with the desired activity for the time of the procedure. This removes some of the potential for calculation errors, although erroneous data entry remains. However, as discussed above, an error at this point merely pauses the procedure until the source of the discrepancy is found and corrected. Again, the operation of the program needs thorough commissioning.

There still remains the possibility that the activity was correctly transferred from the vendor's spreadsheet to the order, but that the time and date assigned to the procedure were incorrect so that the check of the activity either may seem discrepant or, worse, that the procedure would be performed at a different time, meaning the activity would be incorrect due to decay (I). This, and some of the potential failure modes around (J), could be eliminated by instead of using the vendor's spreadsheet, making a database program that would accept the patient information input and as output give possible combinations of times for treatments and available activities. After selecting the most appropriate combination, the program could generate an e-mail message for all principals. This would eliminate transcription errors and omissions of notifications through a combination of procedural changes and technology. During the commissioning of the procedure, all the e-mail addresses would be checked.

An apparent cause for several reported events entailed microspheres adhering to the rubber septum of the vial, possibly resulting from *inversion of the vial* at some point (K). In some of these cases, the nuclear pharmacist and the medical physicist were not aware that such handling could cause a problem. The first step to address this would be to train the persons involved about this problem (a systemic correction). However, training can be *forgotten*, or there may be *inadvertent slips* that result in vial inversion. Procedural changes could minimize the probability of inversion, such as using a cart with a custom holder to transport the vial between the nuclear pharmacy and the procedure room. No matter what the facility does, however, there is always the possibility that the vial was inverted by the manufacturer or during shipping. While this potential failure cannot be eliminated, the vendor recommends gently rocking the vial to rinse microspheres from the septum and tapping the vial on a counter to help drop fluid from the septum before placing it in the delivery box. These steps are in the vendor's checklist and should be followed. Even so, a facility should avoid as much of the potential problem as possible.

Many of the events seem to occur from failures at the time of delivery (L), with several related to *inadequate flow* (M). Many of the progenitor causes entail poor training, such as addressing the effects of clamping the outlet line tube or maintaining sufficient pressure during the injection. As before, the program needs to train all personnel fully before beginning the program. The training should cover all known possible errors that could degrade the treatment and explain how to avoid them. The training should include some form of evaluation. Commissioning the procedure should include establishing the limits of acceptable performance. However, as always, training can be forgotten in the rush of the moment. This potential failure mode also can result from one of many equipment failures. A change in procedures may help mitigate problems of flow. The medical physicist's function during the infusion should be to monitor the flow as best as possible, understanding that little of relevance can be seen. A build-up of microspheres in a tube can be seen, and the failure of the radiation monitor's reading to decrease during the infusion indicates problems. The procedural change could include addressing otherwise unidentified failures for the reading to fall by rocking the vial and trying again. Connections can be tapped and shaken to free hanging microspheres. The physicist potentially needs to have a very active role during this part of the procedure.

The final cause in this set has delivery performed with the *table above the delivery device*, something that is prevented by following the vendor's pretreatment checklist (N). The facility's protocol for these cases should entail using the vendor's checklist or, better yet, a version of the checklist customized for the facility. The facility administration and the facility's radiation safety committee should enforce the checklist protocol after an incident (understanding that one case may happen because of poor understanding, but after a second occurrence, the program should be terminated).

A case may have to be terminated prematurely if the priming line leaks (O). A leaking priming line should pose no hazard because of valves, and it should just contain saline. However, if there is a leak, there may be more serious problems than expected, and the procedure should at least be paused while the medical physicist investigates the issue. Because of this potential problem, the vendor recently changed the preparation procedures by clamping that tube, but clamps can fail. However, a clamp failure will not result in a problem without a valve failure leading to the line. Thus, the whole clamp exercise is actually quality control, and the likelihood of simultaneous valve and clamp failure is extremely low if the clamping forms part of the routine setup. No additional interventions, other than following the checklist, are warranted here, even with this step's high RPN ranking.

Premature termination due to leaking at the catheter connection (P) presents a problem. While one of the causes, *poor training*, can be addressed, a connector failure cannot. The most likely failure would simply be a loose connection that can be fixed at the first sign by simple tightening. A connector failure is a different matter. The connector is tested during the delivery device preparation while assuring flow through the outlet line and out the outlet needle. A bad connector should leak at that point. The connector on the catheter in the patient is used during arterial localization. Since both parts are tested before treatment, the likelihood of failure during treatment is low, unless the connector or catheter is damaged during their connection. In such a case, little can be done since the tubing by then is contaminated with radioactive materials and trying to reseat the connection could contaminate the patient, staff, and room. Thus, this potential failure mode remains unprotected except through use, although it has little probability of failure.

Stasis problems should never happen with this product because of the low number of microspheres used (Q). There should be no need to deliver the microsphere is any manner that would increase the likelihood of stasis, such as extremely slowly out of concern for the patient's condition. The training and changing procedure (actually, just performing the procedures properly) should address this potential failure.

The radiation detector on the box monitors the exposure rate from the source. As the source leaves the vial and flows into the patient, the reading on the monitor decreases. When all the source has been delivered, the reading falls to zero, or thereabout. Having the reading fall to zero prematurely could lead to *termination of the delivery* before deliver of all the microspheres (R). This could happen if the detector fails during delivery and displays zero in the face of a non-zero exposure. Alternatively, reading the detector on the wrong scale could also lead to premature termination if the scale was not sensitive enough to display the exposure rate with only some of the source present. For the reading errors, establishing a relationship between source activity and expected reading provides a QA measure for the reading during the pretreatment setup. Having a second person check the reading protects against erroneous readings.

The last potential failure in this set is equipment failure, not otherwise specified. Unfortunately, without some specification, this potential failure mode will remain unaddressed.

The next failure mode (S) begins three potential causes for *inappropriate subsequent treatments* (T). Before the treatment, the vial is placed on a template and exposure measurements are made at a specific location marked on the template. After the treatment, measurements are made in the same geometry (or as close to the same as feasible) of the residual activity in the waste container, including the used vial, catheter, tubing assembly, and other potentially radioactive wastes. The comparison of the readings before and after determines the activity delivered. This determination is used to verify (or not) satisfaction of the prescription for regulatory compliance (not considered in this analysis) as well as whether the patient received adequate treatment. An erroneous value for the activity delivered could lead to inappropriate subsequent treatments.

As can be seen in the fault tree, there are many possible causes for incorrect activity determinations. The procedure checklist discussed above in dealing with the assay could address neglecting the readings. Having expectations for readings of the background and the pretreatment vial as a function of its activity helps prevent the user from accepting readings that are inappropriately very different. The confidence in the quality of the assessment is greatly increased by taking all the readings and performing the calculations in the tablet computer, and by having the readings checked for consistency with expectations.

The final cause for inappropriate subsequent treatments comes from failure to document the present treatment (U). For the medical physicist, incorrect documentation failure could lead either to excessive or insufficient treatment in the future. Again, following a checklist for the entire procedure reduces the probability of this, particularly if the documentation is generated by the database. Neglecting the documentation is likely self-correcting, particularly if the physician needs the documentation to dictate his or her note or if the documentation is necessary to go with the billing submission. Most hospitals have ways to remind physicians to complete their documentation, usually based on revenue generation.

5.8. Conclusions

In the most fundamental sense, an effective quality management program is based on an understanding of systems approach and the theories of knowledge, learning, and psychology as spelled out by W. Edwards Deming and others (see Section 5.2). This means, among other things, that one must go beyond the traditional way of thinking about quality assurance in radiotherapy involving technology-centered quality assurance through prescriptive tests. Quality management is much more than the routine performance of a set of specific measurements. Indeed, W. Edwards Deming recommends in his 14 points that one should "cease dependence on inspection" (point #3) (Deming 1986). This suggests that the traditional approach of relying on more and more inspections and human QA checks is not fruitful in the long term. A more worthy goal is to design errors out of the system.

Recommendations for prescriptive QA tests, e.g. TG 142 (Klein et al. 2009), and other documents must be understood in this context. A good quality management system will incorporate these recommendations, but it must be appreciated that the work does not stop there. Good quality management goes well beyond a narrow adherence to rules or recommendations. Good quality management incorporates a clear management structure—including regular review of policies and procedures and internal and external audits—all documented appropriately. A good quality management system is adaptable via the plan-do-study-act (PDSA) cycle. A good quality management system employs the tools that are known to work, such as safety barriers, QA tests, and checklists. A quality management system based on these sound principles is the key to improving patient safety and the quality of care.

References

American College of Radiology (ACR). ACR-ASTRO practice guidelines for 3D external-beam radiation planning and conformal therapy. Reston, VA: ACR, 2011. On-line at http://www.acr.org/SecondaryMainMenuCategories/quality_safety/guidelines/ro.aspx. (Accessed Feb. 16, 2013.)

Agency for Healthcare Research and Quality (AHRQ). *Patient Safety Primers: Checklists.* Rockville, MD: ARHQ, 2012. On-line at http://www.webmm.ahrq.gov/primer.aspx?primerID=14. (Accessed Feb. 16, 2013.)

Azmandian, F., D. Kaeli, J.G. Dy, E. Hutchinson, M. Ancukiewicz, A. Niemierko, and S.B. Jiang. (2007). "Towards the development of an error checker for radiotherapy treatment plans: A preliminary study." *Phys Med Biol* 52(21):6511–6524.

Benedict, S.H., K.M. Yenice, D. Followill, J.M. Galvin, W. Hinson, B. Kavanagh, P. Keall, M. Lovelock, S. Meeks, L. Papiez, T. Purdie, R. Sadagopan, M.C. Schell, B. Salter, D.J. Schlesinger, A.S. Shiu, T. Solberg, D.Y. Song, V. Stieber, R. Timmerman, W.A. Tome, D. Verellen, L. Wang, and F.F. Yin. (2010). "Stereotactic body radiation therapy: The report of AAPM Task Group 101." *Med Phys* 37:4078–4101.

Berry, S.L., R.D. Sheu, C.S. Polvorosa, and C.S. Wuu. (2012). "Implementation of EPID transit dosimetry based on a through-air dosimetry algorithm." *Med Phys* 39: 87–98.

Bissonnette, J.P., P.A. Balter, L. Dong, K.M. Langen, D.M. Lovelock, M. Miften, D.J. Moseley, J. Pouliot, J.J. Sonke, and S. Yoo. (2012). "Quality assurance for image-guided radiation therapy utilizing CT-based technologies: A report of the AAPM TG–179." *Med Phys* 39:1946–1963.

Bogdanich, W. "As Technology Surges, Radiation Safeguards Lag." *New York Times*. January 26, 2010. (Accessed Feb. 16, 2013.)

Bogdanich, W. "Radiation offers new cures and new ways to do harm." *New York Times*. January 23, 2010. (Accessed Feb. 16, 2013.)

Boorman, D. (2001). "Today's electronic checklists reduce likelihood of crew errors and help prevent mishaps." *Internat Civil Aviat Org J* 56:17–36.

Bosk, C.L., M. Dixon-Woods, et al. (2009). "Reality check for checklists." *Lancet* 374:444–445.

Boxer, M., D. Forstner, A. Kneebone, G. Delaney, E.S. Koh, M. Fuller, and N. Kaadan. (2009). "Impact of a real-time peer review audit on patient management in a radiation oncology department." *J Med Imaging Radiat Oncol* 53:405–411.

Brundage, M.D., P.F. Dixon, et al. (1999). "A real-time audit of radiation therapy in a regional cancer center." *Int J Radiat Oncol Biol Phys* 43:115–124.

Cafazzo, J.A., P.L. Trbovich, et al. (2009). "Human factors perspectives on a systemic approach to ensuring a safer medication delivery process." *Healthc Q* 12 Spec No Patient:70–74.

Chan, A.J., M.K. Islam, et al. (2010). "The use of human factors methods to identify and mitigate safety issues in radiation therapy." *Radiother Oncol* 97:596–600.

CoxHealth. (2010). "CoxHealth announces some BrainLAB stereotactic radiation therapy patients received increased radiation dose." Company press release. From http://www.coxhealth.com/body.cfm?id=3701. (Accessed Feb. 16, 2013.)

Das, I.J., C.W. Cheng, et al. (2008). "Accelerator beam data commissioning equipment and procedures: report of the TG–106 of the Therapy Physics Committee of the AAPM." *Med Phys* 35:4186–4215.

Degani, A. and E.L. Wiener. (1993). "Cockpit Checklists—Concepts, Design, and Use." Human Factors 35 (2):28–43.

Deming, W.E. *Out of the crisis: quality, productivity, and competitive position*. New York: Cambridge University Press, 1986.

Deming, W.E. *The New Economics: For Industry, Government, Education*. Cambridge, MA: The MIT Press, 1994.

Dieterich, S., C. Cavedon, et al. (2011). "Report of AAPM TG 135: Quality assurance for robotic radiosurgery." *Med Phys* 38:2914–2936.

Dixon-Woods, M., C.L. Bosk, E.L. Aveling, C.A. Goeschel, and P.J. Pronovost. (2011). "Explaining Michigan: Developing an Ex post theory of a quality improvement program." *Milbank Q* 89:167–205.

Ellerbroek, N.A., M. Brenner, P. Hulick, and T. Cushing for the American College of Radiology. (2006). "Practice accreditation for radiation oncology: quality is reality." *J Am Coll Radiol* 3:787–792.

Ely, J.W., M.L. Graber, et al. (2011). "Checklists to reduce diagnostic errors." *Acad Med* 86:307–313.

Essers, M. and B.J. Mijnheer (1999). "In vivo dosimetry during external photon beam radiotherapy." *Int J Radiat Oncol Biol Phys* 43:245–259.

Ezzell, G.A., J.W. Burmeister, N. Dogan, T.J. LoSasso, J.G. Mechalakos, D. Mihailidis, A. Molineu, J.R. Palta, C.R. Ramsey, B.J. Salter, J. Shi, P. Xia, N.J. Yue, and Y. Xiao. (2009). "IMRT commissioning: multiple institution planning and dosimetry comparisons, a report from AAPM Task Group 119." *Med Phys* 36(11):5359–5373.

Ezzell, G.A., J.M. Galvin, D. Low, J.R. Palta, I. Rosen, M.B. Sharpe, P. Xia, Y. Xiao, L. Xing, and C.X. Yu. (2003). "Guidance document on delivery, treatment planning, and clinical implementation of IMRT: Report of the IMRT subcommittee of the AAPM radiation therapy committee." *Med Phys* 30:2089–2115.

Fairchild, A., L. Collette, C.W. Hurkmans, B. Baumert, D.C. Weber, A. Gulyban, and P. Poortmans. (2012). "Do results of the EORTC dummy run predict quality of radiotherapy delivered within multicentre clinical trials?" *Eur J Cancer* 48:3232–3239.

Fiorino, C., D. Corletto, P. Mangili, S. Broggi, A. Bonini, G.M. Cattaneo, R. Parisi, A. Rosso, P. Signorotto, E. Villa, and R. Calandrino. (2000). "Quality assurance by systematic in vivo dosimetry: results on a large cohort of patients." *Radiother Oncol* 56:85–95.

Ford, E.C., L. Fong de Los Santos, T. Pawlicki, S. Sutlief, and P. Dunscombe. (2012). "Consensus recommendations for incident learning database structures in radiation oncology." *Med Phys* 39:7272–7290.

Ford, E.C., R. Gaudette, L. Myers, B. Vanderver, L. Engineer, R. Zellars, D.Y. Song, J. Wong, and T.L. DeWeese. (2009). "Evaluation of Safety in a Radiation Oncology Setting Using Failure Mode and Effects Analysis." *Int J Radiat Oncol Biol Phys* 74:852–858.

Ford, E.C., S. Terezakis, A. Souranis, K. Harris, H. Gay, and S. Mutic. (2012). "Quality control quantification (QCQ): a tool to measure the value of quality control checks in radiation oncology." *Int J Radiat Oncol Biol Phys* 84:e263–269.

Fraass, B., K. Doppke, M. Hunt, G. Kutcher, G. Starkschall, R. Stern, and J. Van Dyke. (1998). "American Association of Physicists in Medicine Radiation Therapy Committee Task Group 53: quality assurance for clinical radiotherapy treatment planning." *Med Phys* 25:1773–1829.

Furhang, E.E., J. Dolan, J.K. Sillanpaa, and L.B. Harrison. (2009). "Automating the initial physics chart-checking process." *J Appl Clin Med Phys* 10:129–135.

Gawande, A. *The Checklist Manifesto: How to Get Things Right.* New York: Metropolitan Books, 2009.

Gawande, A. (2013). "The Checklist Manifesto." On-line at http://gawande.com/the-checklist-manifesto. (Accessed, Feb. 25, 2013.)

Grout, J.R. (2006). "Mistake proofing: Changing designs to reduce error." *Qual Saf Health Care* 15(SUPPL. 1):i44–i49.

Gurses, A.P., K.L. Seidl, V. Vaidya, G. Bochicchio, A.D. Harris, J. Hebden, and Y. Xiao. (2008). "Systems ambiguity and guideline compliance: A qualitative study of how intensive care units follow evidence-based guidelines to reduce healthcare-associated infections." *Qual Saf Health Care* 17:351–359.

Hales, B., M. Terblanche, R. Fowler, and W. Sibbald. (2008). "Development of medical checklists for improved quality of patient care." *Int J Qual Health Care* 20:22–30.

Halvorsen, P.H., I.J. Das, M. Fraser, D.J. Freedman, R.E. Rice 3rd, G.S. Ibbott, E.I. Parsai, T.T. Robin Jr, and B.R. Thomadsen. (2005). "AAPM Task Group 103 report on peer review in clinical radiation oncology physics." *J Appl Clin Med Phys* 6:50–64.

Hart, E.M. and H. Owen (2005). "Errors and omissions in anesthesia: a pilot study using a pilot's checklist." *Anesth Analg* 101:246–250.

Hartford, A.C., M.G. Palisca, T.J. Eichler, D.C. Beyer, V.R. Devineni, G.S. Ibbott, B. Kavanagh, J.S. Kent, S.A. Rosenthal, C.J. Schultz, P. Tripuraneni, and L.E. Gaspar. (2009). "American Society for Therapeutic Radiology and Oncology (ASTRO) and American College of Radiology (ACR) Practice Guidelines for Intensity-Modulated Radiation Therapy (IMRT)." *Int J Radiat Oncol Biol Phys* 73:9–14.

Haynes, A.B., T.G. Weiser, W.R. Berry, S.R. Lipsitz, A.H. Breizat, E.P. Dellinger, T. Herbosa, S. Joseph, P.L. Kibatala, M.C. Lapitan, A.F. Merry, K. Moorthy, R.K. Reznick, B. Taylor, and A.A. Gawande. (2009). "A surgical safety checklist to reduce morbidity and mortality in a global population." *N Engl J Med* 360:491–499.

Herman, M.G., J.M. Balter, D.A. Jaffray, K.P. McGee, P. Munro, S. Shalev, M. Van Herk, and J.W. Wong. (2001). "Clinical use of electronic portal imaging: Report of AAPM radiation therapy committee Task Group 58." *Med Phys* 28:712–737.

Hospital Physicists' Association. (1969). "A code of practice for the dosimetry of 2 to 35 MV x-ray and caesium-137 and cobalt-60 gamma-ray beams." *Phys Med Biol* 14:1–8.

Hunt, M.A., G. Pastrana, H.I. Amols, A. Killen, and K. Alektiar. (2012). "The impact of new technologies on radiation oncology events and trends in the past decades: an institutional experience." *Int J Radiat Oncol Biol Phys* 84:925–931.

Huq, M.S., B.A. Fraass, P.B. Dunscombe, J.P. Gibbons Jr, G.S. Ibbott, P.M. Medin, A. Mundt, S. Mutic, J.R. Palta, B.R. Thomadsen, J.F. Williamson, and E.D. Yorke. (2008). "A Method for Evaluating Quality Assurance Needs in Radiation Therapy." *Int J Radiat Oncol Biol Phys* 71(1 SUPPL.):S170–S173.

Huq, S., B.A. Fraass, P.B. Dunscombe, J.P. Gibbons, G.S. Ibbott, A.J. Mundt, S. Mutic, J.R. Palta, F. Rath, B. Thomadsen, J.F. Williamson, and E.D. Yorke. *Application of Risk Analysis Methods to Radiation Therapy Quality Management: Report of AAPM TG100 of the Therapy Physics Committee of the AAPM.* College Park, MD: AAPM, 2013.

Ibbott, G., C.M. Ma, D.W. Rogers, S.M. Seltzer, and J.F. Williamson. (2008). "Anniversary paper: fifty years of AAPM involvement in radiation dosimetry." *Med Phys* 35:1418–1427.

Ibbott, G.S., D.S. Followill, A. Molineu, J.R. Lowenstein, P.E. Alvarez, and J.E. Roll (2008). "Challenges in credentialing institutions and participants in advanced technology multi-institutional clinical trials." *Int J Radiat Oncol Biol Phys* 71:S71–S75.

Johnstone, A.H. and F. Percival (1976). "Attention Breaks in Lectures." *Education in Chemistry* 13:49–50.

Juran, J.M. and F.M. Gryna. *Quality Planning and Analysis.* New York: McGraw-Hill, 1993.

Kirby, T.H., W.F. Hanson, R.J. Gastorf, C.H. Chu, and R.J. Shalek. (1986). "Mailable TLD system for photon and electron therapy beams." *Int J Radiat Oncol Biol Phys* 12:261–265.

Klein, E.E., J. Hanley, J. Bayouth, F.F. Yin, W. Simon, S. Dresser, C. Serago, F. Aguirre, L. Ma, B. Arjomandy, C. Liu, C. Sandin, and T. Holmes. (2009). "Task group 142 report: Quality assurance of medical accelerators." *Med Phys* 36:4197–4212.

Kolb, A.Y. and D.A. Kolb. (2005). "Learning Styles and Learning Spaces: Enhancing Experiential Learning in Higher Education." *Academy of Management Learning & Education* 4:193–212.

Kubo, H.D., G.P. Glasgow, T.D. Pethel, B.R. Thomadsen, and J.F. Williamson. (1998). "High dose-rate brachytherapy treatment delivery: report of the AAPM Radiation Therapy Committee Task Group No. 59." *Med Phys* 25:375–403.

Kutcher, G.J., L. Coia, M. Gillin, W.F. Hanson, S. Leibel, R.J. Morton, J.R. Palta, J.A. Purdy, L.E. Reinstein, and G.K. Svensson. (1994). "Comprehensive QA for radiation oncology: report of AAPM Radiation Therapy Committee Task Group 40." *Med Phys* 21:581–618.

Langen, K.M., N. Papanikolaou, J. Balog, R. Crilly, D. Followill, S.M. Goddu, W. Grant, 3rd, G. Olivera, C.R. Ramsey, C. Shi, and A.T. Group. (2010). "QA for helical tomotherapy: report of the AAPM Task Group 148." *Med Phys* 37:4817–4853.

Lingard, L., G. Regehr, B. Orser, R. Reznick, G.R. Baker, D. Doran, S. Espin, J. Bohnen, and S. Whyte. (2008). "Evaluation of a preoperative checklist and team briefing among surgeons, nurses, and anesthesiologists to reduce failures in communication." *Arch Surg* 143:12–17; discussion 18.

Litzenberg, D.W., J.M. Moran, and B.A. Fraass. (2002). "Verification of dynamic and segmental IMRT delivery by dynamic log file analysis." *J Appl Clin Med Phys* 3:63–72.

Lorist, M.M., M.A. Boksem, and K.R. Ridderinkhof. (2005). "Impaired cognitive control and reduced cingulate activity during mental fatigue." *Brain Res Cogn Brain Res* 24:199–205.

Low, D.A., J.M. Moran, J.F. Dempsey, L. Dong, and M. Oldham. (2011). "Dosimetry tools and techniques for IMRT." *Med Phys* 38:1313–1338.

Mans, A., P. Remeijer, I. Olaciregui-Ruiz, M. Wendling, J.J. Sonke, B. Mijnheer, M. van Herk, and J.C. Stroom. (2010). "3D Dosimetric verification of volumetric-modulated arc therapy by portal dosimetry." *Radiother Oncol* 94:181–187.

Mans, A., M. Wendling, L.N. McDermott, J.J. Sonke, R. Tielenburg, R. Vijlbrief, B. Mijnheer, M. Van Herk, and J.C. Stroom. (2010). "Catching errors with in vivo EPID dosimetry." *Med Phys* 37:2638–2644.

Marks, L.B., R. Adams, T. Pawlicki, A. Blumberg, D. Hoopes, M.D. Brundage, and B. Fraass. (2013). "Enhancing the Role of Case-Oriented Peer Review to Improve Quality and Safety in Radiation Oncology." *Int J Radiat Oncol Biol Phys* in press.

Mazur, L.M., P.R. Mosaly, M. Jackson, S.X. Chang, K.D. Burkhardt, R.D. Adams, E.L. Jones, L. Hoyle, J. Xu, J. Rockwell, and L.B. Marks. (2012). "Quantitative assessment of workload and stressors in clinical radiation oncology." *Int J Radiat Oncol Biol Phys* 83:e571–576.

McLaughlin, A.C. *What makes a good checklist?* Perspectives on Safety. Washington, D.C.: AHRQ, 2010. On-line at http://www.webmm.ahrq.gov/perspective.aspx?perspectiveID=92. (Accessed Feb. 27, 2013.)

McLeod, P.J., Y. Steinert, T. Meagher, and A. McLeod. (2003). "The ABCs of pedagogy for clinical teachers." *Med Educ* 37:638–644.

Miller, D., P. Block, J. Cunningham, B. Curran, G. Ibbott, D. Jones, S. Jucius, D. Leavitt, R. Mohan, and J. van de Geijn. *Radiation treatment planning dosimetry verification:*

report of AAPM Task Group #23. College Park, MD: American Institute of Physics, 1995.

Miller, G.A. (1956). "The magical number seven plus or minus two: some limits on our capacity for processing information." *Psychol Rev* 63:81–97.

Moran, J.M., M. Dempsey, et al. (2011). "Safety considerations for IMRT: Executive summary." *Pract Radiat Oncol* 1:190–195.

Mutic, S., J.R. Palta, E.K. Butker, I.J. Das, M.S. Huq, L.N. Loo, B.J. Salter, C.H. McCollough, J. Van Dyk. (2003). "Quality assurance for computed-tomography simulators and the computed-tomography-simulation process: report of the AAPM Radiation Therapy Committee Task Group No. 66." *Med Phys* 30:2762–2792.

Norman, D.A. *The Design of Everyday Things.* New York: Basic Books, 1988.

Nuclear Regulatory Commission. (2012). NRC'S Response to Medical Events at the Veterans Affairs Medical Center in Philadelphia, PA. Washington, DC. On-line at http://www.nrc.gov/reading-rm/doc-collections/fact-sheets/fs-response-med-events.html. (Accessed Feb. 25, 2013.)

Palta, J.R., C. Liu, and J.G. Li. (2008). "Current external beam radiation therapy quality assurance guidance: does it meet the challenges of emerging image-guided technologies?" *Int J Radiat Oncol Biol Phys* 71(1 Suppl):S13–17.

Palta, J.R., C. Liu, and J.G. Li. (2008). "Quality Assurance of Intensity-Modulated Radiation Therapy." *Int J Radiat Oncol Biol Phys* 71(1 Supl):S108–S112.

Peiffert, D., J.M. Simon, and F. Eschwege. (2007). "[Epinal radiotherapy accident: passed, present, future]." *Cancer Radiother* 11(6–7):309–312.

Peters, L.J., B. O'Sullivan, J. Giralt, T.J. Fitzgerald, A. Trotti, J. Bernier, J. Bourhis, K. Yuen, R. Fisher, and D. Rischin. (2010). "Critical Impact of Radiotherapy Protocol Compliance and Quality in the Treatment of Advanced Head and Neck Cancer: Results From TROG 02.02." *Jf Clin Oncoly* 28:2996–3001.

Pfeiffer, D., S. Sutlief, W. Feng, H.M. Pierce, and J. Kofler. (2008). "AAPM Task Group 128: quality assurance tests for prostate brachytherapy ultrasound systems." *Med Phys* 35:5471–5489.

Potters, L., L.E. Gaspar, B. Kavanagh, J.M. Galvin, A.C. Hartford, J.M. Hevezi, P.A. Kupelian, N. Mohiden, M.A. Samuels, R. Timmerman, P. Tripuraneni, M.T. Vlachaki, L. Xing, and S.A. Rosenthal. (2010). "American Society for Therapeutic Radiology and Oncology (ASTRO) and American College of Radiology (ACR) Practice Guidelines for Image-Guided Radiation Therapy (IGRT)." *Int J Radiat Oncol Biol Phys* 76:319–325.

Pronovost, P., D. Needham, S. Berenholtz, D. Sinopoli, H. Chu, S. Cosgrove, B. Sexton, R. Hyzy, R. Welsh, G. Roth, J. Bander, J. Kepros, and C. Goeschel. (2006). "An intervention to decrease catheter-related bloodstream infections in the ICU." *N Engl J Med* 355:2725–2732.

Pronovost, P.J. and E. Vohr. *Safe Patients, Smart Hospitals: How One Doctor's Checklist Can Help Us Change Health Care from the Inside Out.* New York: Hudson Street Press, 2010.

Redish, E.F. (1994). "Implications of cognitive studies for teaching physics." *Am J Phys* 62:796–803.

Regehr, G. and G.R. Norman (1996). "Issues in cognitive psychology: implications for professional education." *Acad Med* 71:988–1001.

Shewhart, W.A. and W.E. Deming. *Statistical method from the viewpoint of quality control.* Washington: The Graduate School, The Department of Agriculture, 1939.

Sidney, D. *Patient Safety: A Human Factors Approach.* Oxon, UK: Taylor and Francis CRC Press, 2011.

Siochi, R.A., E.C. Pennington, T.J. Waldron, and J.E. Bayouth. (2009). "Radiation therapy plan checks in a paperless clinic." *J Appl Clin Medl Phys* 10:43–62.

Thomadsen, B., S.W. Lin, P. Laemmrich, T. Waller, A. Cheng, B. Caldwell, R. Rankin, and J. Stitt. (2003). "Analysis of treatment delivery errors in brachytherapy using formal risk analysis techniques." *Int J Radiat Oncol Biol Phys* 57:1492–1508.

Thwaites, D., P. Scalliet, J.W. Leer, and J. Overgaard. (1995). "Quality assurance in radiotherapy. European Society for Therapeutic Radiology and Oncology Advisory Report to the Commission of the European Union for the 'Europe Against Cancer Programme.'" *Radiother Oncol* 35:61–73.

Vaida, A.J. (1999). "Medication error prevention 'toolbox.'" Medication safety alert. On-line at http://www.ismp.org/msaarticles/toolbox.html. (Accessed Feb. 26, 2013.)

Weiser, T.G., A.B. Haynes, A. Lashoher, G. Dziekan, D.J. Boorman, W.R. Berry, and A.A. Gawande. (2010). "Perspectives in quality: designing the WHO Surgical Safety Checklist." *Int J Qual Health Care* 22:365–370.
Williams, M.V. (2007). "Radiotherapy Near Misses, Incidents and Errors: Radiotherapy Incident at Glasgow." *Clin Oncol* 19:1–3.
Willoughby, T., J. Lehmann, J.A. Bencomo, S.K. Jani, L. Santanam, A. Sethi, T.D. Solberg, W.A. Tome, and T.J. Waldron. (2012). "Quality assurance for nonradiographic radiotherapy localization and positioning systems: report of Task Group 147." *Med Phys* 39:1728–1747.
Winters, B.D., A.P. Gurses, H. Lehmann, J.B. Sexton, C.J. Rampersad, and P.J. Pronovost. (2009). "Clinical review: checklists—translating evidence into practice." *Crit Care* 13:210.
Yin, F.F., J. Wong, J. Balter, S. Benedict, J.-P. Bissonnette, T. Craig, L. Dong, D. Jaffray, S. Jiang, S. Kim, C.-M. Ma, M. Murphy, P. Munro, T. Solberg, and Q.J. Wu. *The role of in-room kV x-ray imaging for patient setup and target localization: AAPM Task Group 104.* College Park, MD: American Institute of Physics, 2009.
Yorke, E., R. Alecu, L. Ding, R. Boissone, D. Doracy Fontenla, A. Kalend, D. Kaurin, W.E. Masterson-McGary, G. Marinello, T. Matzen, A. Saini, J. Shi, W. Simon, T. Zhu, X.R. Zhu, G. Rikner, and G. Nilsson. *Diode in vivo dosimetry for patients receiving external beam radiation therapy: AAPM Task Group 62.* College Park, MD: American Institute of Physics, 2005.
Zietman, A.L., J.R. Palta, and M.L. Steinberg. *Safety is No Accident.* Fairfax, VA: American Society for Radiation Oncology, 2012.

Chapter 6

Incident Learning Systems

Steven Sutlief, Ph.D.[1], Barrett Caldwell, Ph.D.[2], Peter Dunscombe, Ph.D.[3], and Sasa Mutic, Ph.D.[4]

[1]Chief Medical Physicist, VA Puget Sound Health Care System, Seattle, Washington
[2]Professor, School of Industrial Engineering, Purdue University, West Lafayette, Indiana
[3]Director of Medical Physics, Tom Baker Cancer Centre, Calgary, Alberta
[4]Professor and Co-director, Department of Radiation Oncology, Washington University School of Medicine, St. Louis, Missouri

6.1 Introduction

This chapter addresses incident learning systems. Such systems rely on an underlying reporting system that can capture the procedural location, severity, and causes of an event, along with other data elements. After an introduction to implementation issues, the processes of incident data collection and root cause analysis are described in depth. Process maps and severity scales are discussed in sections three and four of Chapter 4, respectively. This chapter concludes with a detailed description of incident learning system construction and a comparison between national reporting systems in radiation oncology, in medicine, and in fields outside of medicine.

6.2 Implementation of an Incident Learning System

It has been argued in the opening chapter of this book that quality and safety are not distinct attributes of a system or program. Rather, an unsafe condition can be viewed as an extreme compromise of quality in which the risk of harm is unacceptable. If that argument is accepted, then any improvement in safety would, in general, lead to an improvement in quality. Thus, when we talk about (error) reporting systems, it is worth bearing in mind that such systems, if well constructed, will inevitably lead to an improvement in the quality of the product of the system, whether it be health care or any other system (Williams 2007; Cooke, Dunscombe, and Lee 2007; Clark et al. 2010). As incident learning systems require significant resources to establish—plus an on-going maintenance commitment—it is worth emphasizing the wide-ranging benefits of such a system to administrators.

6.2.1 Characteristics of Incident Learning Systems

There are three general approaches to enhancing the safety (and quality) of a radiotherapy system. They are *prospective*, *reactive*, and *proactive*. The most widely used prospective approach in our field at the present time is Failure Modes and Effects Analysis (FMEA), in which system failures are postulated and their potential effects analyzed. This prospective technique is the basis of the work of the AAPM's Task Group 100 (Huq et al. 2008) and a major focus of this summer school. A reactive approach involves studying failures that have propagated right through the system to affect, in our case, the patient. A reactive approach attempts to ensure that such system failures do not occur again or, at least, that the likelihood of repeat occurrences is reduced. The third approach, proactive, involves observing weaknesses in the system through their effects and strengthening the system to reduce the probability of failures occurring in the first place and, if they do, having steps in place to ensure that errors are discovered and corrected before affecting a patient. It is increasingly recognized that a proactive approach, as described here, can provide major benefits in terms of safety and quality of a radiotherapy system.

Incident learning systems support reactive and proactive approaches to safety and quality improvement (Mutic and Brame 2010). Clearly, if we choose to implement such a system, we would want to achieve the maximum performance to ensure the maximum gain in safety and quality for a given input. Maximum performance in this context means maximizing the input to the incident learning system and maximizing the efficiency of the system to yield the highest gain in terms of safety and quality improvement.

Looking first at the input to an incident learning system, we can ask: what is it that we want to achieve? We want reports of incidents or events from across the complex spectrum of activities taking place within a radiation treatment program. Those activities are performed by a range of professionals, sometimes acting within their own disciplines and sometimes acting in a collaborative fashion across disciplines. For an incident learning system to be fully effective, all members of the department need to be fully engaged (Clark et al. 2012). The system may be initiated and led primarily by one discipline, commonly physics, but ownership of the system should be by a multidisciplinary group reflecting the key activities in the clinical program.

Another important feature of an incident learning system is its scope. Here, opinions vary somewhat. Most systems in use are restricted to clinical incidents, i.e., those incidents that have, or could have, affected patients. However, there is a viewpoint that such a system should be able to handle issues beyond strictly patient safety. Such systems encompass events which negatively impact visitors, staff, and even operational efficiency (Cooke et al. 2006). The argument for a broader approach to incident learning systems is that safety, and indeed quality, cannot and should not be compartmentalized. A safety culture of depth will necessarily acknowledge the importance of the safety of the staff and the facility. The argument against broadening the scope of the system is the incremental resource commitment needed to support a larger system.

Even when the system is restricted to clinical incidents, some guidance must be given to potential reporters on the threshold for reporting. It is apparent that to maximize learning opportunities, the reporting of potential, or close call, incidents must be encouraged. However, whether an incident is actual or potential, some guidance needs to be given to reporters on the level of severity at which it expected that a report will be entered. Judgment is required here, as expecting all incidents, however minor, to be logged could clog the system with trivial non-informative reports. However, being overly prescriptive could result in valuable learning and safety improvement opportunities being lost. The more users understand the uses and benefits of an incident learning system, the better their judgment will be in selecting which incidents to log. This is another reason for full engagement of all team members.

6.2.2 Promoting Staff Participation

With a multidisciplinary system of defined scope, the next challenge is to maximize participation and, hence, safety and quality improvement opportunities. Members of the radiation treatment team have to want to contribute reports of incidents, and further, contribute to the analysis and remediation of system weaknesses. Two environmental conditions need to be in place. First, there should be no deterrents or negative consequences to reporting incidents. Second, it has to be worth the reporter's while to make the effort to report.

Obvious deterrents to reporting include the prospect of disciplinary action against the reporter or a colleague involved in an incident. A feature of some incident learning systems that largely eliminates this deterrent is anonymous reporting. In some environments in which the threat of retribution is real, such an approach may be the only one feasible. However, anonymous reporting does limit the benefits that can be gained from an incident learning system. Frequently, in spite of the best efforts of all concerned, the information submitted in an original report is incomplete or ambiguous. Without knowing who reported the incident, it can be difficult or impossible to follow up and fully investigate the incident—an opportunity lost.

In institutions with a strong and pervasive safety culture, the identification of incident reporters is not an issue. In fact, the willingness of team members to be identified when reporting incidents is a clear measure of the depth of such institutions' safety culture. In any event, systems should be put in place in which the identities of reporters are restricted to the incident investigation team, protecting those reporters from higher levels of administration.

A few years ago, the discussion was about a blame-free culture as a means of encouraging reports of incidents without anonymity. In such a culture, no one could be reprimanded or otherwise disciplined for any behavior, even malicious violations of policies and procedures. The more recently acknowledged reality is that there are members of society who do behave irresponsibly, indulge in substance abuse, and otherwise pose a risk of harm to patients, colleagues, and the integrity of the healthcare system. The notion of a *just* culture has now replaced a blame-free culture, with a just culture requiring individuals to face the consequences of unacceptable behavior. The transition from a blame-free culture to a just culture is not an impediment to incident reporting, provided the culture is truly just and the focus of error-proofing measures is on the system and improving human performance in general.

Removing impediments to reporting is a necessary, but not sufficient, condition for full engagement of staff. The effort involved in logging incidents, when there is a myriad of other clinical tasks vying for priority, has to be justified in the mind of the reporter. If the purpose of the incident learning system is safety and quality improvement, then there needs to be evidence that this is being achieved. In many cases, once the basic causes and contributing factors of an incident, or a series of similar incidents, have been identified, corrective actions become fairly obvious. If these obvious corrective actions are not implemented without good reason, the reporting culture will be seriously undermined. Commitment to an incident learning system will only happen when there is real evidence of change in the clinic and its routine as a result of the analysis of incidents. Experience suggests that a significant number of reported incidents can't always be readily fixed by process or policy changes. Such incidents include those resulting from communication issues, failure to recall the details of some prescribed procedure, and so on. In these cases, measures are taken to make the system as robust as possible, but beyond that, the issue becomes one of reminding the staff on a regular basis of the need for effective, unambiguous communication and diligence in following approved procedures. Whether reporting on corrective actions taken as a result of logged incidents or reminding staff of basic procedural issues, communication with and feedback to staff is absolutely essential if the incident learning system is going to be accepted and realize its full benefit (Clark et al. 2012). Individuals differ in their preferred modes of communication, so it is important to offer a variety of opportunities for staff to learn about safety and quality improvement activities. Such communication modes include e-mails, intranets, and live forums. Some of the new social media may even have a role to play.

6.2.3 Efficiency of the Incident Learning System

Having given the staff a reason to participate in an incident learning system and removed obstacles to participating, the second challenge is to make the reporting, analysis, and learning system as efficient as possible. There is no doubt that incident learning systems consume not insignificant resources that generally have to come from clinical activities. An incident learning system operated off the edge of a desk will not be successful—the dedicated time of skilled professionals should be allocated to the system. The design and operation of an incident learning system must be carefully thought through to ensure resource use is optimized. This topic is discussed in detail later in this chapter, but some of the issues will be introduced here.

The system must appear efficient from the reporter's point of view. Thus the information required to be entered must be known to the reporter at the time of the report. Items such as the patient ID, the step in the process at which the incident was discovered, any safety barriers in operation, and the date and time would be known to the reporter and could be entered into an electronic or paper system very quickly. Whether the reporter should be invited to speculate on where in the process the incident arose and what the basic causes and contributing factors might have been, is a question that needs to be addressed. Synoptic reporting is aided by limiting the number of options for response to a particular question through, for example, drop-down menus (IAEA 2013a). Which dropdowns facilitate reporting and which might conceal critical information? Presumably, a free text description of the incident could be valuable if reporters receive some minimal training on the types of information to include in a free text box. These are issues which deserve due consideration.

In most systems, the investigation of the incident is carried out by someone other than the reporter. Such an approach provides a degree of impartiality when looking for causes and, depending on how the incident learning process is set up, can bring a higher level of expertise to the investigative component of the system. Should there be criteria for deciding which incidents warrant a full investigation and which ones don't? Should investigations only follow incidents that are actually or potentially above some predetermined severity level? Or should one investigate incidents that are less severe but occur frequently? This is a resource issue.

6.2.4 Resource Allocation for Incident Learning

Once causes and contributing factors have been identified, corrective actions need to be considered. In general, these will require additional resources from the clinic's budget. Who has the authority to make the necessary resource re-allocation, and how do they relate to the incident learning system team? As we have noted, if nothing changes following the implementation of an incident learning system, interest in it will soon wane.

There has to be a degree of accountability for the resources committed to the incident learning system. How can it be shown that the system has beneficial effects on the safety and quality of the clinical program? More incidents reported? An increasing ratio of potential to actual incidents? Staff engagement?

As a community, we continue to gain experience with incident learning systems. The questions posed above may not have definitive answers that suit all organizational structures and cultures. However, in the rest of this chapter, these and other issues will be addressed, and some options for solutions will be offered.

6.3 Root Cause Analysis

6.3.1 Purpose of Root Cause Analysis (RCA)

Root cause analysis is an important tool in retrospective event analysis and quality improvement. Although it is frequently described as a single methodology, RCA actually includes a number of tools that can be used both to identify causes of a prior event and to focus proactive attention on how to prevent future events. RCA has been developed and applied in a number of civilian and military settings since the 1940s to improve processes, safety, and production. In fact, the roots of RCA can be seen in the work of Frank and Lillian Gilbreth in their efforts to minimize error, fatigue, and waste in analyses of human performance (Gilbreth 1924; Boring 2010).

The goal of RCA and other incident or systems engineering analysis tools (including FMEA) is to help improve our understanding of the design, interactions, and performance of complex systems. RCA is intended to focus on different types of causal factors that influence adverse outcomes in a complex system—not to affix blame, but to evaluate contributing factors so that their effects can be minimized in the future. Despite criticisms of RCA, it remains an important framework for quality improvement by systematically identifying causes, generating corrective actions, and formalizing monitoring plans.

6.3.1.1 Three Motivations

Many and motivations and responses can arise when something goes wrong or nearly goes wrong. Victims, participants, bystanders, and others may experience:

- a desire to place blame on the individuals or group responsible for the event;
- a desire to learn from the event so that corrective measures can be taken to prevent the event, or events like it, from happening again; and
- a desire to place causes of the event in categories so that deeper patterns can be identified and more effective measures taken to address vulnerabilities in the system.

It is hoped that the focus on blame will progress to a more effective focus on learning from events and understanding deeper causes to prevent similar occurrences in the future. To learn directly from an event in isolation is to treat it in an episodic manner. The specific combination of details in any given event may include unique or otherwise tangential factors that may or may not play a major role in other, similar cases. A purely episodic approach to individual events increases the chance that corrective "improvements" will themselves have unanticipated effects or focus on irrelevant elements. For events with serious effects, or the possibility of serious effects, prompt and decisive action is essential, but an optimal solution will take account of similar situations which had similar conditions that help to identify broader patterns of causal factors (also known as event clustering), or even situations where the event could have happened but did not (also known as a near miss or close call). This is the systemic approach, and it requires an incident learning system in which enough "close call" data is collected that trends become apparent.

When events are dealt with in isolation (episodically), the general process is investigation to determine causation, followed by the generation of corrective actions, implementation of those actions, and review of their efficacy. In this context, a list of all possible cause types will help ensure the investigators select the best set of basic causes to explain why the event occurred. Such a list is called a *causal taxonomy*. While the use of a causal taxonomy is helpful when investigating events in isolation, it becomes essential when working with a large set of incident and close call reports in an attempt to determine event clusters. A vibrant close call reporting system will supply sufficient data to allow events with similar basic causes to be grouped into categories (systemically) so that program weaknesses are brought out more clearly.

6.3.1.2 RCA and Event Frequency

Root cause analysis can arise in the following contexts:

- A significant event can be identified, perhaps through submission to an incident reporting system local to the department or hospital.
- An event may require reporting to an outside agency (e.g., NRC, The Joint Commission) which requires investigation using root cause analysis.
- A pattern of non-critical close call events can be identified, perhaps through submissions to an incident reporting system within the department or hospital.

In the first two cases above, the RCA is performed in response to a particular event. This is the traditional and standard context in which RCA is used. However, general RCA techniques of probing deeper, assigning causes, and developing corrective actions also can be applied to less significant events reported to an incident reporting system.

This latter type of RCA use requires an active incident reporting system, whether formal or informal. One institution's reporting system encouraged reports of all potential process deviations, and they observed a reporting rate of about 0.6 reports per patient (Mutic et al. 2010). At this rate, a clinic treating 500 patients per year would receive about 26 incident reports each month. This is the size of a department with two linear accelerators with a moderate patient load. Compare this with a facility treating 1600 patients per year that collected 688 reports over three years (Arnold et al. 2010). They report receiving an average of 19 reports per month for their much larger program. This rate, when scaled to a department treating 500 patients per year, gives a rate of six reports per month. The difference in report frequency underscores the variation of close call reporting practices where reporting criteria and participation levels may differ greatly between institutions. The minor events won't trigger their own RCA investigations, but by applying RCA concepts informally, underlying causes can be assigned to each event, from which trends can be identified and corrective actions implemented before a more serious event occurs. This is one of the main uses of a root cause taxonomy. The Veterans Health Administration National Center for Patient Safety (VHA NCPS) calls this "aggregate review" (VHA 2011).

6.3.1.3 Areas of Application

The modern use of RCA has its roots in systems engineering analyses done for process engineering companies and large military systems. Growing in use during the 1940s and 1950s, FMEA and RCA tools were used in tandem as components of reliability analysis. The combination of these tools allowed system operators to do prospective and retrospective analyses of failures and design improvements in military systems and in the development of nuclear weapons to improve their operational reliability. By the 1960s and 1970s, systematic and quantitative techniques for FMEA and RCA—including determinations of the sources of operator, equipment, and procedural errors—were applied in nuclear power, aviation, and other industries. The Nuclear Regulatory Commission has requirements for RCA and corrective action (NRC 2012).

The National Transportation Safety Board (NTSB) was established in 1967 to conduct independent investigations of all civil aviation accidents in the United States and major accidents in the other modes of transportation. The NTSB only uses the phrase "probable cause." These independent investigations represent some of the most thorough and systematic retrospective analyses of major accidents, and they focus on multiple immediate, proximal, and distal causes of those accidents.

Other examples of incident reporting and system analysis tools include the Aviation Safety Reporting System (ASRS) developed by the Federal Aviation Administration (FAA) in 1975 (National Aeronautics and Space Administration 2013a). Compared to NTSB investigations, the ASRS represents a more informal system that includes capabilities for self-reporting of close call situations and not just identified accidents. A critical element of the ASRS is that it specifically solicits reports from pilots in a blame-free setting.

After the close of World War II, the Toyota Production System was developed in Japan. This system had many of the properties now referred to as Lean management. One of these properties was the "Five Whys" technique. This technique is the process of asking what cause led directly to the incident, asking what indirect cause led to the direct cause, and repeating the process until the underlying causes are revealed. The technique of causal diagrams created by Kaoru Ishikawa in the 1960s represents a more formalized approach. Causal perspectives were also integrated in the Six Sigma approach to quality.

An early application of RCA in health care is the work done on the PRISMA model at Eindhoven University in the Netherlands (Van Vuuren et al. 1997). The system encompasses a causal tree incident description method, the Eindhoven Classification Model (ECM), of system failure and a classification/action matrix. Other projects were initiated to develop RCA and improvement tools in U.S. healthcare settings. See, for example, the work of Caldwell and associates (Caldwell and Anderson 1996; Caldwell and Kapp 1996). The Rancho Mirage conference was seen as one of the first events drawing specific attention to the concerns of error and process improvement in health care, with an eye toward the use of RCA and other reliability analysis tools. The 1999 publication of *To Err is Human* spurred additional emphasis on quality improvement in U.S. health care. This document focuses on the need for both mandatory and voluntary reporting structures, and alludes to cause identification and corrective action.

In the mid-1990s, The Joint Commission defined criteria for sentinel events which would trigger the requirement for a facility to conduct a root cause analysis. The Joint Commission provides guidance documentation for how to conduct a root cause analysis (The Joint Commission 2013). The Veterans Health Administration began a special focus on patient safety improvement in 1997 (VHA 2011) which led to the creation of the National Center for Patient Safety (NCPS). The NPSC has developed freely available materials on RCA, FMEA, Safety Culture, and other patient safety topics.

6.3.1.4 Limitations of Causal Analysis

One drawback of causation investigation is that it can lead one to see linear relationships between causes and events. This may lead the investigator to focus on the most apparent cause of an event rather than identifying multiple contributing causes. The investigator may also be hampered by hindsight bias, which can prevent one from addressing additional factors that led the operator to think he or she was doing the right thing at the time (Paradies 2010). As described above, the episodic approach to causal analysis may underestimate or confuse the effects of common causal factors and overestimate unique (or even spurious) contributing factors.

6.3.2 Perspectives on RCA

6.3.2.1 RCA for Process Control

One way to think of error is as a deviation from the set of expected processes. That is, if one is operating in state X, and conditions Y prevail, after performing a particular action, one should end up in state Z. In an ideal, well-known operational environment, one would never arrive at an undefined state. Of course, this perfect knowledge of states and conditions is not achievable. The function of incident learning is to identify weaknesses in defined processes that could lead to undefined or undesirable outcomes. Although there are some circumstances where undesirable outcomes (in the medical context, these may be defined as *adverse events*) are due to negligence, willful negative intent, or criminal behavior, one need not assume that these motivations drive all human contributions to adverse events. In fact, many of the deviations from written policy may be more appropriately viewed as employee-driven innovation with their own logic, or even as understandable workarounds to overcome the limitations of poorly designed or implemented operational procedures. Unless one is able to place one's self in the operator's perspective, it will not be possible to understand the situation which led the operator to believe he or she was making the most appropriate choice at the time (Dekker 2006).

6.3.2.2 RCA for Error Reduction

RCA is often a component of a larger quality improvement program whose goal is the reduction of errors. This concept is most applicable in high-volume, high-repetition disciplines like manufacturing. For large-volume radiation therapy departments, a well-structured incident reporting system will enable event rates to be tracked. One recent study showed how the total incident rate dropped from 1.1% to 0.7% to 0.3% over successive years in a three-year study (Arnold et al. 2010). Note that the focus of RCA is not simply on the determination of causal factors contributing to errors or unsafe outcomes. Investigation is followed by corrective actions that improve system design so that errors are less likely, severe, or hard to recover from.

6.3.2.3 RCA for System Understanding

Efficiency and safety—as well as cost-effective productivity—are seen as hallmarks of effective quality management programs. RCA provides an opportunity to reflect on the process facets that have led to adverse events and close call situations. In addition to making a system safer (avoiding adverse events and destructive outcomes), corrective actions may also improve the overall efficiency of the system by identifying procedural simplifications which eliminate certain error pathways or sources of system degradation. In a similar sense, causal analysis can be used in an incident reporting system to generate statistics on the rates of certain error types, which can then inform better-decision making when using prospective safety tools such as FMEA and fault tree analysis.

6.3.2.4 Industry Differences

Many stakeholders have become involved in the implementation of incident learning within particular contexts: safety professionals, regulatory agencies (NRC, NPSA, etc.), oversight organizations (The Joint Commission, ECRI, NIH, etc.), and professional bodies (ASTRO, AAPM, nursing associations, etc.). In the case of regulatory agencies and oversight organizations, the oversight can be general to a field (aviation, nuclear power, medicine, etc.); a discipline within a field (such as radiation oncology, cardiology, etc.); or to a role (occupational safety, patient safety, nursing, radioactive materials handling and use, etc.). Combining oversight (incident collection) and regulatory (punitive) roles in one RCA organization can lead to a reluctance to report or the desire to omit critical incident details. For the ASRS, the collection agency (NASA) does not have regulatory (i.e., FAA) authorization to punish pilots involved in adverse events. In fact, pilots who do report close call events to ASRS prior to event investigation can obtain some immunity through confidential reporting receipts. This issue should be noted in healthcare organizations, where RCA and incident analysis investigation team members may also be involved in personnel performance reviews or decisions regarding departmental resource allocations.

6.3.3 Design and Use of RCA Incident Analysis Tools

6.3.3.1 Taxonomy Structure

A causal taxonomy is just one of many tools for categorization of incidents. Other such tools include process steps (Ford et al. 2012) and incident-type classifications (Arnold et al. 2010). Table 6.1 compares several published taxonomies that are applicable to health care or to radiation therapy in particular. The causal elements are divided into the following considerations: organizational, technical, human, and other factors. As can be seen, with a sufficient amount of reorganization, it is possible to map the elements of one taxonomy onto another. A comparison of the first three classification schemes (Itoh, Omata, and Andersen 2009) emphasized the importance of latent conditions (risk factors) in contributing to events and de-emphasized active failures, thus encouraging the use of corrective actions with greater leverage. In contrast, a focus on active failures leads to corrective actions such as "always do it this way" and "never do it that way," which does little to assure an event will not recur and does nothing to mitigate indirectly related fault paths. Table 1 utilizes the following classification frameworks:

1. "PRISMA-Medical: A Brief Description." (Eindhoven University Taxonomy) (Van der Schaaf and Habraken 2005), Table 1. This is also cited in (Snijders et al. 2009).
2. "The JCAHO Patient Safety Event Taxonomy." (Chang et al. 2005), Figure 4.
3. "Towards an International Classification for Patient Safety: Key Concepts and Terms." (Sherman et al 2009, WHO 2009b).
4. "A Reference Guide for Learning from Incidents in Radiation Treatment." (HTAU: the Health Technology Assessment Unit of the Alberta Heritage Foundation for Medical Research.) (Cooke et al. 2006), Table 11.
5. "Taxonomic Guidance for Remedial Actions." (Thomadsen and Lin 2005), Figures 1, 2 and 3.
6. "The AAPM Working Group for the Prevention of Errors Causal Taxonomy." (AAPM WGPE). (Ford et al. 2012).

It is apparent from Table 6–1 that there is substantial overlap between the various published causal taxonomies. Of particular interest are the gray boxes, which represent elements not translatable to the given taxonomy. For example, the Eindhoven taxonomy does not include intentional bad behavior, and the HTAU taxonomy does not explicitly address external factors. Judicious limitation of the causal choices is an important tool to prevent users from identifying superficial causes or from arriving at different nodes that equally describe the cause sought for.

Unlike the other taxonomies shown here, the AAPM WGPE taxonomy includes a section devoted to more formal human factors concepts, such as problem detection, interpretation, rule selection, response planning, and plan execution. This section is intended to be a resource for those with training in human factors analysis. The AAPM WGPE taxonomy is shown in full in Appendix 6.1.

6.3.3.2 Single Versus Multiple Causal Factors

Regardless of the taxonomy selected, there will be the possibility of identifying multiple causal factors, ranging from individual human performance to organizational policies to fundamental design flaws in equipment. Relatively simple, coarse-grained taxonomy distinctions will result in smaller numbers of distinct causal factors; more fine-grained taxonomies allow for a greater distinction in causal contributors. An active incident reporting system will increase the ability to identify event clusters sharing multiple causal factors.

In incident analysis and causal determination, the structure of the process itself can affect what conclusions are ultimately reached. Many RCA tools are implemented based on attempts to determine a primary causal factor. However, the concept of "primary" is frequently an artifact of the order of questions presented in the incident investigation tool. This can be described as a "pinball" model of RCA incident investigation (Caldwell 2008): one proceeds down the list of possible causal factors as defined by the taxonomy structure and incident investigation tool until a single, identifiable causal factor is selected. The pinball metaphor comes from the behavior of a single ball's path through a pinball machine until it disappears down the ball return. At this point, the determination of causal contributions to the incident is completed. If the incident investigation tool was ordered differently, a different "primary" cause would have been selected.

By contrast, a "pachinko" model of incident investigation suggests that the determination of causal contributors continue until *all* reasonable causal factors have been identified. In contrast to the pinball model, this metaphor reflects the pachinko game, where many balls travel down multiple paths to accumulate in several bottom collection areas. This technique allows for multiple contributing factors, without the need to impose biases or reporting system artifacts to select a specific "primary" factor. As the number of incident reports increases, the pachinko model has greater statistical and organizational power to identify patterns of causal contributors to adverse events (Caldwell 2008). Determining multiple patterns of contributors can help improve the ability to cluster causal factors due to work flow processes, common causes of failure in equipment, or other situational factors (such as accelerator beam variations due to interference from electrical power fluctuations at meal times, as determined in one incident analysis).

Table 6–1. Comparison of classification schemes

Eindhoven van der Schaaf & Habaken 2005			JCAHO Chang et al. 2005			WHO Sherman 2009; WHO 2009b	
Organizational		Management priorities	Systems	Organizational	Management	Organizational/ Service Factors	Organization of Teams
		Culture			Organization and Culture		Organizational decisions, Culture
		Protocols			Protocols, Processes		Protocols, Policies, Procedures, Processes
		Transfer of knowledge			Transfer of knowledge		Resources, Workload
		External			External		External[6]: Services, Systems, Policies
Technical		External		Technical	External	Work[1]/ Environme	External[6]: Products, Technology, Infrastructure
		Design Construction Materials			Facilities		Physical environment, Distance from service, Risk & safety…
Human		External (human factors in other departments)	Intentional human behavior	Error	External	(hatched)	(hatched)
		Knowledge-based behavior			Practitioner	Staff Factors[2]	Cognitive > Knowledge based
	Rule-based behavior	Qualifications					Performance > Rule based: misapplication of good rules, application of bad rules
		Coordination					
		Verification					
		Intervention					
		Monitoring					
	Skill-based	Slips[4]					Performance > Technical > Slips/Lapse/error
		Tripping[5]					
Other		Patient related factor			Patient	Patient factors: Cognitive, performance, …[3]	
		(hatched)	Negligence			Behavior > Risky Behavior	
			Recklessness			Behavior > Reckless Behavior	
		(hatched)	Intentional rules violations			Behavior > Sabotage/ Criminal act	
		Unclassifiable	(hatched)			External[6]: Natural Environ.	

The superscripts indicate items requiring further explanation:

1. The WHO classification of Work/Environment Factors include: Physical Environment/Infrastructure; Remote/Long Distance from Service; Environmental Risk Assessment/Safety Evaluation; and Current Code/Specifications/Regulations.
2. The WHO classification of Staff Factors (page 92) is extensive: cognitive factors, performance factors, behavior, communication factors, patho-physiologic/disease related factors, emotional factors, social factors. Only those elements that mesh with the other systems have been shown.
3. The WHO classification of Patient Factors (page 92) is extensive: cognitive factors, performance factors, behavior, communication factors, patho-physiologic/disease-related factors, emotional factors, and social factors.
4. Slips = "Failures in performance of highly developed skills."

Continued on next page

Table 6–1, cont. Comparison of classification schemes

HTA		Thomadsen		AAPM WGPE	
Cooke et al. 2006		Thomadsen & Lin 2005		Ford et al 2012	
Systemic/Management	5. Communication	Organizational subpath	Communication system	Organizational Manage.	Communication
	4. Planning		Leadership, Competition for attention		Program planning
	1. Standards, Policies, Procedures				Policies, Procedures, Regulations
	6. Knowledge, Skill		Training Programs – Knowledge transfer		Training; Transmitting knowledge and skills
					Physical environment
					Leadership & external
Job Factors		External		Technical	Environment (within the facility)
	3. Design	Environmental subpath: Tangible elements			Acc.test & Commiss.
	2. Materials, Tools, Equipment				Equipment design
		Hardware/software subpath			Equipment maintenance
Personal Factors		Human Error subpath[8]		Human Behavior	
			Knowledge based: Information problem		*See training*
			Rule Based: Mistaken choices (recall), Interference		Acting outside scope
	7. Capabilities: physical, sensory, substance sensitivities		Skill Base: Physical coordinations; slips, trips		Slip causing physical error
Judgment	8. Judgment: Failure to address issues, Extreme demands…[7]			Patient related circumstances	
			Lack of vigilance	Hum. Beh.	Negligence
					Poor judgment
	8.5 Criminal intent		Intentional violation		Intentional rules violations
Natural factors: 9. Hazards		Environmental subpath		External factors	

Continued from previous page

5. Tripping = "Failures in whole body movements."

6. The table on page 95 of the referenced document (WHO 2009b) for "Contributing Factors/Hazards – External Factors" is divided into the three separate subsections used here.

7. The referenced document (Cooke et al. 2006) addresses judgment issues differently than JCAHO or the WHO by listing causes rather than simply attributing the cause to negligence or recklessness.

8. Figure 3 of the referenced document (Thomadsen and Lin 2005) breaks down its "Human Error sub-subpath" by both functional categories and Rasmussen's model for human actions based on the level of intellectual consciousness required for the work: whether the action is skill-based, rule-based, or knowledge-based.

6.3.4 The Conduct of RCA

The process of incident investigation is discussed in depth in publications such as the NCRP's "Investigation of Radiological Incidents" (NCRP 2012) and the *Aviation Investigation Manual* of the National Transportation Safety Board (NTSB 2002).

6.3.4.1 Determination of the Need for RCA

Formal RCA is generally triggered when an incident occurs satisfying pre-established criteria, such as The Joint Commission's sentinel event definition or the NRC's medical event definition. When an RCA is triggered by a pattern of near-miss events captured by an incident reporting system, the formal methods described in this section are largely inapplicable.

6.3.4.2 Preserving the Scene

In most applications of RCA outside health care, it is important to control the incident scene so that no information is lost and no physical items are removed because these would limit the gathering of information related to the incident. Although this is equally desirous in the context of health care, it is seldom possible.

6.3.4.3 Team Formation

While incidents of low to moderate consequence may require only a single investigator, incidents of higher consequence or complexity will require representative subject matter experts to be on the investigation team. The team may require non-team member subject matter experts for additional consultation. Incident investigation teams are generally limited to five or six participants.

6.3.4.4 Charter

The investigation charter functions as an agreement between management and the investigation team in order to specify the problem statement as well as the investigation's scope, anticipated duration, and team membership. The problem statement should describe the gap between the desired situation and the current situation. It should do so in a way that does not bias the possible findings.

6.3.4.5 Investigation

Formal investigations will usually start with an entrance meeting, at which time the terms of the charter are presented. This is followed by a physical inspection of the areas of concern, the collection of testimony, a review of documents, and daily briefings. At the close of the investigation, the investigation team will meet with management to present preliminary information about the factual findings and, if possible, the possible causes that contributed to the event.

6.3.4.6 Report

After the investigation phase is complete, a written report is generated. The body of the report should focus on affected program activity and equipment, sequence of events, barrier analysis, inappropriate actions or equipment failures, and cause analysis (NCRP 2012).

6.3.5 Conclusions

The true purpose of RCA and other incident investigation or failure modes analysis tools is not to assign blame, but to improve system reliability and performance. True RCA within the health care industry may involve multiple tools, ranging from informal incident and close call descriptions through exhaustive analyses of critical adverse events resulting in catastrophic outcomes. An effective RCA moves beyond a focus on anger and blame to emphasize systematic methods for quality improvement and error pathway elimination. Multiple RCA taxonomies exist; most include causal contributing factors involving human task performers, organizational policies and procedures, technical equipment and maintenance, work processes, and environmental conditions. In order to effectively use RCA for process improvement, the incident analysis process should consider mechanisms to:

- identify multiple potential causal contributors,
- avoid biases and artifacts that lead to specific emphases on provider fault isolated from other conditions, and
- empower investigation teams to focus on opportunities for system improvement apart from punitive examinations of individuals and departments.

6.4 Data Elements

The reported incidents can be classified and used in various ways. On one side of the spectrum are the high-severity, infrequent incidents, and on the other side are the low-severity, high-frequency events. The incident type and severity will determine whether the event should be analyzed promptly or grouped with similar incidents to be processed together. For example, a wrong patient treatment is an isolated and potentially high-severity incident that would require a full RCA as soon as practical. Contrast this with a situation in which a significant fraction of patients have planning computed tomography (CT) volumes longer than necessary, creating a situation where patients receive small amounts of unnecessary dose, and the additional CT images create unnecessary work and require extra time for processing. The first incident would be readily obvious, and all of the necessary data could be available in a short span of time. However, it would take some time to realize that there is a systematic problem in the second type of incident, and it would take even more time to collect a significant number of events to determine all of the contributing factors. The data elements contained in the reported events have to be extensive enough to enable processing of a high-severity, infrequent event through an RCA, as well as provide enough information so a trend or a pattern can be identified from low-severity, high-frequency events.

The broader benefit of an incident learning system is that the collected data enables individual clinics to learn from incidents that occurred elsewhere and to potentially benchmark their performance of certain aspects of their individual practice against the outside standard. Whether the learning and benchmarking happens between two or three small practices who all belong to a same medical group or the data comes from a national reporting system, there is a need for communication of event data so the receiving facility can consume it efficiently and maximize the benefit from the details contained in the incident description.

The incident report and the related metadata need to be organized in data elements which would enable efficient and effective processing within an institution or allow for distribution to a broader reporting system (a chain of medical practices, state or national reporting system, etc.). The AAPM Working Group for Prevention of Errors (Ford et al. 2012) has reported recommendations for incident database structures in radiation oncology. This report details the data elements which should be submitted by the initial reporter, as well as a variety of data elements which can be subsequently collected to process an incident report within an institution (whether individually or in aggregate with other similar events) or where the incident can be passed to a broader reporting body.

The World Health Organization's *Draft Guidelines for Adverse Event Reporting and Learning Systems* offers additional rationales for creating incident report taxonomies, recommendations for organizing data in elements, and some other useful information for creation of incident learning systems (World Health Organization 2005).

Based on the AAPM working group recommendations (Ford et al. 2012), data should be organized in three levels: report, analysis, and response. Table 6–2 is reproduced from this document and lists the data elements for each of the report levels. It is important to note that while at the first glance these lists seem quite extensive, a number of the individual elements can be collected automatically from the patient data or are a part of a facility's demographics. The data in Table 6–2 are sufficiently comprehensive to be used for local incident learning purposes or to be shared with an outside reporting system. If the initial implementation of a local system seems too daunting if all of these elements are to be collected, it is reasonable to evaluate which elements are truly needed for the purposes of the initial implementation. Ideally, all of the elements shown should be collected, and an institution would be able to utilize the benefits of a comprehensive incident learning system early on. However, if implementing a comprehensive reporting system would prevent the implementation of *any* reporting system, it is likely better to start with a narrowed version and expand as the organization becomes more comfortable with operating the reporting system.

The data elements are just a part of a reporting system, and the use of data is another consideration. It is important to remember that an incident reporting system is a tool, and ultimately the benefits that can be gained from it will depend on how that tool is used. Periodic evaluation of the data elements that are routinely collected and their ability to provide insight to individual event analysis or trend analysis will determine if the full benefits of the incident learning system are realized or if certain improvements need to be made in order to provide more meaningful data and better insight into processes. As it is expected that the data elements will change over time, forward and backward compatibility of the collected data and the scalability of data elements is a consideration when creating an incident learning

Table 6–2. Data elements for 1) a reporter's form, 2) an analyst's form, and 3) a responder's form (Appendix E of Ford et al. 2012)

1. Reporter's form

#	Data element	Level	Format	Encrypt	Definition
1.1	Date of incident(s)	Required	YYYY-MM-DD	N	The date that the incident occurred. If the incident occurred on more than one date, this is the date of the first occurrence
1.2	Time of incident(s)	Optional	HH:mm 24-hr format UTC	N	The time of day that the incident occurred. If the incident occurred on more than one time this is the time of the first occurrence
1.3	Date of report	Recommend	YYYY-MM-DD	N	The date that the report is logged
1.4	Date of discovery	Optional	YYYY-MM-DD	N	The date that the incident was first discovered
1.5	Incident type	Required	Pulldown	N	Near miss or actual event
	Options: 1: actual incident 2: near-miss; no-one affected				
1.6	Person affected	Required	Pulldown	N	The person(s) affected in the incident
	Options: 0: no one affected 1: one patient affected 2: several patients affected (give number) 3: staff 4: other (specify) 5: unknown				
1.7	Number of fractions delivered incorrectly	Required	Integer	N	The number of radiation treatment fractions delivered incorrectly
1.8	Incident description	Required	Text	N	Description of the incident provided by the person reporting. This should be a detailed description that can be read and understood by any radiation therapy professional in any clinic
1.9	Where found	Recommend	Pulldown	N	The point in the workflow where the incident was initially discovered
	Options: refer to detailed process maps (Secs. 2 and 3 of Appendix B, WGPE process maps)				
1.10	How discovered	Recommend	Text	N	A brief description of how the event was discovered
1.11	Patient's first name	Recommend	Text	Y	The first name of the patient affected. "Multiple" if multiple patients were affected
1.12	Patient's last name	Recommend	Text	Y	The last name of the patient affected. "Multiple" if multiple patients were affected
1.13	Patient's medical record number	Recommend	Text	Y	The medical record number of the patient affected. "Multiple" if multiple patients were affected
1.14	Treatment modality	Recommend	Pulldown	N	The radiation therapy modality used or planning to be used at the time of the incident
	Options: 1: photons 2: electrons 3: particles 4: brachytherapy 5: radioisotope 6: other (specify) 7: not applicable				
1.15	Person reporting first name	Optional	Text	Strip	The first name of the person filing the initial report
1.16	Person reporting last name	Optional	Text	Strip	The last name of the person filing the initial report
1.17	Person reporting 's role	Optional	Pulldown	Strip	The departmental role of the person filing the initial report
	Options: 1: attending radiation oncologist 2: resident radiation oncologist 3: other physician 4: radiation therapist 5: dosimetrist 6: physicist 7: nurse, NP or PA 8: administrator 9: patient 10: other (specify)				
1.18	Action taken by reporter	Optional	Text	N	The immediate response of the person reporting. This information is provided so the evaluator has it up front for follow up

2. Analyst's form

#	Data Element	Level	Format	Encrypt	Definition
2.1	Location	Recommend	Text	Y	City and state/province where the incident occurred
2.2	Facility profile	Recommend	TBD	Y	Information about the facility where the incident occurred including number of linacs, patients on treatment, etc.
2.3	Patient's age at the time of the incident	Recommend	Integer	Y	The age of the patient when the incident occurred
2.4	Patient's race	Recommend	Pulldown	N	The ethnic race of the patient.
	Options: 1: American Indian or Alaska Native 2: Asian 3: Black or African American 4: Hispanic 5: Native Hawaiian or Other Pacific Islander 6: White 7: Some other race 8: More than one race 9: Unknown 10: not applicable				
2.5	Gender	Recommend	M/F/unknown	N	The sex of the patient
2.6	Patient notification	Recommend	Y/N/unknown	N	Were patient, patient's family or guardian notified?

Table 6–2, cont. Data elements for 1) a reporter's form, 2) an analyst's form, and 3) a responder's form (Appendix E of Ford et al. 2012)

#	Data Element	Level	Format	Encrypt	Definition
2.7	Treating radiation oncologist	Recommend	Pulldown	Y	The radiation oncologist of record at the time of the incident
	Options: for individual clinic systems the pulldown consists of a list of physician provides for a national system the list would derived from a database of provider IDs				
2.8	Treating radiation oncologist notified	Recommend	Y/N/unknown	N	Was treating radiation oncologist notified?
2.9	Referring physician(s) notified	Recommend	Y/N/unknown	N	Was referring physician(s) notified?
2.10	Disease being treated	Recommend	Text	N	The primary disease being treated. Use ICD-9 code if available
2.11	Disease stage	Optional	Text	N	The stage of the disease at the time of the incident. Use TNM staging descriptors
2.12	Intended treatment site	Recommend	Text	N	Intended anatomical site of treatment. For multiple sites indicate the most appropriate site
2.13	Treatment intent	Optional	Pulldown	N	Immediate intent of radiation treatment
	Options: 1: curative 2: palliative 3: unknown 4: not applicable				
2.14	Total prescribed dose	Recommend	Real number	N	Total prescribed radiation dose for the full treatment being delivered at the time of the incident in units of Gy
2.15	Dose per fraction	Recommend	Real number	N	Total prescribed radiation dose per fraction at the time of the incident in units of Gy
2.16	Course	Recommend	Pulldown	N	Radiation course at the time of the incident
	Options: 1: primary phase 2: boost phase 3: modification during treatment 4: other (specify) 5: not applicable				
2.17	Treatment technique	Recommend	Pulldown	N	The radiation therapy technique used or planning to be used at the time of the incident
	Options (select all that apply): 1: simple 2: 3D conformal 3: IMRT 4: SRT/SRS cranial 5: SBRT 6: modulated arc therapy 7: intracranial, intraluminal, intravascular or surface 8: interstitial 9: LDR, PDR 10: HDR 11: temporary implant 12: permanent implant 13: orthovoltage 14: other (specify) 15: not applicable				
2.18	Image guidance	Optional	Pulldown	N	The image-guidance technique(s) used or planning to be used at the time of the incident
	Options (select all that apply): 1: kV or MV radiographs 2: IGRT: kV cone-beam CT 3: IGRT: MV cone-beam CT 4: MV CT 6: stereo camera tracking 7: other (specify) 8: none 9: not applicable				
2.19	Person discovering first name	Optional	Text	Strip	The first name of the person who initially discovered the incident
2.20	Person discovering last name	Optional	Text	Strip	The last name of the person who initially discovered the incident
2.21	Person discovering 's role	Optional	Pulldown	Strip	The departmental role of the person who initially discovered the incident. Note may be different than the person reporting (element 1.15, 1.16, and 1.17)
	Options: 1: attending radiation oncologist 2: resident radiation oncologist 3: other physician 4: radiation therapist 5: dosimetrist 6: physicist 7: nurse, NP or PA 8: administrator 9: patient 10: other (specify)				
2.22	Treatment unit manufacturer(s)	Recommend	Pulldown	N	The manufacturer of the hardware used to deliver treatment in the incident
2.23	Treatment unit model(s)	Recommend	Pulldown	N	The model of the hardware used to deliver treatment in the incident
2.24	Treatment planning system manufacturer	Recommend	Pulldown	N	The manufacturer of the software used to plan treatment in the incident
2.25	Treatment planning system model	Recommend	Pulldown	N	The model of the software used to plan treatment in the incident
2.26	Record and verify system manufacturer	Recommend	Pulldown	N	The manufacturer of the record and verify system in the incident
2.27	Record and verify system model	Recommend	Pulldown	N	The model of the record and verify system in the incident
2.28	Third-party ancillary device manufacturer	Recommend	Pulldown	N	The manufacturer of any third-party ancilliary devices being used or planned for use in the incident.
2.29	Third-party ancillary device model	Recommend	Pulldown	N	The model of any third-party ancilliary devices being used or planned for use in the incident
2.30	Number of patients on-treatment	Optional	Integer	N	The number of patients being treated on the day of the incident on the treatment unit where the incident occurred
2.31	Number of radiation therapists present	Optional	Integer	N	The number of RTTs present at the time when the incident occurred. Leave blank if incident did not occur on treatment unit

Table 6–2, cont. Data elements for 1) a reporter's form, 2) an analyst's form, and 3) a responder's form (Appendix E of Ford et al. 2012)

#	Data Element	Level	Format	Encrypt	Definition
2.32	Number of other staff present	Optional	Integer	N	The number of other staff present at the time when the incident occurred. Leave blank if incident did not occur on treatment unit
2.33	Staff involved	Optional	Pulldown	Y	Names and roles of staff present when the incident occurred
2.34	Error type	Optional	Pulldown	N	The type of error that occurred or potentially could occur
	Options (select all that apply): 1: wrong patient or potentially wrong patient 2: wrong anatomical site or potentially wrong site 3: wrong laterality or potentially wrong laterality 4: wrong dose to all or part of the tumor or normal tissue or potentially wrong dose 5: wrong modality or energy or potentially wrong modality or energy 6: other (specify) 7: not applicable				
2.35	Dosimetric severity	Optional	Pulldown	N	See Appendix C
	Options: refer to Appendix C, WGPE severity scales				
2.36	Reportable event	Optional	Y/N/unknown	N	Was the event reportable to the relevant regulatory agency (state, NRC)?
2.37	Medical severity	Required	Pulldown	N	See Appendix C. For potential incidents (near-misses) assume a worst-case scenario if the incident were actually to occur
	Options: refer to Appendix C, WGPE severity scales				
2.38	Severity assessment type	Required	Pulldown	N	The method used for assessing the severity of harm
	Options: 1: estimated severity 2: actual observed severity 3: not applicable				
2.39	When severity assessed	Recommend	Pulldown	N	The time after the incident at which the severity of harm was observed or at which it is estimated to occur
	Options: 1: before 24 hours 2: after 24 hours but before 3 days 3: three days or later 4: unknown				
2.40	Summary title	Optional	Text	N	A very brief title phrase describing the incident
2.41	Incident analysis (from evaluator)	Optional	Text	N	Description of the incident provided by the person evaluating the incident in the clinic. This should provide very detailed information from which one can infer root causes
2.42	Where originated	Recommend	Pulldown	N	The point in the workflow where the incident originated
	Options: refer to detailed process maps (Secs. 2 and 3 of Appendix B, WGPE process maps)				
2.43	Attached files	Optional	File	N	Files (e.g., screenshots, pdf reports, etc.) that may help clarify the details of the incident
2.44	Causes	Required	Pulldown	N	List all the causes and contributing factors at work in the incident. For potential incidents assume a worst-case scenario if the incident had actually occurred. See Appendix D
	Options (select all that apply): Refer to Appendix D, WGPE causality table				

3. Responder's form

#	Data Element	Level	Format	Encrypt	Definition
3.1	Clinical action scale	Optional	Pulldown	N	Priority scale to guide the follow-up actions of the clinic
	Options: A,B, C, D (exact meaning of each to be determined by the individual clinic)				
3.2	Intervention	Recommend	Pulldown	N	Was any intervention attempted in order to "rescue" the patient (i.e., to prevent, minimize or reverse harm)?
	Options (select all that apply): 1: treatment stopped 2: treatment modified 3: added additional treatment fractions 4: patient's stay in hospital extended 5: intervention pending 6: no intervention 7: unknown 8: other (specify) 9: not applicable				
3.3	Safety barriers	Recommend	Pulldown	N	The safety barriers or quality control measures in place which prevented or could prevent the incident
	Options: (select all that apply). Refer to detailed process maps (Secs. 2 and 3 of Appendix B, WGPE process maps)				
3.4	Corrective action	Optional	Text	N	Corrective action(s) taken to mitigate the harm for this particular patient
3.5	Preventive action	Optional	Text	N	Preventive action taken to insure that a similar incident will not happen to a patient in the future
3.6	Learning actions	Optional	Text	N	The learning activities that were undertaken in response to the incident, for example, presentation at rounds, etc.
3.7	Incident closed	Optional	Y/N	N	Has the incident been closed by the evaluator responsible for incident handling? This includes the implementation and evaluation of corrective, preventive and learning actions. Quality control of the data should also be performed

system. Most of the modern databases can be structured where addition or subtraction of data elements does not affect the longitudinal use of the data.

6.5 Reporting System Applications

In this section we address how the principles of good incident reporting system design are applied. Incident reporting systems may be set up within a department, a hospital, a network of hospitals, the state, federally, or internationally. This chapter primarily addresses departmental systems designed to catch close calls with the intent to prevent more serious incidents from occurring. This is dependent on completing the cycle of reporting, analysis (perhaps root cause analysis), identifying corrective actions, and monitoring the implementation of corrective actions.

In discussions of incident learning systems, the following purposes are frequently cited:

1. Generate immediate corrective actions based on root and contributing causes of individual events.
2. Gather statistics on the relative frequency of categories of close calls.
   A. Compare year to year and machine to machine.
   B. Trend data with time (on the time scale of days and weeks).

The first purpose hinges on the causal factors, while the second hinges on the breadth, depth, consistency, and electronic availability of data collected. Note also that the first purpose deals with events on an episodic level and can obscure the relative significance of events and corrective actions, especially when viewed by someone without specialization in radiation therapy. In contrast, the second purpose offers a wider vantage on the significance and evolution of recorded events and the culture in which our processes are embedded.

6.5.1 Implementation Format

6.5.1.1 Dividing Reporting and Analysis Tasks

The processes of reporting and analysis differ in their required expertise, required time, and vantage point. Incident reporting should include immediately available factual information, such as the name of the patient, the reporter, and the device involved (if it is to be included), as well as a narrative description of the event. The incident description will be critical for understanding the context of the event. Later review of the event can easily fall into the traps of hindsight bias and counterfactuals—seeing the event from outside the "tunnel" containing the event participants, to use Sidney Dekker's metaphor (Dekker 2006). Thus it is important to be able to reconstruct not only the process deviations that led to the event, but also the circumstances that would have led the operator to think his or her action was appropriate at the time. This is the human factors perspective, which will address the fact that different people in a similar circumstance could be led to perform a similar action. Unfortunately, it is unlikely that the operator will have the time to include a full account of the circumstance in the initial report, which means that the broader narrative won't be collected until later, if at all.

After the completion of the initial incident report, it should be reviewed by a higher-level reviewer for the following purposes:

1. Mandatory reporting to the facility or an outside agency (which may involve disclosure to the patient among other considerations).
2. Immediate corrective action to prevent repeat occurrences of the incident.
3. Follow up analysis (e.g., RCA) for the purpose of later corrective actions.
4. Data trending.

The first pair of tasks forms threshold considerations requiring immediate action. The second pair of actions requires that time be set aside later for completion. The higher-level report reviewer should know the external requirements for event reporting. Requirements are set by regulatory agencies, such as the state and the Nuclear Regulatory Commission; non-governmental outside agencies, such as The Joint Commission (sentinel event reporting); and by the facility. Communication outside the department should generally be authorized by a designated person and performed by a designated person. The threshold for immediate corrective action should be based on consideration of severity and may lead to suspending a process if resources for completing necessary corrective actions are not available. This circumstance would arise if either the allocation of resources (repair funds, additional staff, equipment investment, or training) cannot be completed prior to scheduled treatments, or if deeper analysis is needed to delineate the scope of the problem and the appropriate corrective actions required prior to resuming the process.

The second pair of considerations (analysis and data trending) does not require or allow immediate action. In particular, formal RCA requires a sequence of steps involving investigation team selection, interviewing, analysis, and report generation. Less formal analysis should also include data gathering through discussions with those closest to the incident. It is important to minimize the delay before gathering information, since it is essential to understand the sequence of events from the perspective of the operator at the time, especially with regard to how competing issues influenced one's choices and decisions. The second of these activities is the documentation of the event. Which event data will be queried from an event database also influences the time and staffing required to maintain the database. For example, a paper-based system may be hand-audited to gather statistics for a small department. In a larger department, it may be preferable to use a hybrid system in which reporting forms are scanned and some of the information is captured into an electronic database, such as a spreadsheet or a relational database. In a large department, for which a large number of events are compiled, the use of a fully electronic reporting system is essential.

6.5.1.2 Paper Versus Electronic Submission

The choice to use a paper-based versus electronic submission system depends on software availability, software expertise, and availability of computer access. If the

Table 6–3. Incident reporting system characteristics as a function of scope

Scope	Immediate Corrective Action	Disclosure to Patient	Initiate Formal Analysis	Trending	Report & Data Access
Within the department	Possibly	No, for close calls	Yes	Yes	Department Leadership
Within the medical facility	No	Yes, when warranted	Yes	Yes	Safety Officer
Within a cluster of medical facilities under a single organization	No	Maybe	Yes	Yes	Safety Officer
Externally through an agency independent of the facility	No	No	Yes	Yes	Identified Regulators or Officials

department is limited in one or more of these areas, then a paper system is likely to be preferable. It should be noted that the initial handwritten report cannot be considered fully anonymous, since handwriting is recognizable. If anonymous reporting is a goal, then the possibility of identifying grammar and handwriting may be problematic. Thus the recorded narrative should be written in different words. Examples are available of both paper-based reporting forms (Cooke et al. 2006) and electronic forms (Cunningham 2011). Elements in these forms were discussed above in Section 4.

6.5.1.3 Closing the Loop

A healthy clinical culture requires belief that each employee's time is valued and that their work is of value to the department. It is important that the reporting system have positive impact in the clinic and thus be deserving of the time it requires. This impact can occur through immediate corrective actions for those incidents whose potential severity warrants immediate changes. It can also occur through the careful analysis of a given event or a more subtle trend inferred from an aggregate of events. Without the clinic benefiting from the reporting system, the support of the frontline staff, who are expected to initiate reports, will quickly erode.

6.5.1.4 How the Information is Used

As mentioned before, incident reporting systems exist at the following levels: within the department, within the medical facility, within a cluster of medical facilities under a single organization, and externally through an agency independent of the facility, such as a national voluntary or mandatory reporting system. The characteristics of an incident learning system depend on its scope, as shown in Table 6–3.

6.5.1.5 Corrective Actions

A crucial measure of effectiveness for an incident reporting system is the perceived value of the departmental process improvements it leads to. This is inherently subjective. The more obvious measure of effectiveness is the reduction in the rate of events addressed by the corrective action. However, this will apply only to the most obvious trends.

Corrective actions may be immediate, based on the response to single episodes of an event type, or they may be in response to a trend in an event type. Corrective actions may take the form of a process modification (so that the error pathway is no longer possible), in the form of an additional safety barrier added onto the existing process (so that there is an additional opportunity to catch errors of the kind addressed), or in the form of reinforcement of the existing policy, either through additional training, direct competency assessment, or punitive action. These categories of corrective action have been listed from strongest to weakest in terms of effectiveness.

6.5.1.6 Follow-up

The incident reporting system should be run within a more comprehensive quality improvement program administered by designated individuals. This higher-level group should meet periodically, and at that time the status of recent corrective actions should be discussed. It is important to have benchmark goals for corrective actions. These goals should be monitored over a period of time, with the staff being aware of the monitoring process.

6.5.2 Implementation Decisions

An incident reporting system will generate a number of documents that must be controlled in terms of access (keeping it from the wrong people), availability (getting it to the right people), and proper archiving (not losing it). These documents include the initial event report, any initial reports submitted beyond the department, investigation notes or transcripts, and final reports, which may include a causal analysis and statistical data and analyses.

6.5.2.1 Who Writes the Event Report

As a general approach to human factors analysis and investigation, it is important to consider not only the perspective of the person or people involved in the incident, but also the perspective from a detached vantage point. This means the initial narrative should be collected in writing or verbally from someone close to the event, while the subsequent analysis should be done by someone not involved in the event. Ideally, the person performing the follow-up analysis should be knowledgeable about causal analysis and have experience with investigatory techniques.

6.5.2.2 Integration with Hospital-wide Event Reporting

For those radiation therapy departments within a hospital system, it is important to meet both the needs of the department and the needs of the facility. It is common to have a low threshold for close call reporting for internal departmental incident reports. These incident reports are then filtered through a higher-threshold medical event reporting process to determine whether to post to the hospital reporting system.

The secondary reporting may need to happen within a prescribed time interval. Alternatively, it may be permissible to wait until an analysis has been performed.

The departmental reporting system should capture all the information required by the hospital reporting system.

6.5.2.3 Who Has Access to Reports

Access to the reports should be limited to authorized personnel. This will typically be the department's or hospital's quality committee. Those with access should be properly trained so that handling of incident reports—which are considered quality improvement documentation—is performed in a manner compliant with HIPAA regulations and prevents unintended legal discovery.

6.5.2.4 Reporter-identifiable Information

Sidney Dekker (Dekker 2006) points out that it is preferable to submit incident reports to a patient safety committee rather than to an individual within the clinical department, since it is conceivable that a supervisor may behave adversely toward the reporter, depending on the content of the report. This is a question of safety culture, and it should be assessed before making the decision of whether to route incident reports through the safety office, or to allow anonymous report submission.

Without trace-back, incident reports are of limited value. For this reason, it would be preferable to enlist the participation of the safety office rather than allow anonymous reporting. However, this introduces a new set of issues, such as the reportability criteria used by the reporting office, the data collected on their form (which may likely be insufficient for radiotherapy analysis), access to submitted reports for aggregate analysis, and the safety office's own criteria for required follow-up, which may result in a minor close call garnering greater scrutiny than one would wish.

6.5.2.5 Patient-identifiable Information

For an incident reporting system internal to a department, the opportunity for quality improvement is optimal when identifiable information is left intact. However, from a privacy and legal perspective, it is imperative to comply with all HIPPA requirements and to protect such reports from legal discovery. There are states where quality improvement and discoverability has not been tested, and the requirements for protection vary. Therefore the departmental quality improvement team should confer with its risk management or legal group on this issue of identification protection. The following remarks are offered as general guidance only.

Several steps must be taken to protect the reports from legal discovery. First, it is best to label each page with a header that states that the document is for quality improvement and is protected under the appropriate legal code. Second, the protected quality documentation must be secured so that it is only accessible by authorized personnel. Third, direct reference to these reports should only be in documents which are also dedicated to quality improvement (QI) and handled in the same way. No unprotected e-mail or non-QI-designated document should make reference to individual incident reports. Lastly, QI documentation should only talk about relevant QI concerns and should not cross into discussions of legal concerns.

In regard to HIPPA requirements, since the Privacy Rule includes quality improvement in the definition of healthcare operations, a covered entity may use an individual's private health information for its own QI activities without first obtaining the individual's authorization.

6.5.3 Conclusions

1. Departmental incident reporting systems can help to identify highly consequential circumstances for error that warrant immediate correction. However, by keeping and reviewing complete records of incidents, it is also possible to identify trends in close call reports by which one can revise processes to prevent errors from occurring.
2. Incident reporting serves no benefit without setting aside time and resources for a quality improvement team to collect incident reports, perform follow-up, analyze the reports, generate corrective actions, and monitor their workability. The quality improvement team must have the authority or the support of management to ensure cooperation from the staff.
3. It is important to record a narrative description of each incident so that the situation can be seen from the perspective of the people involved at the time. Anonymous incident reporting has the following benefits: extra precaution against accidental legal discovery (potential HIPPA violation) and less fear of recrimination among the staff. However, anonymous incident reporting has the following shortcomings: the inability to clarify ambiguities in the incident report and the inability to trend on the kinds of data which could potentially identify staff members (such as on which linear accelerator the event occurred).

6.6 National Reporting Systems

A major push for national incident reporting in medicine was triggered by the publication of *To Err is Human* in 2000 (Institute of Medicine 2000). This book famously states that as many as 98,000 Americans die each year as a result of medical errors.

Radiation therapy fell under the spotlight due to a series of articles by Walt Bogdanich for the *New York Times* (Bogdanich 2009–2010; Bogdanich and Rebelo 2010a–b; Bogdanich and Ruiz 2010). These articles underscored the severity of errors possible in radiation therapy and the challenge of detecting errors before harm is done.

In the years following the Bogdanich stories, a series of recommendations were made within the radiotherapy community. These included the ASTRO Six Point Action Plan and the 20 recommendations identified at the meeting "Safety in Radiation Therapy: A Call to Action" held in 2010 (Hendee and Herman 2011). The ASTRO Six Point Action Plan includes a recommendation for the "creation of an anonymous national database for event reporting." The summary recommendations from the "Call to Action" meeting speak more generally about the need to develop error-reporting systems. They mention the interest of the FDA, NRC, and CRCPD in creating such a system, as well as the International Atomic Energy Agency (IAEA) voluntary system called SAFRON.

More recently, ASTRO has begun work on a safety initiative to implement a voluntary national radiation therapy reporting system handled through a certified patient safety organization. The role of patient safety organizations per the Patient Safety and Quality Improvement Act of 2005 is detailed in the next section.

6.6.1 Patient Safety Organizations in the U.S.

The Patient Safety and Quality Improvement Act of 2005 (Patient Safety Act) was enacted to provide for the improvement of patient safety and to reduce the incidence of events that adversely affect patient safety. It facilitates the creation of patient safety organizations which perform the collection and analysis of submitted documents and disseminates information such as "recommendations, protocols, or information regarding best practices." The act's "Network of Patient Safety Databases" section includes the following two items:

> (a) In General. – The Secretary shall facilitate the creation of, and maintain, a network of patient safety databases that provides an interactive evidence-based management resource for providers, patient safety organizations, and other entities. The network of databases shall have the capacity to accept, aggregate across the network, and analyze nonidentifiable patient safety work product voluntarily reported by patient safety organizations, providers, or other entities. The Secretary shall assess the feasibility of providing for a single point of access to the network for qualified researchers for information aggregated across the network and, if feasible, provide for implementation.…
>
> (c) Use of Information. – Information reported to and among the network of patient safety databases under subsection (a) shall be used to analyze national and regional statistics, including trends and patterns of health care errors. The information resulting from such analyses shall be made available to the public and included in the annual quality reports prepared under section 913(b)(2).

The key words here are "voluntary" and "nonidentifiable." Nonidentifiable means the identities of any provider that is a subject of the work product, individually identifiable health information, and reporting individuals (page 119, stat. 425, (2) and (3) in (Department of Health and Human Services, sec. 921–926)).

The database itself is not directly assessable in any form, so it cannot be used to drill down to individual submissions or to get current information, since information is only available in statistical form and only when published after analysis. This makes the tool very limited compared to the ROSIS/SAFRON or FAA systems, which allow search and analysis using individual events.

The Patient Safety and Quality Improvement Act of 2005 focuses on event aggregation, which is made explicit in the discussion of regulations to implement the Act: "Patient safety experts have long recognized that the underlying causes of risks and hazards in patient care can best be recognized through the aggregation of significant numbers of individual events; in some cases, it may require the aggregation of thousands of individual patient safety events before underlying patterns are apparent. It is hoped that this proposed rule will foster routine reporting to PSOs of data on patient safety events in sufficient numbers for valid and reliable analyses" (Federal Register 2008). The Patient Safety Act states that each Patient Safety Organization (PSO) bears responsibility for data collection, analysis, and reporting. There are currently over 80 PSOs.

To implement this legislation, the U.S. Department of Health and Human Services (HHS) issued the Patient Safety and Quality Improvement Final Rule (Patient Safety Rule) in 2009. It sets up the framework for voluntary reporting to PSOs and outlines the requirements for PSOs. It gives responsibility for certifying and listing PSOs to the Agency for Healthcare Research and Quality (AHRQ), which is a part HHS. More broadly, the AHRQ is responsible for health services research, including quality improvement and patient safety.

The main impetus for restrictions on data access stems from legal concerns. In particular, reporting documentation (formally known as the "patient safety work product") is privileged and not subject to legal discovery, disclosure, or admission (Sec. 922 (a) in (Department of Health and Human Services, sec. 921–926)).

6.6.2 Reporting System Attributes

A number of choices need to be made in the design of a health incident reporting system: voluntary versus mandatory, close call versus harmful event, protected versus public, and identifiable versus unidentifiable. A comparison of recommendations coming from several agencies is shown in Table 6–4.

6.6.2.1 Required Elements

Anticipating the need for guidance on national incident reporting system construction, the AAPM Working Group for the Prevention of Errors has published a set of recommendations (Ford et al. 2012). Five topics are addressed: definitions, process maps, severity scales, causality taxonomy, and data elements. An incident learning system should prompt the reporter for where in the process the incident occurred, the severity or potential severity of the incident, and basic causes leading to the incident. The sets of definitions and data elements facilitate both recommendations for consistent terminology and complete incident data collection, as well as greater opportunities to aggregate data between systems.

6.6.2.2 Voluntary Versus Mandatory

As discussed in the section about patient safety organizations, voluntary reporting is appropriate when anonymous close call data is collected, while mandatory reporting is suitable when collection is limited to medical events, in which case the submission is stored in an identifiable format.

Table 6–4. Comparison of reporting system recommendations and implementations

	To Err is Human (pp. 87–88)	CRCPD H–38	ASTRO	SAFRON	NRC
Adverse events	Yes (5.1)	Yes	Yes	Yes	Yes
Near misses	Yes (5.2)	No	Yes	Yes	No
Mandatory	Yes (5.1)	Yes	No	No	Yes
Voluntary	Yes (5.2)	No	Yes	Yes	No
Protection from Legal Discovery	Yes (p. 102)	No	Yes	Yes	No
Administer Through States	Yes (5.1)	Yes	No	No	* Only through agreement states

6.6.2.3 Reportability Threshold

The reportability threshold is a major limitation of statistical assessments based on collected data. Either the threshold is set high, as in the case of mandatory reporting, for which only harmful events are collected, or the threshold is set low, and the focus shifts to close call events.

In high-threshold cases, the scope of submitted incident reports will depend on the phrasing of the criteria and on their interpretation. For example, the U.S. NRC has a brachytherapy reporting requirement which includes a threshold for non-target organs with dose volumes exceeding 150% of the expected dose. The problem is that most treatment planning systems for prostate brachytherapy don't give any simple way to compute this dose volume in the periprostatic tissues. Thus the rate of reporting of such incidents will depend on the initiative and ingenuity of the facility to make this assessment.

In low-threshold cases, close call incident reports are collected through voluntary submission. Depending on the level of training and experience, what one staff member considers to be outside the bounds of expected practice, another staff member may consider to be within the bounds of expected practice and, hence, not reportable. Thus the reporting threshold may, in practice, differ between staff members.

Due to the inherent biases introduced by reporting thresholds, it is important to avoid criteria that flag situations still within the realm of sound clinical judgment. For example, the NRC currently requires treatments whose dose to target differs by more than 20% from the prescribed dose to be reported. In low dose rate prostate brachytherapy, there is clinical evidence that insufficient dose reduces the likelihood of cure, but there is little data to show that excess dose, except to certain normal structures, is harmful. Thus, practitioners may differ in whether they choose to avoid insufficient clinical dose by loading a greater number of seeds, or whether they instead choose to avoid crossing the regulatory threshold by limiting the number of seeds implanted. In either case, the high dose threshold is not clinically motivated and can become a regulatory impediment to optimal patient care.

Table 6–5. A sampling of reporting systems outside of health care

	NTSB/ FAA	ASRS NASA	NFF NMRS NASA	FRA Railroad	C3RS NASA	NRC/ REACTS	IAEA NEWS
Industry	Aviation	Aviation	Fire	Rail	Rail	Nuclear	Nuclear
Status	Active	Active	Active	Active	Pilot	Active	Active
Criteria	Injury Death Damage	Near Miss	Near Miss	Injury Death Damage	Near Miss	Defined by Regulation	Near Miss
Mandatory?	Mandatory	Volun.	Volun.	Mandatory	Volun.	Mandatory	No
Confidential	Yes	Yes	Yes	No	Yes	Public	Yes
Adverse Events?	Yes	Yes	No	Yes	No	Yes	Yes
Near Misses?	No	Yes	Yes	No	Yes	No	Yes
Searchable?	Yes	Yes	Yes	In Person	Data Sent to Participants	Yes	Browse Only
Data Collected in 2012	~1700	~72,000	~290	>1700	~340	~1600	~22

References: FAA (FAA 2013), NRC/REACTS (NRC 2013), NFF NMRS NASA (National Aeronautics and Space Administration (NASA 2013b), C3RS NASA (National Aeronautics and Space Administration (NASA 2013c), ASRS NASA (National Aeronautics and Space Administration (NASA 2013a), FRA Railroad (FRA 2013), IAEA (IAEA 2013b).

6.6.2.4 Anonymity

Anonymity involves four parties: the patient, the reporters and staff, the organization, and identifiable equipment information. The scrubbing of data to ensure full anonymity is a laborious process requiring careful attention to detail. One organization facilitates this through a two-step process: the scrubber removes all identifiable information and then the reviewer looks over the anonymized copy to ensure that no identifiable information remains.

It is also necessary to obtain appropriate permissions, preferably in writing, from your hospital safety office or executive staff before engaging in an external incident reporting system. Your facility must ensure that the processes used by the incident reporting system do not run afoul of HIPPA regulation or other privacy concerns.

6.6.3 Examples of National Reporting Systems

The national reporting systems for various fields have originated from the government, manufacturers, and professional organizations. As can be seen in the following lists, the NASA incident reporting system developed for the aviation industry (ASRS) has been adapted to health care, fire fighting, and the railroads, respectively called the "Patient Safety Reporting System," the "National Fire Fighters Near-Miss Reporting System," and the "Confidential Close Call Reporting System" (National Aeronautics and Space Administration (NASA) 2013a–d). These applications of this innovative NASA system are discussed below.

6.6.3.1 Non-medical Reporting Systems

The following list contains prominent incident reporting systems outside of health care. A comparison of some of their characteristics is shown in Table 6–5.

1. **Aviation**

   NTSB/FAA Accident and Incident Data. This is a mandatory system under the U.S. National Transportation Safety Board. It rapidly shares preliminary information on aviation accidents and incidents (FAA 2013).

   Aviation Safety Reporting System (ASRS, NASA). This NASA-developed system is characterized by ample narratives describing the circumstances of each incident. This is a voluntary system (National Aeronautics and Space Administration (NASA) 2013a).

2. **Fire**

   National Fire Incident Reporting System (NFIRS). According to NFIRS, "participating fire departments report 23,000,000 incidents and 1,000,000 fires every year." It claimed to be "the world's largest, national, annual database of fire incident information." Participation is voluntary (FEMA/NFIRS 2013).

   The National Fire Fighters Near-Miss Reporting System (NASA). This reporting project follows the template of the other NASA reporting projects in that reports are submitted voluntarily. They are reviewed, and all identifying information is removed. The reports are then posted on the website in a searchable database (National Aeronautics and Space Administration (NASA) 2013b).

3. **Railroads**

   Confidential Close Call Reporting System (C3RS, NASA). This system is focused on near misses. Although incidents involving harm could be entered, this use is not encouraged because in those cases the reporter loses protection from discoverability. Note also that data is not readily available for review (National Aeronautics and Space Administration (NASA) 2013c).

   U.S. Federal Railroad Administration Office of Safety Analysis. This reporting system tracks those incidents resulting in "injury or death to an individual or damage to equipment or a roadbed." Accident data is not freely available, but can be reviewed in person or through a Freedom of Information Act (FOIA) request (FRA 2013).

4. **Nuclear**

   U.S. NRC. The U.S. Nuclear Regulatory Commission requires that incident reports be submitted for those events meeting the reporting criteria. Event summaries are publicly posted on their website (NRC 2013).

   IAEA. Since nuclear power incidents may have the potential to escalate into an emergency response situation, the current IAEA approach is to couple the response networks for incidents and emergencies through its Incident and Emergency Centre (IEC). The intent is to handle event information receipt, validation, and dissemination in a unified way. Different systems are used to provide government representatives reporting tools for nuclear accidents (USIE), events (NEWS), and illicit trafficking (ITDB) (IAEA 2013b).

6.6.3.2 Medical Reporting Systems

The following list contains prominent incident reporting systems within health care, but not limited to radiation therapy. A comparison of some of their characteristics is shown in Table 6–6.

1. **UK Voluntary System.** National Patient Safety Agency (NPSA) National Reporting and Learning System (NRLS): a voluntary reporting system to collect and learn from patient safety incidents. The NPSA is a part of the National Health System in the United Kingdom. A trigger code is used in the incident reports to identify radiotherapy incidents (NPSA 2010). This is a voluntary reporting system which includes near misses and "other non-conformances." The NPSA system is independent of, and does not supersede, the IR(ME)R system described below (NPSA 2013).
2. **UK Mandatory System.** Regulation of health care, adult social care services, and primary dental care services in England is performed by the Care Quality Commission (CQC) based on the Ionising Radiation (Medical Exposure) Regulations (IR(ME)R).The CQC manages a mandatory reporting system for radiation exposures "much greater than intended," for which they provide a technical definition. In 2011 the CQC received 538 notifications, of which 68 were from radiotherapy (CQC 2012). Of the radiotherapy incidents, two thirds were considered treatment errors and one fourth were considered planning errors (CQC 2013).
3. **U.S. FDA.** According to their website, "The FDA Adverse Event Reporting System (FAERS) is a database that contains information on adverse event and medication error reports submitted to FDA. The database is designed to support the FDA's post-marketing safety surveillance program for drug and therapeutic biologic products." Furthermore, "Reporting of adverse events and medication errors by healthcare professionals and consumers is voluntary in the United States" (FDA FAERS 2013). Serious adverse events are reported through MedWatch, which provides a definition for reportability (FDA MedWatch 2013).
4. **U.S. NRC.** The U.S. Nuclear Regulatory Commission requires event reports for those incidents satisfying their definition of "medical event." These reports are posted on their website (NRC 2013).
5. **The Joint Commission.** The Joint Commission requires reporting of those incidents satisfying their definition of a "sentinel event" (The Joint Commission 2013).

6.6.3.3 Radiation Therapy Reporting Systems

The following list contains prominent incident reporting systems limited to radiation therapy. Several of these systems (SAFRON, CRCPD H38, and the ASTRO project) are in the pilot stage. SAFRON is administered by the IAEA. A comparison of some of their characteristics is shown in Table 6–6.

1. **ROSIS.** ROSIS was originally developed with support from ESTRO. This database has collected over 1000 voluntarily contributed incident reports as of 2011 (Cunningham et al. 2010; IAEA 2013c).
2. **SAFRON.** As of 2012, SAFRON is undergoing a pilot study. It is supported by the IAEA, in contrast to the largely volunteer ROSIS program. In other respects, SAFRON is similar to ROSIS (IAEA 2013d).
3. **CRCPD H38 Pilot Project.** The Conference of Radiation Control Program Directors (CRCPD) is investigating an incident reporting system for radiation therapy. As of January 2013, this project is still in the pilot stage. It will be a mandatory reporting structure for radiation oncology medical events in agreement states (i.e., those not directly under the NRC) within the U.S. (Elee 2013).
4. **ASTRO/AAPM National PSO System.** As of 2012, this project is nearing the pilot stage. It is based on the utilization of a patient safety organization to manage and protect data. The reporting structure will include near miss reporting (Tomlinson 2012).

6.6.4 Conclusions

The following conclusions can be drawn about national reporting systems.

1. Many occupational fields (health care in the UK, railroads, fire fighters, aviation, non-U.S. nuclear power) have both a mandatory reporting system for events meeting some predefined criteria and a voluntary reporting system for close calls.
2. Searchable event databases and limited-access event databases serve very different purposes. A searchable event database provides current data, specific equipment information, detailed narratives, free exploration, and the opportunity for expert trend review, but at the risk of potential discoverability. A limited-access event database—such as would be managed by a patient safety organization—would protect anonymity while allowing expert review and trend reporting, but at the expense of the other benefits of an open access system.
3. Mandatory reporting systems contain skewed data because what is reported is constrained by the criteria used for reportable events. Voluntary reporting system contain skewed data because what is reported is locally dependent on the culture of reporting. Both cases are in contrast to an exhaustive, retrospective review of practice, such as was done in the Harvard Study of 1991 (Leape et al. 1991).

Table 6–6. Prominent reporting systems for medical use with or without radiation

	UK RPSA NRSL	TJC (local, not database	PSRS NASA	FAERS/ MedWatch	ROSIS/SPHRON	CRCPD H38	ASTRO/AAPM
Industry	Health	Health	Health	Devices	Rad. Therapy	Rad. Therapy	Rad. Therapy
Status	Active	Active	On Hold	Active	Active	Pilot	Pending
Criteria	Near Miss/ Serious	Sentinel Events	Near Miss+	Near Miss/ Serious Adv. Event	Near Miss	Serious Defined by State	Near Miss
Manda-tory?	Voluntary	Mandatory	Voluntary	Voluntary	Voluntary	Mandatory	Voluntary
Confiden-tial?	Yes	N/A	Confidential	Confidential	Anonymous	Public / via States	Anonymous
Adverse Events?	Yes	Yes	Yes	MedWatch	Yes	Yes	Yes
Near Misses?	Yes	No	Yes	FAERS	Yes	No	Yes
Search-able?	No	N/A	No	Yes	Yes	Pilot	Pending
Data Collected	~1,400,000 in 2012	N/A	N/A	~760,000 in 2010	>1000 from 2003–2008	Pilot	Pending

References: UK NPSA NRSL (NHS 2013), TJC (The Joint Commission 2013), PSRS NASA (National Aeronautics and Space Administration (NASA) 2013d), FAERS FDA (FDA FAERS 2013), ROSIS/ SPHRON (IAEA 2013c), CRCPD H38 (Elee 2013), ASTRO/AAPM (Tomlinson 2012).

6.7 Summary

- Incident learning systems serve as reactive and proactive tools to improve safety and quality improvements by either reducing the chance that a failure is repeated or by acting on reported weaknesses in the system to reduce the chance of related failures from occurring.
- As much as any other quality improvement tool, incident learning systems require the broad support of staff and management in order to be sustaining and of optimal use.
- At a minimum, an incident learning system consists of incident report collection, incident report analysis, implementation of corrective actions, and follow-up on the efficacy of the corrective actions.
- Root cause analysis is an industry standard approach to incident investigation and corrective action. Its success is contingent on the makeup of the team, the availability of time and resources, and appropriate training.
- Root cause analysis can be hampered by factors such as hindsight bias, insistence on finding a single root cause or causal origin, inability to see events from the perspective of the operators at the time of the event, and

reducing a complex multifocal situation to an artificially linear trajectory of events.

- When designing an incident learning system, one must consider who will have access to the collected information, whether anonymity will be preserved, whether reporting is mandatory or voluntary, whether the system will be paper-based or electronic, whether close call events are reported, what the reportability criteria will be, how follow-up of corrective actions will be handled, and to what extent the focus will be on incident trends versus responding to each incident episodically. These considerations are not mutually independent.
- Incident reporting systems have been implemented in many disciplines, including aviation, fire fighting, rail transportation, nuclear power, and medicine (including radiation therapy). These implementations vary widely with regard to the design choices mentioned above.
- In general, the quality of an incident learning system is dependent on the quality of the reports and the corrective actions which, in turn, are dependent on the care and competence of those who design, administer, and report into the system.

References

Arnold, A., G.P. Delaney, L. Cassapi, and M. Barton. (2010). "The use of categorized time-trend reporting of radiation oncology incidents: a proactive analytical approach to improving quality and safety over time." *Intl J Radiat Oncol Biol Phys* 78:1548–1554.

Bogdanich, W. "As Technology Surges, Radiation Safeguards Lag." *The New York Times*, January 26, 2010.

Bogdanich, W. "At V.A. Hospital, a Rogue Cancer Unit." *The New York Times*, June 21, 2009.

Bogdanich, W. "Case Studies: When Medical Radiation Goes Awry." *The New York Times*, January 27, 2010.

Bogdanich, W. "F.D.A. Toughens Process for Radiation Equipment." *The New York Times*, April 8, 2010.

Bogdanich, W. "Oncologist Defends His Work at a V.A. Hospital." *The New York Times*, June 30, 2009.

Bogdanich, W. "Radiation Offers New Cures, and Ways to Do Harm." *The New York Times*, January 24, 2010.

Bogdanich, W. "Safety Features Planned for Radiation Machines." *The New York Times*, June 9, 2010.

Bogdanich, W. "V.A. Is Fined over Errors in Radiation at Hospital." *The New York Times*, March 17, 2010.

Bogdanich, W. and K. Rebelo. "The Radiation Boom—Stereotactic Radiosurgery Overdoses Harm Patients." *The New York Times*, December 28, 2010.

Bogdanich, W. and K. Rebelo. "They Check the Medical Equipment, But Who Is Checking Up on Them?" *The New York Times*, January 27, 2010.

Bogdanich, W. and R.R. Ruiz. "Radiation Errors Reported in Missouri." *The New York Times*, February 24, 2010.

Boring, R.L. "How Many Performance Shaping Factors Are Necessary for Human Reliability Analysis?" Idaho Falls, ID: Idaho National Laboratory, 2010. Available at http://www.inl.gov/technicalpublications/Documents/4814133.pdf.

Caldwell, B.S. (2008). "Tools for developing a quality management program: human factors and systems engineering tools." *Intl J Radiat Oncol Biol Phy* 71:S191–S194.

Caldwell, B.S. and E.A. Kapp. "Domains of performance enhancement, accident reduction, and error recovery in complex systems." In *Human Factors in Organizational Design and Management–V: Proceedings of the Fifth International Symposium on Human Factors in Organizational Design and Management,* O. Brown, Jr. and H.W. Hendrick, eds. Amsterdam: North-Holland, 1996.

Caldwell, B.S. and T.J. Anderson. "Design and active phases of medical process improvement systems." In *Examining Errors in Health Care: Developing a Prevention, Education and Research Agenda.* Rancho Mirage, CA: Annenberg Center for Health Sciences, 1996.

Care Quality Commission (CQC). (2012.) "Ionising Radiation (Medical Exposure) Regulations 2000: A Report on Regulatory Activity in 2011." Accessed Jan. 7, 2013. Available at http://www.cqc.org.uk/organisations-we-regulate/special-reviews-and-inspection-programmes/ionising-radiation/key-findings-.

Care Quality Commission (CQC). (2013.) "Care Quality Commission–Reporting Incidents." Accessed Jan. 7, 2013. Available at http://www.cqc.org.uk/organisations-we-regulate/special-reviews-and-inspection-programmes/ionising-radiation/reporting-inc.

Chang, A., P.M. Schyve, R.J. Croteau, D.S. O'Leary, J.M. Loeb. (2005). "The JCAHO patient safety event taxonomy: a standardized terminology and classification schema for near misses and adverse events." *Int J Qual Health Care* 17:95–105.

Clark, B.G., R.J. Brown, J.L. Ploquin, A.L. Kind, L. Grimard. (2010). "The Management of Radiation Treatment Error Through Incident Learning." *Radiother Oncol* 95:344–349.

Clark, B.G., R.J. Brown, J.L. Ploquin, P.B. Dunscombe. (2012). "Patient Safety Improvements in Radiation Treatment Through 5 Years of Incident Learning." *Radiother Oncol.* Forthcoming.

Cooke, D., M. Dubertz, R. Heshmati, S. Iftody, E. McKimmon, J. Powers, R. Lee, and P.B. Dunscombe. *A Reference Guide for Learning from Incidents in Radiation Treatment.* Health Technology Assessment Unit #22. Edmonton, Canada: Alberta Heritage Foundation for Medical Research, 2006. Available at http://www.ihe.ca/publications/library/archived/a-reference-guide-for-learning-from-incidents-in-radiation-treatment/

Cooke, D.L., P.B. Dunscombe, R.C. Lee. (2007). "Using a Survey of Incident Reporting and Learning Practices to Improve Organisational Learning at a Cancer Care Centre." *Qual Saf Health Care* 16:342–348.

Cunningham, J. "Radiation Oncology Safety Information System (ROSIS)—A Reporting and Learning System for Radiation Oncology." Dublin: University of Dublin, Trinity College, 2011. Available at www.rosis.info/docs/Joanne_Cunningham_Thesis_ROSIS.pdf.

Cunningham, J., M. Coffey, T. Knöös, O. Holmberg. (2010). "Radiation Oncology Safety Information System (ROSIS)—Profiles of participants and the first 1074 incident reports." *Radiother Oncol* 97:601–607.

Dekker, S. *The Field Guide to Understanding Human Error.* Aldershot, England: Ashgate, 2006.

Department of Health and Human Services. Patient Safety and Quality Improvement Act of 2005. Available at http://www.gpo.gov/fdsys/pkg/PLAW-109publ41/html/PLAW-109publ41.htm.

Elee, J. (2013). Personal communication.

Federal Aviation Administration (FAA). "Federal Aviation Administration—Accident & Incident Data." 2013. Accessed January 7, 2013. Available at http://www.faa.gov/data_research/accident_incident/.

Federal Emergency Management Agency (FEMA). National Fire Incident Reporting System (NFIRS) home webpage. Accessed March 6, 2013. Available at http://www.nfirs.fema.gov/.

Federal Railroad Administration (FRA). Office of Safety Analysis. "Federal Railroad Administration Office of Safety Analysis." (2013.) Accessed January 7, 2013. Available at http://safetydata.fra.dot.gov/OfficeofSafety/default.aspx.

Federal Register. 2008. "Proposed Rules" 74, No. 29, February 12: 8113.

Food and Drug Administration (FDA). "FDA Adverse Event Reporting System (FAERS) (formerly AERS)." (2012.) Accessed January 7, 2013. Available at http://www.fda.gov/Drugs/GuidanceComplianceRegulatoryInformation/Surveillance/AdverseDrugEffects/default.htm.

Food and Drug Administration (FDA). MedWatch. "Reporting Serious Problems to FDA: What Is a Serious Adverse Event?" Accessed January 8, 2013. Available at http://www.fda.gov/Safety/MedWatch/HowToReport/ucm053087.htm.

Ford, E.C., L.F. de Los Santos, T.L. Deweese, S. Bisgaard, P. Aspden. (2012). "Consensus Recommendations for Incident Learning Database Structures in Radiation Oncology." *Med Phys* 39:7272–7290.

Gilbreth, F.B. (1924). "The Classification of Work." Proceedings of the 11th National Convention of the Society of Industrial Engineers, Buffalo, NY, May 1–2. *Soc Ind Engrs* 7:105–120.

Hendee, W.R., M.G. Herman. (2011). "Improving patient safety in radiation oncology." *Med Phys* 38:78–82.

Huq, M.S., B.A. Fraass, P.B. Dunscombe, J.P. Gibbons, Jr., G.S. Ibbott, P.M. Medin, A. Mundt, S. Mutic, J.R. Palta, B.R. Thomadsen, J.F. Williamson, E.D. Yorke. (2008). "A method for evaluating quality assurance needs in radiation therapy." *Int J Radiat Oncol Biol Phys* 71:S17–3.

Institute of Medicine. *To Err is Human: Building a Safer Health System.* Washington, D.C.: National Academy Press, 1999.

International Atomic Energy Agency (IAEA). (2013a). "Improving Patient Safety in Radiotherapy." Accessed March 6, 2013. Available at rpop.iaea.org/RPOP/RPoP/Content/News/safron.htm.

International Atomic Energy Agency (IAEA). (2013b). "International Atomic Energy Agency – Nuclear Safety & Security – Incident Reporting." Accessed March 6, 2013. Available at http://www-ns.iaea.org/tech-areas/emergency/incident-reporting-mechanisms.asp.

International Atomic Energy Agency (IAEA). (2013c). "Radiation Oncology Safety Information System." Accessed March 6, 2013. Available at http://www.rosis.info/index.php.

International Atomic Energy Agency (IAEA). (2013d). "Recently Raised Issues of Radiation Risks to Patients." Accessed March 6, 2013. Available at https://rpop.iaea.org/RPOP/RPoP/Content/ArchivedNews/recent-issues-radiation-risks.htm.

Itoh, K., N. Omata, H.B. Andersen. (2009). "A human error taxonomy for analysing healthcare incident reports: assessing reporting culture and its effects on safety performance." *J Risk Res* 12: 485–511.

Leape, L.L., T.A. Brennan, N. Laird, A.G. Lawthers, A.R. Localio, B.A. Barnes, L. Hebert, J.P. Newhouse, P.C. Weiler, H. Hiatt. (1991). "The nature of adverse events in hospitalized patients. Results of the Harvard Medical Practice Study II." *N Engl J Med* 324:377–84.

Mutic, S. and S. Brame. "Error and near miss reporting: a view from North America." In *Quality and Safety in Radiotherapy*, T. Pawlicki, P. Dunscombe, A.J. Mundt, and P. Scalliet, eds. New York: Taylor & Francis, 2010.

Mutic, S., R.S. Brame, S. Oddiraju, P. Parikh, M.A. Westfall, M.L. Hopkins, A.D. Medina, J.C. Danieley, J.M. Michalski, I.M. El Naqa, D.A. Low, B. Wu. (2010). "Event (error and near-miss) reporting and learning system for process improvement in radiation oncology." *Med Phys* 37:5027–36.

National Aeronautics Space Administration (NASA). "Aviation Safety Reporting System." (2012.) Accessed March 6, 2013. Available at http://asrs.arc.nasa.gov/.

National Aeronautics Space Administration (NASA). "Confidential Close Call Reporting System (C3RS)." (2013.) Accessed March 6, 2013. Available at http://www.closecall-srail.org/.

National Aeronautics Space Administration (NASA). "National Fire Fighter Near-Miss Reporting System." (2013.) Accessed March 6, 2013. Available at http://www.firefighternearmiss.com/.

National Aeronautics Space Administration (NASA). "The Patient Safety Reporting System." (2013.) Accessed March 6, 2013. Available at http://www.psrs.arc.nasa.gov/.

National Council on Radiation Protection and Measurements (NCRP). *Report 173—Investigation of Radiological Incidents*. Bethesda, MD: National Council on Radiation Protection and Measurements, 2012.

National Patient Safety Agency (NPSA). (2010.) "Implementing Towards Safer Radiotherapy: Guidance on Reporting Radiotherapy Errors and Near Misses." Accessed March 6, 2013. Available at http://www.nrls.npsa.nhs.uk/resources/?EntryId45=75033.

National Patient Safety Agency (NPSA). (2013.) "Patient Safety—Patient Safety Homepage." Accessed March 6, 2013. Available at http://www.nrls.npsa.nhs.uk/.
National Transportation Safety Board (NTSB). "Aviation Investigation Manual, Major Team Investigations." (2002.) Accessed March 6, 2013. Available at http://www.ntsb.gov/doclib/manuals/majorinvestigationsmanual.pdf.
Paradies, M. (2010). "Under scrutiny." *Qual Prog* 43:32–37.
Runciman, W., P. Hibbert, R. Thomson, T. Van der Schaaf, H. Sherman, P. Lewalle. (2009). "Towards an international classification for patient safety: key concepts and terms." *Int J Qual Health Care* 21:18–26.
Sherman, H; G. Castro, M. Fletcher, M. Hatlie, P. Hibbert, R. Jakob, R. Koss, P. Lewalle, J. Loeb, T. Perneger, W. Runciman, R. Thomson, T. Van Der Schaaf, M. Virtanen. (2009). "Towards an International Classification for Patient Safety: the conceptual framework." *Int J Qual Health C* 21:2–8.
Snijders, C., A. Molendijk, H. Klip, R.A. Van Lingen, W.P.F. Fetter. (2009). "Feasibility and reliability of PRISMA-Medical for specialty-based incident analysis." *Qual. Saf. Health Care* 18:486–491.
The Joint Commission. "Root Cause Analysis and Action Plan Framework." (2013.) Accessed January 7, 2013. Available at http://www.jointcommission.org/Framework_-for_Conducting_a_Root_Cause_Analysis_and_Action_Plan/.
The Joint Commission. "The Joint Commission—Sentinel Event." (2013.) Accessed January 7, 2013. Available at http://www.jointcommission.org/sentinel_event.aspx.
Thomadsen, B. and S.-W. Lin. "Taxonomic Guidance for Remedial Actions." In *Advances in Patient Safety: From Research to Implementation*, vol. 2. Bethesda, MD: National Institutes of Health, 2005. Available at http://www.ncbi.nlm.nih.gov/bookshelf/br.fcgi?book=aps2&part=A1796.
Tomlinson, C. (2012). "ASTRO to Develop Radiation Oncology Error Reporting System." *ASTROnews* 15:59. Accessed March 6, 2013. Available at https://www.astro.org/News-and-Media/ASTROnews/Annual-Meeting-Guide-2012/ASTRO-to-develop-radiation-oncology-error-reporting-system.aspx.
U.S. Nuclear Regulatory Commission (NRC). (2012.) 10 CFR 50, Appendix B, Criterion XVI. Available at http://www.nrc.gov/reading-rm/doc-collections/cfr/part050/part050-appb.html.
U.S. Nuclear Regulatory Commission (NRC). (2013.) "NRC: Reports Associated with Events." Accessed January 7. Available at http://www.nrc.gov/reading-rm/doc-collections/event-status/.
Van der Schaaf, T.W., M.M.P. Habraken. (2005.) "PRISMA-Medical: A Brief Description." Eindhoven, the Netherlands: Technische Universiteit Eindhoven. Available at http://www.who.int/patientsafety/taxonomy/PRISMA_Medical.pdf.
Van Vuuren, W., C.E Shea, and T.W. van der Schaaf. *The Development of an Incident Analysis Tool for the Medical Field*. Eindhoven, The Netherlands: Eindhoven University of Technology, Faculty of Technology Management, 1997.
Veterans Health Administration (VHA). "VHA National Patient Safety Improvement Handbook". Washington, DC: Veterans Health Administration, 2011. Available at http://www1.va.gov/vhapublications/ViewPublication.asp?pub_ID=2389.
Williams, M.V. (2007). "Improving Patient Safety in Radiotherapy by Learning from Near Misses, Incidents and Errors." *Br J Radiol* 80:297–301.
World Health Organization (WHO). World Alliance for Patient Safety. *Conceptual Framework for the International Classification for Patient Safety*. Geneva, Switzerland: World Health Organization, 2009. Available at: http://www.who.int/patientsafety/implementation/taxonomy/icps_technical_report_en.pdf.
World Health Organization (WHO). World Alliance for Patient Safety. *WHO Draft Guidelines for Adverse Event Reporting and Learning Systems: From Information to Action*. Geneva, Switzerland: World Health Organization, 2005.

Appendix 6.1: The AAPM WGPE Causal Taxonomy

1) Organizational management
   a) Planning for program maintenance or expansion
      i) Inadequate human resources
         (1) Staffing not consistent with professional clinical recommendations
         (2) Staffing not consistent with vendor specs, or
         (3) Staffing not consistent with regulations
         (4) No provision for incremental increases in activities (reports, additional duties, committee participation)
      ii) Inadequate capital resources
         (1) Inadequate budget for equipment
         (2) Inadequate support/service contracts
         (3) Inadequate training support
         (4) Insufficient educational budgets
         (5) Insufficient IT infrastructure
         (6) Inappropriate or inadequate equipment
      iii) Admin/contractor negotiations for support or staff
   b) Policies, procedures, regulations
      i) Relevant policy nonexistent
      ii) Policy not implemented
      iii) Policy inadequate
      iv) Policy not followed
      v) External regulation not followed
      vi) Conflicting policies
   c) Training; acquiring, and transmitting knowledge and skills
      i) Appropriate skills not acquired from facility training
      ii) Appropriate skills not acquired from vendor provided training
      iii) Inadequate periodic assessment of staff competencies
      iv) Lack of continuing education
   d) Communication
      i) Poor, incomplete, unclear or missing documentation
      ii) Inadequate communication patterns designed
      iii) Inappropriate or misdirected communication
      iv) Failure to request needed information
      v) Outside medical records old/incorrect/incomplete/absent
      vi) Lack of timeliness
      vii) External factors
      viii) Verbal instructions not supported by written documentation
   e) Physical environment
      i) Physical environment inadequate (poor lighting, excessive sound etc.)
      ii) Distracting environment
      iii) Interruptions

f) Leadership and external issues
   i) Inadequate safety culture
   ii) Failure to remedy past known shortcomings
   iii) Environment not conducive to safety
   iv) Hostile work environment
   v) Inadequate supervision
   vi) Lack of peer review
   vii) Leaders not fluent in the discipline
   viii) Outdated practices

2) Technical
   a) Proper acceptance testing and commissioning of new equipment
      i) Not following explicit referral to best-practice documents (AAPM TG reports, ASTRO, ACR, IPEM, COMP, etc.)
      ii) Lack of independent review
      iii) Lack of review of preexisting reports
      iv) Lack of effective documentation (vendor or self)
   b) Equipment design and construction issues
      i) Inadequate policies and procedures for quality assurance and quality control
      ii) Poor human factors engineering
      iii) Interoperability problems
      iv) Networking problems (IT)
      v) Software operation failure
      vi) Hardware failure
      vii) Poor construction (physical)
   c) Equipment maintenance issues
      i) Failure to report problems to vendor
      ii) Failure to follow vendor notices (field change orders)
      iii) Failure to provide adequate preventive maintenance
      iv) Failure on the vendor's part to share failure/safety issues in a timely manner
      v) Unavailability of local and field support as needed
   d) Environment (within the facility)
      i) Ergonomics (room layout and equipment setup)
      ii) Machine collision issues (room specific)
      iii) Environment (water, HVAC, electrical, gas)
      iv) IT infrastructure and networking issues (including compliance to expected security and capacity standards)
      v) Delay in corrective actions for facility problems

3) Human behavior involving staff
   a) Acting outside one's scope of practice
   b) Slip causing physical error (failure in performance of highly developed skills as intended or maintained)
   c) Poor judgment (e.g., failure to carry out quality control on a patient due to time limitation)
   d) Language and comprehension issues

  e) Intentional rules violations (sabotage/criminal acts, criminal intent, intentional violation)
  f) Negligence (risky behavior, poor judgment in failure to address issues or extreme demands, lack of vigilance; recklessness)
4) Patient-related circumstances
  a) Misleading representation
  b) Cognitive performance issues
  c) Non-compliance
  d) Language issues and comprehension
  e) Patient medical condition (inability to be positioned or remain still)
5) External factors (beyond facility control)
  a) Natural environment
  b) Hazards
6) Procedural issues
  a) Failure to detect a developing problem
    i) Environmental masking
    ii) Distraction
    iii) Loss of attention
    iv) Lack of information
  b) Failure to interpret the nature of the developing problem
    i) Inadequate search
    ii) Missing information
    iii) Incorrect information
    iv) Expectation bias
  c) Failure to select the correct rule to address problem
    i) Incomplete or faulty rule
    ii) Old or invalid rule
    iii) Misapplication of a rule
      (1) Similarity bias/stereotype fixation
      (2) Familiar pattern not recognized
      (3) Familiar association short-cut
  d) Failure to develop an effective plan
    i) Information not seen or sought
    ii) Inappropriate assumptions
    iii) Information misinterpreted
    iv) Side effects not adequately considered
    v) Mistaken options
  e) Failure to execute the planned action
    i) Stereotype take-over/faulty triggering
    ii) Plan forgotten in progress
    iii) Plan misinterpreted
    iv) Plan too complicated (bounded reality)
7) Other

Appendix 6.2: List of Acronyms

AAPM American Association of Physicists in Medicine
ACR American College of Radiology
AHRQ Agency for Healthcare Research and Quality
ASRS Aviation Safety Reporting System
ASTRO American Society of Radiation Oncology
C3RS Confidential Close Call Reporting System
COMP Canadian Organization of Medical Physicists
CQC Care Quality Commission
CRCPD Conference of Radiation Control Program Directors
CT Computed tomography
ECM Eindhoven Classification Model
ECRI Emergency Care Research Institute
ESTRO European Society for Radiotherapy and Oncology
FAA Federal Aviation Administration
FAERS The FDA Adverse Event Reporting System
FDA Federal Drug Administration
FMEA Failure Mode and Effects Analysis
FOIA Freedom of Information Act
FRA Federal Railroad Administration
HHS Department of Health and Human Services
HIPAA Health Insurance Portability and Accountability Act
HTAU The Health Technology Assessment Unit of the Alberta Heritage Foundation for Medical Research Group
HVAC Heating, ventilation, and air conditioning
IAEA International Atomic Energy Agency
ID Identification
IEC IAEA Incident and Emergency Centre
IPEM Institute of Physics and Engineering in Medicine
IR(ME)R Ionising Radiation (Medical Exposure) Regulations
IT Information technology
ITDB IAEA Illicit Trafficking Database
JCAHO Joint Commission on Accreditation of Healthcare Organizations
NASA National Aeronautics and Space Administration
NCPS VHA National Center for Patient Safety
NCRP National Council on Radiation Protection & Measurements
NEWs IAEA Nuclear Event Web-based System
NFIRS National Fire Incident Reporting System
NIH National Institutes for Health
NPLS National Reporting and Learning System
NPSA National Patient Safety Agency
NRC Nuclear Regulatory Commission
NTSB National Transportation Safety Board
PRISMA Prevention and Recovery System for Monitoring and Analysis
PSO Patient Safety Organization
QI Quality improvement
RCA Root Cause Analysis
ROSIS Radiation Oncology Safety Information System
SAFRON Safety in Radiation Therapy
TG AAPM Task Group
UK United Kingdom
USIE IAEA Unified System for Information Exchange
VHA Veterans Health Administration
WGPE AAPM Working Group for the Prevention of Errors

Chapter 7

Quality Improvement Paradigms

Steven Sutlief, Ph.D.[1], Luis Fond de los Santos, Ph.D.,[2] and Todd Pawlicki, Ph.D.[3]

[1]Chief Medical Physicist, VA Puget Sound Health Care System, Seattle, Washington
[2]Department of Radiation Oncology, Mayo Clinic, Rochester, Minnesota
[3]Professor and Vice Chair of Medical Physics and Technology, Department of Radiation Medicine and Applied Sciences, University of California–San Diego, La Jolla, California

7.1 Introduction

The quality improvement movement (or "quality movement" for short) embodies several conceptual threads that span the twentieth century and cross several continents. The movement originated in manufacturing and spread to other fields, including health care. A narrative of the quality movement is given in this chapter,

Table 7–1. Milestones in the quality movement

Year	Event	Key Ideas
1861	**Ignaz Semmelweis** publishes *The etiology, concept, and prophylaxis of childbed fever*	Statistical analysis in medicine
1911	**Frederick Taylor** publishes *The Principles of Scientific Management*	Scientific management, manager driven, inspection as a QI tool
1917	**Ernest Codman** publishes *A Study in Hospital Efficiency*	Outcome-based health care, morbidity and mortality conferences, founded precursor to The Joint Commission
1931	**Walter Shewhart** publishes *Economic Control of Quality of Manufactured Product*	Behavior charts, the distinction of systematic and random errors, foundation of statistical process control
1948	**Toyota Production System** begins to be developed in Japan	Participatory management, just-in-time production, continuous improvement, design out overburden and inconsistency, eliminate waste
1950	**W. Edward Deming** lectures in Japan	PDCA, 14 points for management
1951	**The Joint Commission** formed (called JCAH at that time and then JCAHO)	Continuous improvement, National Patient Safety Goals
1954	**Jospeh M. Juran** consults in Japan	Pareto principle, change management, cost of poor quality
1957	**Genichi Taguchi** publishes Vol. 1 of his two-volume book on *Design of Experiments*	Taguchi loss function, off-line quality control, design of experiments using orthogonal arrays
1962	**Kaoru Ishikawa** develops quality circles	Fishbone diagrams, quality circles, internal customer
1966	**Avedis Donabedian** publishes *Evaluating the Quality of Medical Care*	Quality in health care, outcomes research, systems management
1982	**ISO 9000** standard developed	Quality management system standards, process improvement, procedures, and monitoring documentation
1986	**Total Quality Management** movement launched when W.E. Deming publishes *Out of the Crisis* which states his 14 points	Customer focus, employee participation, continuous improvements, builds on work of Deming and Juran, focusing on long-term success through customer satisfaction
1987	**Six Sigma Quality Program** launched at Motorola	DMIAC, quality management tools (many), focus on reducing defects by minimizing variation in processes
1987	**Malcolm Baldridge Award** initiated by U.S. Congress	Results focus, organization-wide goal alignment, draws on ideas of Total Quality Management
1990	**Lean manufacturing** described in the book *The Machine that Changed the World*	Derived from the Toyota Production System, focus on process flow and waste

coupled with examples drawn from radiation therapy. Following this introduction, the second section describes the origin of the quality movement in manufacturing, where a key concept is the introduction of quantitative assessment in order to make "scientific" decisions. The third section describes a parallel focus on quantitative assessment within health care, starting with the early work of Semmelweis and Codman and continuing with the work of Donabedian. The fourth section describes particular frameworks applicable to radiation therapy departments, and the final section summarizes recent research in the application of quality processes in radiotherapy.

The quality community has an extensive internal language that we will survey in this chapter. Here is a short list of the expansive and bewildering set of terms with overlapping meanings:

- *Prominent frameworks:* Continuous Quality Improvement, ISO 9000, ISO 9001, The Joint Commission, Lean, Malcolm Baldridge Award, Six

Sigma, Statistical Process Control, Total Quality Management, and Toyota Production System.

- *Prominent tools:* Cause-and-Effect Diagram, Check Sheet, Control Chart, DMAIC, Fishbone Diagram, Histogram, Pareto Chart, PDCA, Quality Circles, Scatter Diagram, Taguchi Loss Function, Taguchi Orthogonal Arrays, and Value Stream Mapping.
- *Prominent names (outside medicine):* W. Edwards Deming, Kaoru Ishikawa, Josph M. Juran, Walter A. Shewhart, Genichi Taguchi, and Frederick Taylor.
- *Prominent names (in medicine):* Ernest Codman, Avedis Donabedian, and Ignaz Semmelweis.

The relationship of many of these names and concepts over time is shown in Table 7–1.

7.2 Quality Improvement Paradigms Originating Outside Health Care

Quality improvement has four different guises for us to consider:

- quality as a set of production improvement tools,
- quality as a set of standards for demonstrating the implementation of certain tools,
- quality as an approach to equipment QA and QA design, and
- quality as a test of notions about radiation therapy standards of practice.

A theme in this chapter will be the role of change culture when introducing quality improvement in the first of the senses listed above. Change culture refers to the buy-in by employees of new ideas and strategies introduced by management. W. Edward Deming using the following analogy: "A usual stumbling block [in improving quality] is management's supposition that quality control is something that you install, like…a new carpet" (Deming 1986). This metaphor emphasizes the prevalent attitude that quality improvement is a compartmentalized task rather than a toolset requiring a system-wide mindset.

Because of the importance of mindset as just mentioned, it is important to distinguish between the underlying tools (such as process charts and Pareto diagrams), the organizing frameworks (such as PDCA and DMIAC), and the overall label (such as Total Quality Management, Lean, and Six Sigma). The underlying tools are easily described and uncontroversial, as are the organizing frameworks, but what tools one places under each overall label varies between authors. For example, Total Quality Management (TQM) has received substantial criticism, but that criticism often focuses on what has been described as the *soft model* of TQM with its human relations emphasis (Yong and Wilkinson 2001).

7.2.1 Taylorism and Scientific Management

Frederick Winslow Taylor began looking at how to increase human and machine productivity while working as a machinist foreman. He published his ideas in a 1911 book titled *The Principles of Scientific Management* (Taylor 1967). His four "principles" include using scientific study of tasks to replace ad hoc rules, emphasizing training, making each employee responsible for discrete tasks, and making managers responsible for the application of scientific principles while workers are responsible for carrying out the tasks. These notions were precursors to notions of quality in production. Taylor is credited with changing the "craftsman" model of creating products by his emphasis on separating the engineering of a new product from its production and the management of the entire system.

In reviewing Taylor's influence on production, it is important to note that the responsibility and opportunity for innovation were squarely in the realm of management, while the worker's role was to work hard and benefit from the higher wages afforded by the greater efficiencies provided by scientific management. This distribution of roles is very different in later systems, where the worker is seen to have an interest in the final product and is given the opportunity to suggest innovations.

7.2.2 Shewhart and Statistical Process Control

Statistical process control (SPC) is a quality improvement tool (or set of tools) that is supported by established research and development. Walter A. Shewhart is known as the father of SPC, although he did not coin the term as we know it today. Process control has been around since the mid-1920s, but it really became known in 1931 with the publication of Walter Shewhart's textbook on the topic (Shewhart 1931). Dr. Shewhart's innovations include the distinction between common causes and special causes, plus the subsequent development of control charts. Common causes represent the inherent variation or noise in a process. Special causes represent an extrinsic change or signal in a process. While SPC has been around for a long time, it is new to radiation therapy and will be described below.

To describe SPC, we first note that no two things are exactly the same. This is quickly learned in any introductory experimental physics course. When something is repeatedly measured, one obtains an average value of the thing being measured. Many values are created by repeatedly running an experiment, and then a spread of data around the average value is obtained. The monitoring of mass production processes is essentially a repeated physics experiment for each part that is being produced. If the process of manufacturing products changes—e.g., equipment deteriorates, new procedures are used, or workers get tired—then the parts will not meet the engineering specifications, and the final product will be of poor quality. The main question in manufacturing is: how does one know if the process is changing and needs to be fixed? Control charts provide an answer to this question.

The paradigm of manufacturing quality assurance in the United States has traditionally been to: 1) fabricate, 2) inspect, 3) rework or scrap, and 4) ship (Wheeler and Chambers 1992). For example, consider a company that makes multi-leaf collimator (MLC) leaves. After each leaf is completed, it is inspected to ensure that it meets the engineering specifications. Then, one of two things happens. Either the leaf passes the inspection and proceeds to the consumer or the leaf fails the inspection and is rejected (either reworked or discarded outright). The manufacturer can only accumulate data on how many passes and fails are tallied. Medical physicists routinely use this approach (fabricate, inspect, rework or scrap, and ship) with quality assurance. Physicists take a measurement and determine if it is outside some preset limits. If it is out of limits, one usually takes another measurement to check it again. Assuming the measurement is still out of limits, the physicist proceeds to identify what the underlying cause is. At this point, different things may happen, depending on the magnitude of the error.

To discuss the different things that can happen, let's consider the case of measuring linear accelerator output. If the output is 10% high, then the decision is easy—no treatments until you find out what the problem is. If the output is 0.2% high, then the decision is also easy—fill out the quality assurance documentation and allow patients to be treated. However, what if the output is 2.1% high? Now there is a decision to make. Should the physicist let it go and measure again tomorrow? Should the physicist adjust the potentiometers to reduce the output? Or, is this within the stability limits of your linear accelerator output and nothing should be done? This last question is of the kind that the techniques of SPC can provide a statistically based answer. As physicists, we deal with these types of questions all the time in radiation therapy. Usually a physicist is left to use his or her own clinical judgment to determine whether or not to accept the difference and proceed with treatment.

Clinical judgment is not completely on the shoulders of the individual physicist in making the decision. The medical physics community has guidelines that give the physicist some idea of what to do. We know that final delivered dose should be with about ±5% of the prescribed dose, and Task Groups 40 (Kutcher et al. 1994) and 142 (Klein et al. 2009) of the American Association of Medical Physicists (AAPM) provide action limits for linear accelerator quality assurance to achieve this level of dose accuracy.

It is desirable for process variation to remain within acceptable levels and not cross beyond the action limits. A long-held tenet within industrial and systems engineering is that optimal quality is defined as process performance being on target with minimum variance (see the Taguchi Loss Function in Section 7.2.5). Therefore, even variations within acceptable limits are detrimental to quality improvement. Another reason is because increasing amounts of process variation make it difficult to understand the causes of a poor result. Control charts were created specifically to provide workers with a tool to achieve that definition of optimal quality. Once process variation is well within action limits, there likely are diminishing returns to further quality improvement efforts for that process. Continued expenditure of large amounts of effort to bring process variance toward zero may have opportunity costs that could be detrimental to improvement of the global system. For example, effort applied for small reductions in already-low variances might be better spent addressing other processes that have larger, or more meaningful, variances.

Control charts show process variation over time. The quantity to be measured will determine the process, or part of a process, that will be investigated. Consecutive instances of the quantity being measured are then batched into groups which could be as small as a single instance ($n = 1$) or as large as necessary to make sense for the process being measured ($n > 1$). Take patient-specific IMRT dose QA as an example. One can take each IMRT case as an instance of the quantity of interest ($n = 1$), or cases can be grouped by a certain number of cases ($n > 1$). In SPC vernacular, these are called subgroups. Using $n = 1$ versus $n > 1$ will highlight different aspects of the IMRT QA process. To have the best chance of identifying special causes, subgroups should be as homogeneous as possible, meaning that a subgroup should consist of the same things. Thus, in our IMRT QA example, it is not ideal to consider different types of IMRT cases in the same subgroup. Head and neck cases can reasonably be expected to show more case-to-case variation in dose QA than prostate cases, for example. We will provide the equations for three different strategies for creating control charts depending on whether the subgroup size n is $n = 1$, $1 < n \leq 10$, or $n > 10$. Note that this will be an oversimplification of the decisions that need to be made when using control charts, but it provides a good starting point.

If $n = 1$ is used, then a control chart is created with upper and lower control limits (UCL and LCL) as,

$$\begin{aligned} UCL &= \bar{x} + 2.66\ \overline{MR} \\ LCL &= \bar{x} - 2.66\ \overline{MR} \end{aligned} \tag{7.1}$$

and center line (CL), CL = $\bar{x}$ for which $\bar{x}$ is the mean value of the measurements $\{x_1, x_2, ...\}$ and $\overline{MR}$ is the average of the moving ranges of two consecutive observations $\overline{MR} = | x_i - x_{i-1} |$. (For definitions of all the quantities involved, see this chapter's appendix.) The general case shown next will clarify why the control limits are set with these values.

Control charts use control limits customarily defined as the distance three standard deviations away from the mean measurement value. However, the standard deviation is not known except through an estimation based on measurements. To increase specificity of the tool, the sequence of measurements is divided into subgroups, each containing n measurements. The upper and lower control limits for the control chart can then be obtained by using either the range or the standard deviation of the subgroups to estimate the overall standard deviation. A set of chronologically labeled measurements $\{x_1, x_2, ...\}$ are divided into subgroups of size n: $\{x_1, ..., x_n\}$, $\{x_{n+1}, ..., x_{2n}\}$, ... For each subgroup we can extract the measured mean $\bar{x} = (x_1 + ... + x_n) / n$, the range $R = \max(x_1, ..., x_n) - \min(x_1, ..., x_n)$, and the measured standard deviation

$$s = \sqrt{\left((x_1 - \bar{x})^2 + \cdots + (x_n - \bar{x})^2\right) / (n-1)} \tag{7.2}$$

which is an estimate of the standard, which is an estimate of the standard deviation $\sigma_{\bar{x}}$ within each subgroup. The standard deviation $\sigma_{\bar{x}}$ within each subgroup is related to the overall standard deviation σ_x by $\sigma_{\bar{x}} = \sigma_x / \sqrt{n}$, which follows from the Central Limit Theorem. The grand mean based on the mean of m preliminary subgroups is

$$\bar{\bar{x}} = \left(\bar{x}_1 + \cdots + \bar{x}_m\right) / m \tag{7.3}$$

and the mean range over m preliminary subgroups is

$$\bar{R} = \left(R_1 + \cdots + R_m\right) / m. \tag{7.4}$$

If the subgroup size is small ($n \leq 10$), then it is sufficiently accurate to establish the control limits using the subgroup range R rather than the estimated subgroup standard deviation s. Before the widespread availability of computers, it was more practical to use R than s, which requires greater computation. By defining d_2 to be $\bar{R}$, the mean of the range in the sample, divided by the overall standard deviation σ_x, we can write $\sigma_x = \bar{R}/\mathrm{d}_2$ and thus upper and lower control limits can be written in terms of the mean subgroup range as

$$\begin{gathered} UCL(x)_{\bar{R}} = \bar{\bar{x}} + 3\sigma_{\bar{x}} = \bar{\bar{x}} + 3\sigma_x / \sqrt{n} = \bar{\bar{x}} + 3\bar{R} / \left(d_2\sqrt{n}\right) \\ LCL(x)_{\bar{R}} = \bar{\bar{x}} - 3\sigma_{\bar{x}} = \bar{\bar{x}} - 3\sigma_x / \sqrt{n} = \bar{\bar{x}} - 3\bar{R} / \left(d_2\sqrt{n}\right) \\ CL = \bar{\bar{x}} \end{gathered} \tag{7.5}$$

The constant d_2 serves as a bias correction factor depending only on the sample size n. The d_2 can be computed using Tippett's integral formulas (Barbosa, Gneri, and Meneguetti 2013) and are found to be $d_2 = 1.128$, 1.693, 2.059, and 2.326 for $n = 2$, 3, 4, and 5 respectively. In the example below we will use $d_2 = 3.078$ for $n = 10$. Note that if the subgroup size is $n = 1$, the method of moving ranges (described above) is used, effectively creating a sample size of two, such that $3/\left(d_2\sqrt{n}\right) = 2.66$, in which case the control limits are those shown above for the $n = 1$ case.

As the subgroup size increases (e.g., $n > 10$) the range-based approach becomes progressively less accurate because it discards all information except for the range boundaries and is thus more sensitive to low-probability outliers. The standard deviation $\sigma_{\bar{x}}$ characterizes the variation within the subgroups. Thus for larger subgroups ($n > 10$) it is preferable to use

$$\begin{gathered} UCL(x)_{\bar{s}} = \bar{\bar{x}} + 3\sigma_{\bar{x}} = \bar{\bar{x}} + 3\sigma_x / \sqrt{n} = \bar{\bar{x}} + 3\bar{s} / \left(c_4\sqrt{n}\right) \\ LCL(x)_{\bar{s}} = \bar{\bar{x}} - 3\sigma_{\bar{x}} = \bar{\bar{x}} - 3\sigma_x / \sqrt{n} = \bar{\bar{x}} - 3\bar{s} / \left(c_4\sqrt{n}\right) \\ CL = \bar{\bar{x}} \end{gathered} \tag{7.6}$$

The mean measured standard deviation $\bar{s}$ is the mean of the standard deviations for all m subgroups. The constant c_4 is a bias correction factor to account for the fact that the sample standard deviation(s) will underestimate the population standard deviation. It is dependent on the sample size n and is defined to be the proportionality between the sample standard deviation s and the population standard deviation σ_x: $s = c_4\sigma_x$. The first few values are found to be $c_4 = 0.7979$, 0.8862, 0.9213, and 0.9400 for $n = 2, 3, 4$, and 5 respectively. In the example below, we will use $c_4 = 0.9717$ for $n = 10$. The values for c_4 and d_2 (as well as d_3 defined below) are tabulated in most quality engineering textbooks (Wheeler and Chambers 1992; Montgomery 2012).

Note that the use of control charts usually includes the analysis of a chart for subgroup variation (R or s) as well as subgroup mean—the x-chart—which was shown above. The reason for two charts is because special causes process variation can manifest itself by a change in the process mean or a change in the subgroup variation. While the x-chart depicts the variation between subgroups, if we want to see the variation within a subgroup, we must turn to either an R-chart or an s-chart. Like d_2 defined above, we define d_3 to be the standard deviation of the range $\bar{R}$ in the sample divided by the standard deviation σ_x. The constant depends only on the subgroup size n and is computable (Barbosa, Gneri, and Meneguetti 2013). As an example, the parameter d_3 is 0.880 and 0.797 for subgroup sizes $n = 4$ and $n = 10$, respectively. We can now write $\sigma_R = d_3\sigma_x = d_3\bar{R}/d_2$ and thus the upper and lower control limits of an R-chart can be written in terms of the mean subgroup range $\bar{R}$ as

$$\begin{aligned} UCL(R)_{\bar{R}} &= \bar{R} + 3\sigma_R = \bar{R} + 3(d_3/d_2)\bar{R} \\ LCL(R)_{\bar{R}} &= \bar{R} - 3\sigma_R = \bar{R} - 3(d_3/d_2)\bar{R} \\ CL(R) &= \bar{R} \end{aligned} \tag{7.7}$$

The s-chart depicts the standard deviation of the subgroups. The control limits for the sample standard deviation s are harder to derive. Using c_4 as defined above, the standard deviation of an estimate of the sample standard deviation s can be shown to be $\sigma_{\bar{s}} = \sigma_x\sqrt{1-c_4^2}$, which allows us to write these control limits for s:

$$\begin{aligned} UCL(s)_{\bar{s}} &= \bar{s} + 3\sigma_{\bar{s}} = \bar{s} + (3/c_4)\bar{s}\sqrt{1-c_4^2} \\ LCL(s)_{\bar{s}} &= \bar{s} - 3\sigma_{\bar{s}} = \bar{s} - (3/c_4)\bar{s}\sqrt{1-c_4^2} \\ CL(s) &= \bar{s} \end{aligned} \tag{7.8}$$

The most important aspect of these derivations to estimate process variation is the subgrouping of the data, which distinguishes the decisions based on these charts from a decision one would make based simply on calculating the sample standard deviation from a stream of data from the process. Furthermore, once control chart limits are determined (and no out-of-control process behavior is detected), then one should compare the process capability to the process requirements.

An example using simulated data is shown in Table 7–2 and figures 7–1, 7–2, and 7–3. In this example, 80 measurements have been taken using normally distrib-

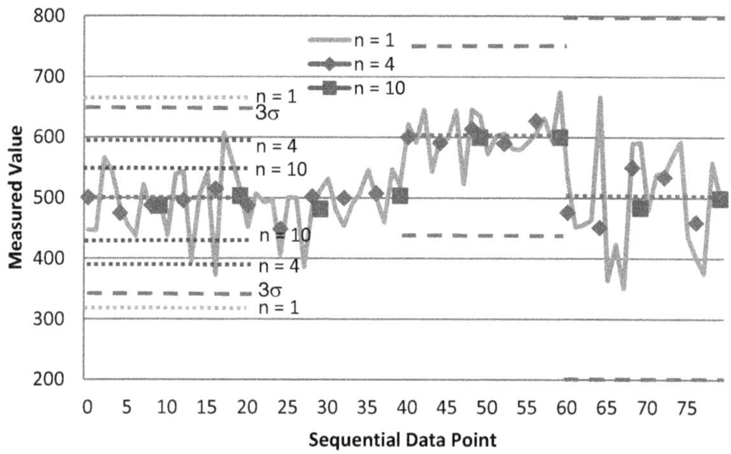

Figure 7–1. The x-chart for the example in the text. The horizontal axis is the sequential number of the data point and the vertical axis is the data point value. The continuous line is the raw data, randomly generated for this example as described in the text. The dotted line (unlabeled) represents the mean, and the dashed lines represent the 3σ limits used to generate the data, which would not be known in a real-world situation. The diamonds are the mean values of the data taken in groups of four. The squares are the mean values of the data taken in groups of 10. There are three labeled pairs of dotted lines that represent the three kinds of control limits that can be applied directly to the data. The dotted line represents the control limits with subgroup sizes as labeled: n = 1, in which case a moving range calculation is used, and n = 4 and n = 10 using the R-values for subgroups with four and 10 elements each, respectively. For each of these cases, the control lines are based on the first 20 measurement points, which is why the dotted lines are only drawn to point 20. See the text for further description.

uted simulated data. Measurements 1 through 40 have a mean of 500 and a standard deviation of 50. Measurements 41 through 60 have a mean of 600 and a standard deviation of 50. Measurements 61 through 80 have a mean of 500 and a standard deviation of 100. These values were chosen to illustrate how the use of control limits on *x*-, *R*-, and *s*-charts is much more capable of detecting uncontrolled variation than could be accomplished through visual inspection of the data alone. Sufficient information has been provided here to allow the reader to replicate these computations and plots.

Figure 7–1 represents the *x*-chart in this example, where control limits are established by the first 20 points and are used to detect process variation in the remaining data points. The raw data is represented by the solid line. The control limits determined by the moving range method (appropriate for subgroup size n = 1) using the first 20 data points are represented by the dotted line "n = 1." As can been seen, these control limits are larger than the actual 3σ limits shown with dashed lines, which are intrinsic to the process being measured, but which can only

Table 7–2. Data and computed values for the statistical process control example described in the text (See note on next page)

#	Data in pairs	n = 4			n = 10			Control Limits (UCL, CL, LCL)				
		$\bar{x}$	*R*	σ	$\bar{x}$	*R*	σ	X(MR)	X(R)	X(s)	R	s
2	447 446	501	120	63	487	129	45	hi=673 mid=493 lo=312	n=4: hi=595 mid=495 lo=394 n=10: hi=551 mid=495 lo=439	n=4: 597 495 393 n=10: 553 495 436	n=4: 314 138 0 n=10: 322 182 41	n=4: 142 63 0 n=10: 104 60 16
4	566 543											
6	485 455	475	85	37								
8	437 522											
10	476 498	488	101	42								
12	438 539				503	234	75					
14	543 395	496	148	70								
16	504 541											
18	373 607	515	234	101								
20	562 516											
22	452 507	487	55	24	482	145	47	hi=622 mid=495 lo=369	n=4: hi=547 mid=489 lo=430 n=10: hi=529 mid=493 lo=456	n=4: 549 489 429 n=10: 532 493 454	n=4: 183 80 0 n=10: 27 119 211	n=4: 83 37 0 n=10: 69 40 11
24	493 497											
26	406 501	448	115	61								
28	500 386											
30	486 512	502	52	24								
32	531 479				504	93	33					
34	454 489	500	92	38								
36	509 546											
38	502 460	507	87	37								
40	547 520											
42	621 591	600	102	44	599	122	44	hi=725 mid=601 lo=476	n=4: hi=663 mid=604 lo=545 n=10: hi=640 mid=600 lo=559	n=4: 661 604 547 n=10: 639 600 560	n=4: 184 81 0 n=10: 235 132 29	n=4: 80 35 0 n=10: 70 40 11
44	645 543											
46	597 598	591	121	50								
48	644 523											
50	645 635	614	73	33								
52	572 602				600	142	37					
54	606 581	590	27	12								
56	579 592											
58	606 631	627	81	36								
60	594 675											
62	533 452	476	81	38	484	315	102	hi=732 mid=494 lo=256	n=4: hi=617 mid=494 lo=371 n=10: hi=573 mid=491 lo=410	n=4: 622 494 366 n=10: 577 491 405	n=4: 387 169 0 n=10: 472 266 59	n=4: 178 78 0 n=10: 152 88 24
64	456 464											
66	666 364	451	315	147								
68	423 351											
70	590 591	550	111	53								
72	480 539				499	216	74					
74	540 569	534	158	70								
76	592 434											
78	402 376	459	182	85								
80	558 501											

TABLE 7–2 NOTE: Data and computed values for the statistical process control example described in the text. To aid the layout, two data points have been shown per line. Columns 3 to 5 show the values of $\bar{x}$, *R*, and σ when the sample size is n = 4. Columns 6 through 8 show the corresponding values when n = 10. The remaining columns show the control limit triplets (upper control limit, center line, and lower control limit), based on the data batched in groups of 20 points, for the chart types described above. Column 9 shows the control limits for the mean based on the moving range (i.e., n = 1). Columns 10 and 11 show the control limits for the mean using the R-values and the s-values respectively. Columns 12 and 13 show the control limits for the R-chart and s-chart respectively.

be estimated and are not directly knowable. Depending on the random fluctuation of the data, these moving range method control limits may overestimate or underestimate the underlying variation. Later measured values (e.g., points 21 to 40) lie well within the control limits, and we expect less than one in a hundred measurements to be an outlier, since the control limits were based on a 3σ criterion. This figure illustrates the problem of analysis on ungrouped data: even when a change occurs leading to a 2σ shift in the mean (points 41 to 60), the control limit may not be exceeded with sufficient frequency to call attention to the shift. By clustering the data into groups of four—as shown by the diamonds for the data and the dotted lines labeled "n = 4" for the control limits—the control chart becomes a much more sensitive tool. In this case, the mean values of the $n = 4$ subgroups are seen to routinely exceed the original control limits (shown by the dotted lines based on points 1 through 20). When the subgroup size is made even larger—as shown by the squares and dotted lines for the case $n = 10$—the control chart becomes more specific. However, the use of the range data itself to determine the control limits becomes less reliable because the ranges will be dominated by infrequent outliers. This is why the *s*-value should be used for larger sample sizes, since it is less sensitive to outliers.

As can been seen from Table 7–2, for the $n = 4$ subgroup size for the first 20 points, using the s-value to compute the control limits for x, which are 393 and 597, does not differ significantly from the control limits derived from the *R*-values, which are 394 and 595. For the $n = 10$ subgroup size, the control limits based on *R*-values and s-values are also similar. Note also that points 61 through 80 use the same mean, but now the standard deviation is twice as large, which causes the width between the control limits based on the R-values to expand from 204 to 246 and the width of the control limits based on s-values to expand from 201 to 256 for the $n = 4$ subgroup size in both cases. This interval represents where the result of the control subgroup means should fall 99.7% of the time (that is, within 3σ of the mean) if the data are drawn from a normal distribution. The important point to note is that by combining consecutive measurements into subgroups, for this data the control limits have only increased by about 27% (from 201 to 256) even though the

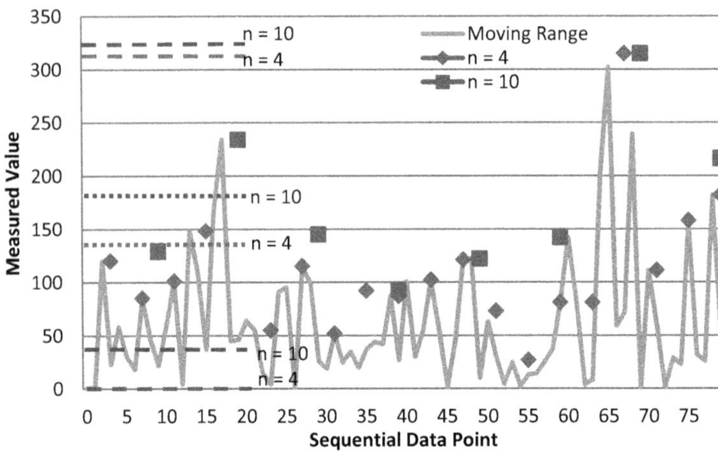

Figure 7–2. R-chart for the example in the text. The continuous line is the moving range for the data. The diamonds are the R-values of the data taken in groups of four. The squares are the R-values of the data taken in groups of 10. There are two other sets of lines that represent the control limits depending on the sample size. The dotted lines are the mean R-values and the dashed lines are the upper and lower control limits for the cases n = 4 and n = 10 as labeled. For each of the cases, the control lines are based on the first 20 measurement points. See the text for further description.

standard deviation has doubled, so by grouping we have a more sensitive indicator of process change, even when the standard deviation is large.

Figure 7–2 represents an *R*-chart for this example. Although the moving range data is shown for convenience, we are interested in the *R*-values for the $n = 4$ and $n = 10$ subgroups, shown by the diamonds and the squares, respectively. As can be seen, the R-values for the $n = 10$ subgroups are always larger than those for the $n = 4$ subgroups. In particular, the data point at 20 is anomalously high and skews the $n = 10$ mean to be large. This sensitivity to outliers is why the *R*-chart is less efficient for larger subgroup sizes. The purpose of the *R*-chart is to detect changes in the variation of the process which, in this example, has been simulated by doubling the size of sigma for points 61 through 80. Even in the four R-values for subgroup size $n = 4$ and the two *R*-values for subgroup $n = 10$ size, there is one point that has approached the original 3σ limit, shown by the upper dotted line, which should only occur at a rate of 1 in 370 cases. Thus the *R*-chart is able to quickly signal a shift in the internal variation of the process.

Figure 7–3 represents an *s*-chart for this example. This chart is qualitatively similar to the *R*-chart, except for the following characteristics. The *s*-values for the $n = 10$ subgroups do not necessarily exceed those of the $n = 4$ subgroups. The control limits vary with the sample group size (becoming smaller for larger groups) in a more stable way than the control limits for the *R*-chart. Like the *R*-chart, the *s*-chart is very sensitive to changes in the internal variation of the samples.

Control charts are generally considered as a pair: the control chart for mean plus either the control chart for range or the control chart for standard deviation. So what do we do when a control limit is exceeded on the mean chart or the range/standard deviation chart? Montgomery recommends that a process should be investigated when the data trend shows a rising or falling run (typically eight or more points), a cyclic pattern, or any of the criteria from the "Western Electric guidelines":

- When one point plots exceeds the 3σ control limits,
- When two out of three consecutive points fall above the upper 2σ limit or below the lower 2σ limit,
- When four out of five consecutive points exceeds the 1σ limit, or
- When eight consecutive points fall on one side of the center line (Montgomery 2012).

These guidelines have the effect of increasing the sensitivity of the charts to process changes. The goal is to understand the expected distribution of measurements so that any changes in the measurements can be addressed, possibly through re-commissioning or re-engineering of the process, if necessary, or else through the establishment of revised control limits if the new values are still within the required tolerances (e.g., as specified in TG 142).

There is sometimes a question about the relationship between control charts, statistical process control, and statistical quality control (SQC). There is no single definition of SPC and SQC, but some generalities can be made. In 1974, Kaoru Ishikawa outlined seven main tools to be used in quality improvement. These were the cause-and-effect diagram, check sheet, control chart, histogram, Pareto chart, scatter diagram, and stratification (Tague 2005). In 1976, these were later supplemented by seven other tools more designed for management, which included: the affinity diagram, relations diagram, tree diagram, matrix diagram, matrix data analysis, arrow diagram, and process decision program chart (Tague 2005). The use of these 14 tools is generally thought of as SPC and SQC. The difference between SPC and SQC is that SPC is used to monitor process inputs (independent variables) and SQC is use to monitor process outputs (dependent variables). However, a dependent variable is usually used in another process and can be thought of as an independent variable, so the distinction between SPC and SQC is not that important; the 14 tools of quality improvement are the important aspect.

7.2.3 Deming and the PDCA Approach

During the 1940s, W. Edward Deming, a physicist by training, promoted new ideas about quality in manufacturing. These ideas included statistical process control and process improvement. Although U.S. industry was not receptive to Deming's ideas at first, Ishikawa took an interest and invited Deming to Japan to instruct their industrial leaders in his ideas. A core tenant of Deming's approach is that the

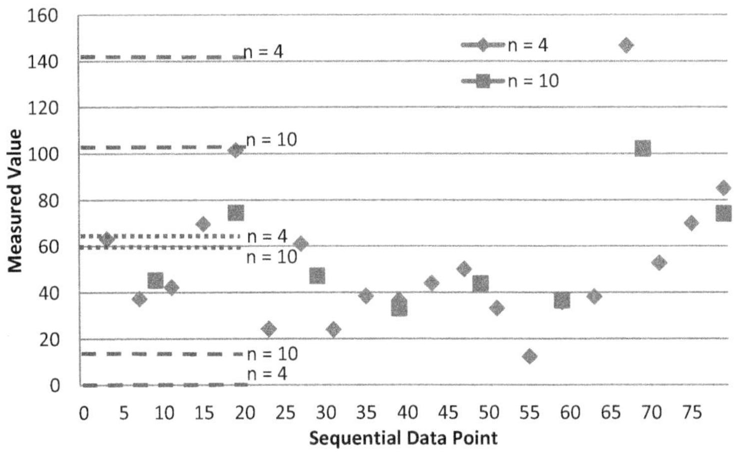

Figure 7–3. The s-chart for the example in the text. The diamonds are the s-values of the data taken in groups of four. The squares are the s-values of the data taken in groups of 10. There are two other sets of lines that represent the control limits depending on the sample size. The dotted lines are the mean *s*-values and the dashed lines are the upper and lower control limits for the cases $n = 4$ and $n = 10$ as labeled. For both the $n = 4$ and $n = 10$ cases, the control lines are based on the first 20 measurement points. See text for further description.

worker is the one who actually controls the production process. This led to the Deming Cycle: Plan, Do, Check, Act (PDCA), which Deming credited to Walter Shewhart. The success of the quality movement in Japan during the 1960s led to its adoption in many other countries.

PDCA, as depicted in Figure 7–4, is thought of as a cyclic process, which fits the "continuous" aspect of continuous quality improvement. One notion originating with Deming and repeated by others is the recommendation to perform the PDCA cycle on a small scale before adopting wider implementation of a process modification (Scholtes, Joiner, and Streibel 2003; Massoud et al. 2001). PDCA has these components:

Plan. Formulate how a process will be modified. Then establish what needs to be done, who will do it, and by when it needs to be done. Collect baseline data by which to compare measurements after the change has been initiated.

Do. Implement the change. Collect data from the revised process.

Check (or Study). Assess the collected data in light of the baseline data to determine whether there was an improvement. Study the results to see what worked and what didn't, which may include determining whether the work team needs additional knowledge, training, or resources to make the change successful.

Act. If the change was successful, implement the revised process as a standard procedure. If the change was less than successful, then proceed to the planning step in order to develop an alternative.

Deming is also known for his 14 principles of transforming business effectiveness and for his seven diseases that plague American management. Since the seven diseases pertain to management and human resources in a way only indirectly bearing on quality, they are not addressed here. Of the 14 principles for transforming business effectiveness (Deming 1986), the following are most relevant when translated to health care:

- ***#6: Institute effective training***—Training provides opportunities to provide adequate rehearsal for the worker to gain competency. It also provides an underlying understanding of processes so that dangerous deviations can be recognized in time.
- ***#7: Institute effective leadership***—Modeling behaviors to support the desired culture and supporting supervisors and workers to implement and maintain the desired culture.
- ***#8: Eliminate fear***—This refers to what one can say, what one can ask, and the risks one can take.
- ***#12: Eliminate barriers to pride of workmanship***—An important principle in quality management (and, in fact, for safety as well) is that staff members come to work wanting to do a good job. Just criticism and evaluations are important so as not to undermine the desire for quality and improvement.

7.2.4 Juran and Managing for Quality

While Deming focused on the use of statistical process control, Joseph M. Juran focused on managing for quality, which includes issues such as cross-functional

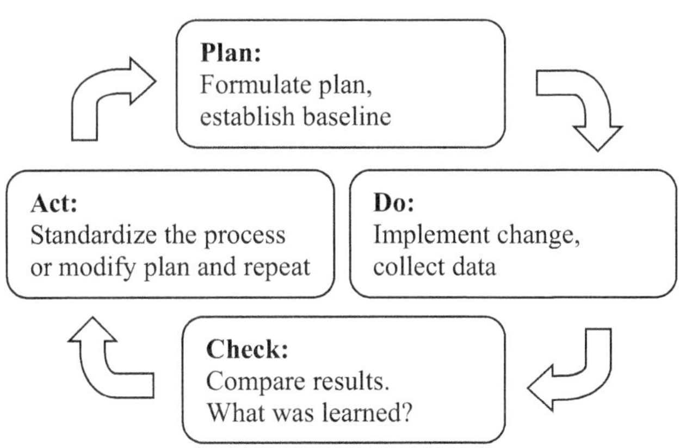

Figure 7–4. The Plan-Do-Check-Act cycle.

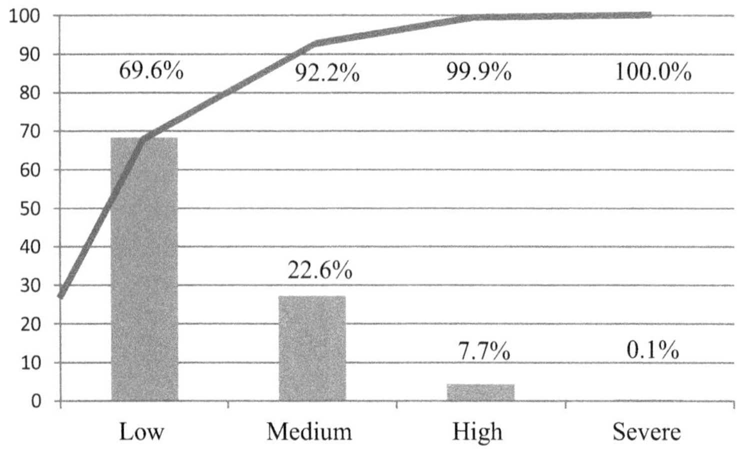

Figure 7–5. An example of a Pareto chart using results from (Mutic et al. 2010). Their implementation of an incident reporting system included a qualitative assessment of severity to the patient for all events involving potential deviations in the treatment planning and delivery process.

management and resistance to change. He popularized the Pareto Principle, the importance of resistance to change, and the cost of poor quality.

The Pareto Principle is the 80/20 rule that the distribution of most things (wealth, production, errors, and so on) is concentrated among "the vital few." In the healthcare industry, this situation would arise if a practice could prevent 80% of the possibly harmful situations by addressing the most significant 20% of the hazards—"the vital few." The Pareto Principle can be displayed schematically through a Pareto diagram, as shown in Figure 7–5, which shows the relative number of events of a given potential severity captured by an incident reporting system, as well as the sum of severities (Mutic et al. 2010). From the curve in the figure it is immediately apparent that 92.2% of the events had no worse that moderate potential severity. It is interesting that the potential severity of events reported into their system (69.6%, 22.6%, 7.7%, and 0.1% for low, medium, high, and severe respectively) closely match the percent area under a bell curve (68.3%, 27.2%, 4.3%, and 0.3% from 0 to 1σ, 1σ to 2σ, 2σ to 3σ, and greater than 3σ respectively). Both examples exemplify the Pareto principle that a small group (only the low potential severity events, only data within 1σ of the mean) generates most of the effect (number of incident reports, area under the bell curve). This kind of diagram works for any data that can be classified into unique groups that add up to a total of 100%, and it can be used to assess the significance of the first few data categories relative to the total of the data. A Pareto chart would not work for a root cause classification, for example, since a single event may have multiple basic causes.

Juran's views on change management center on the role of staff culture in building alignment for change as described in his book *Managerial Breakthrough* (Juran 1964). His view on the cost of poor quality is contained in his "quality trilogy" of managerial processes, composed of quality planning to prepare to meet quality goals, quality control of processes to meet the goals, and quality improvement to reach new levels of performance. While working as an engineer at Western Electric's Hawthorne factory, Juran developed his focus on managing for quality—to broaden quality improvement from a purely statistical effort to an approach combining statistical and managerial strategies. Although Juran advocated for organizations to adopt the tools of quality, he later became critical of quality paradigms, such as Six Sigma and ISO 9000, as ways to hype quality while not necessarily producing long-term quality improvement, since broad and deep knowledge is not always stressed in quality paradigms (Phillips-Donaldson 2004).

7.2.5 Taguchi's Loss Function and Design of Experiments

Genichi Taguchi was a Japanese statistician who made several important contributions to the quality movement in manufacturing, including the Taguchi loss functions and orthogonal arrays for design experiments (Taguchi and Yokoyama 1993; Ross 1988). An important underlying concept of Taguchi's work is the quantification of quality, which he characterizes in terms of quality loss. He defines quality loss as "the amount of functional variation of products plus all possible negative effects, such as environmental damages and operational costs" (Taguchi 1993). Thus Taguchi's concept of quality represents the sum of cost to the manufacturer, the customer, and society itself.

A typical problem would be how a power supply manufacturer chooses the output voltage tolerance that triggers in-factory repair so as to reduce the cost of out-of-factory repair of power supplies under warranty. Exceeding the tolerance for output voltage deviation will lead to one cost if repaired before leaving the factory, but will lead to a higher cost if the product must be repaired under warranty. In this instance, the Taguchi loss function gives a quantitative way to determine the optimal in-house repair threshold so as to minimize the overall expense.

The Taguchi loss function is based on the idea that the quality loss L, as a function of some parameter y, can be written in terms of a Taylor expansion about the nominal (optimal) value m of the parameter. The Taylor expansion will have a constant term, a linear term, a quadratic term, and higher order terms. We consider the loss function L to be zero when the parameter y is at its design value m. The linear term must be zero since we are expanding around the design value, which will be a minimum of the quality function where the first derivative is zero. Near the minimum point m the higher order terms are presumed to be much less than the quadratic term, which leaves the simple formula

$$L(y) = k(y - m)^2 \tag{7.9}$$

where k is a constant based on the cost of action when the parameter y reaches its tolerance threshold. For example, consider our example of the manufacture of power supplies. If the tolerance interval for power supply voltage is 4.95 V to 5.05

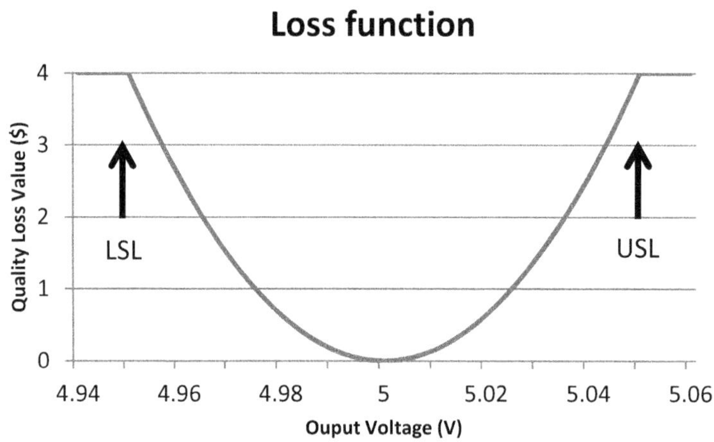

Figure 7–6. Loss function for the example described in the text of output voltages from a batch of power supplies. The replacement cost for a power supply is taken to be $4.00, and the loss function is modeled as a quadratic curve with a value of zero at the design voltage (5 V). The tolerance interval is 0.1 V, the lower specification limit (LSL) is 4.95 V, and the upper specification limit (USL) is 5.05 V.

V (with design value m = 5.00 V) and the cost for replacement $L(5.05)$ is $4.00, then the constant k for the formula given above will be

$$k = L(5.05)/(5.05 - 5.00)^2 = \$4.00/(50 \text{ mV})^2 = \$1600/\text{V}^2. \quad (7.10)$$

This loss function is shown in Figure 7–6. Note that since the mean square deviation from the target σ^2 is equal to the mean value of $(y - m)^2$, we can write the mean value of the quality loss function as

$$\overline{L} = k\sigma^2 \quad (7.11)$$

As described in (Taguchi 1993), at the end of the 1970s, industry was focused on differences between Japanese and U.S. manufacturing practices. A television unit was manufactured to identical specifications in the U.S. and in Japan, yet most American consumers preferred the model manufactured in Japan. It was found that the U.S. manufacturing plant produced televisions with color density distributed uniformly across most of the tolerance interval. In contrast, the Japanese manufacturer of the same model produced units with a normal distribution filling the tolerance interval. In Figure 7–7 this situation has been recast to fit our power supply example.

To compare these manufacturing distributions using the quality loss function, we need more statistical facts. Since 99.7% of a normal distribution lies within three standard deviations on either side of its mean value, the length of one standard deviation will be about one sixth the width of a normal distribution that fits within

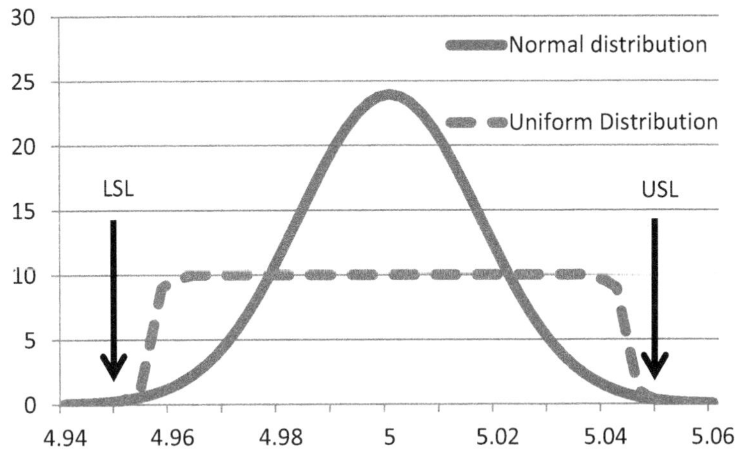

Figure 7–7. Distribution of output voltages from a batch of power supplies. The design voltage is 5 V, the tolerance interval is 0.1 V, the lower specification limit (LSL) is 4.95 V, and the upper specification limit (USL) is 5.05 V. The dashed curve represents a uniform distribution nearly spanning the tolerance interval. The solid curve represents a normal distribution such that the 3σ distance extends to the specification limit on either side. The curves are normalized to an area of 1 V^{-1}.

the tolerance interval. In contrast, the standard deviation for a uniform distribution is the width of the tolerance interval divided by the square root of 12 (Taguchi 1993). Apply this to the power supply example, given a uniform distribution and a normal distribution of voltages across the tolerance interval of width 0.1 V:

Normal distribution of the output voltage in the tolerance interval	Uniform distribution of the output voltage in the tolerance interval
$= \$1600 \times 0.1^2 / 6^2 = \$0.44$	$= \$1600 \times (0.1)^2 / 12 = \$1.33$

Thus, if it was reasonable to model the loss in quality by a quadratic function scaled to equal the corrective cost when the deviation of a parameter reaches the tolerance limit, manufacture of the power supply with a uniform distribution of output voltages yields a loss in quality about three times greater than that for a normal distribution—even though about 0.3% of units produced in the latter case exceed the tolerance. This was the case with the manufacturing of television sets cited above. Thus, in a quantitative assessment of quality, it is not sufficient to remain within tolerances. Instead, it is more important to strive toward the optimal value.

This quantification of quality can be compared with efforts within radiation therapy to achieve accurate external beam dose output. In 1976, *ICRU Report 24* stated that "the available evidence for certain types of tumour points to the need for an accuracy of ±5% in the delivery of an absorbed dose to a target volume if the eradication of the primary tumour is sought" (International Commission on Radiation Units and Measurements 1976), where the uncertainties were estimated at the 95% confidence level, corresponding to a 2σ variation. This accuracy level is referenced in each of the three AAPM reports on external beam quality assurance (listed below). In keeping with the analysis given above, if quality assurance is run such that controllable parameters are allowed to vary uniformly within the tolerance interval, then it is not accurate to model the overall uncertainty as a sum of uncertainties in quadrature. This concern is recognized in AAPM TG 24 (Svensson et al. 1984), TG 40 (Kutcher et al. 1994), and TG 142 (Klein et al. 2009), each on external beam quality assurance. For example, TG 142 states that "if certain baseline parameters barely satisfy the tolerance value repeatedly, an appropriate action should be taken to correct the equipment." This is formalized by a stratification of actions into three levels:

- *Level 1 (inspection level)*—triggered when there is an unexpected change in a routinely measured parameter, but which remains within tolerance after the change; the change should be investigated during routine inspection. This presupposes the user is reviewing trends in the data, which is essentially the purpose of Shewhart's control charts, described previously.
- *Level 2 (scheduled action)*—triggered when consecutive measurements approach the tolerance limit or a single measurement slightly exceeds the limit; the parameter should be investigated or adjusted during maintenance to be scheduled within one or two working days.
- Level 3 *(immediate action)*—triggered when a pass/fail parameter (such as a door interlock) fails or when there is an excessive error in a dosimetry parameter; treatment with the device must not resume until it operational within the allowed tolerances.

Thus, even though the AAPM task group reports on external beam quality assurance do not explicitly use Taguchi's framework, they adopt an implementation framework designed to meet the same ends for sensible quality assurance.

The Taguchi loss function is related to statistical process control through its management of statistical variation, in particular through the consideration of predictable and non-predictable variation. This is echoed in the uncertainty classifications used by the IAEA, described as follows. The estimate of uncertainties contributing to dose accuracy involves parameters that can be modeled from a set of measurements (e.g., matching a normal distribution). These are called Type A uncertainties. However, there are also parameters that cannot be characterized from a set of measurements (e.g., a parameter which has an uncharacterized dependence on another parameter). These are called Type B uncertainties, as discussed in the IAEA publication *Absorbed Dose Determination in External Beam Radiotherapy* (International Atomic Energy Agency 2000). For example, ionization chamber calibration factors in the kV range depend on both the kilovoltage generating potential and the filtration (HVL), and thus a Type B uncertainty is applied when estimating the dependence on HVL alone. This labeling of uncertainties is used instead of the

Table 7–3. An orthogonal array OA_9 for testing four parameters (A, B, C, and D), each with three levels (1,2, and 3)

Run	A	B	C	D
1	1	1	1	1
2	1	2	2	2
3	1	3	3	3
4	2	1	2	3
5	2	2	3	1
6	2	3	1	2
7	3	1	3	2
8	3	2	1	3
9	3	3	2	1

classification of uncertainties into random and systematic contributions "because classifying the components instead of the method of evaluation is prone to ambiguities" (International Atomic Energy Agency 2000).

The quantitative valuation of quality has also been applied to radiation therapy treatment efficacy. Harry, Whitaker, and Pawlicki apply the Taguchi loss function to model the effect of dose accuracy and precision on the tumor control probability and normal tissue complication probability (Harry, Whitaker, and Pawlicki 2012). In this way, they estimate the failure probability of a given treatment plan.

Taguchi's other major contribution to process engineering is the use of orthogonal arrays in experimental design, which is associated with a strategy for quality improvement known as Design of Experiments (DOE). Vlachogiannis applied orthogonal arrays to determine the settings for optimal efficiency of a radiographic screen, which is quantified by the ratio of the energy of the optical signal absorbed by the optical sensor to the energy produced in the x-ray tube (Vlachogiannis 2003). Four parameters are adjusted: the peak voltage of x-ray tube, the product of tube current and exposure time, the surface thickness of the phosphor screen, and the spectral matching factor. For the experiment, three settings are used for each of these four parameters. The relative impact of each of these parameters can then be tested using a nine-item orthogonal array (Taguchi 1986), as shown in Table 7–3. Each pair of columns includes all possible pairings of parameters an equal number of times. Thus, the columns are mutually orthogonal. In this study, two trials were performed: one presuming a 20 cm body thickness and the other presuming a 40 cm body thickness. Measured or simulated data for the optical signal is then obtained for each of the nine experimental configurations, and the mean values and signal-to-noise ratios are calculated. In this way, Vlachogiannis was able to quantify how the optical sensitivity is more dependent on the peak voltage of the x-ray tube and the product of tube current and exposure time than on the other factors studied. The author then goes on to perform an ANOVA statistical analysis to assess the contribution of each factor to the variation in screen performance ratio.

7.2.6 Ishikawa's Causal Diagrams and Quality Circles

Several of Kaoru Ishikawa's contributions to quality have already been mentioned. Ishikawa was a quality management innovator in Japan. He developed the use of causal diagrams, which bear his name, and of quality circles. Causal diagrams, also known as Ishikawa diagrams or fishbone diagrams, are used to categorize possible causes leading to an event. In the diagram, the stem extending from the event branches into categories, with minor branches showing the individual factors. An example of a causal diagram is shown in Figure 7–8.

Another major innovation was the use of quality circles. Despite their success in Japan, the adoption of quality circles in the U.S. in the 1980s provides an example of a less than fully successful quality tool. Ishikawa originated the idea in Japan in 1962 (Hutchins 1999). A quality circle is volunteer group of employees who meet regularly (at least once a week) and are trained to find solutions to work-related problems and present them to management (Ishikawa 1985). While highly successful in Japan and other Asian countries, they failed to meet expectations in the U.S., and by the late 1980s they were largely abandoned (Hindle 2008). One explanation for this was the lack of supporting total quality management concepts embraced by U.S. companies at that time (Juran 1988). Japanese work culture was more participatory, while U.S. work culture was more individualistic, such that the identification of problems in the U.S. was often followed by placing blame.

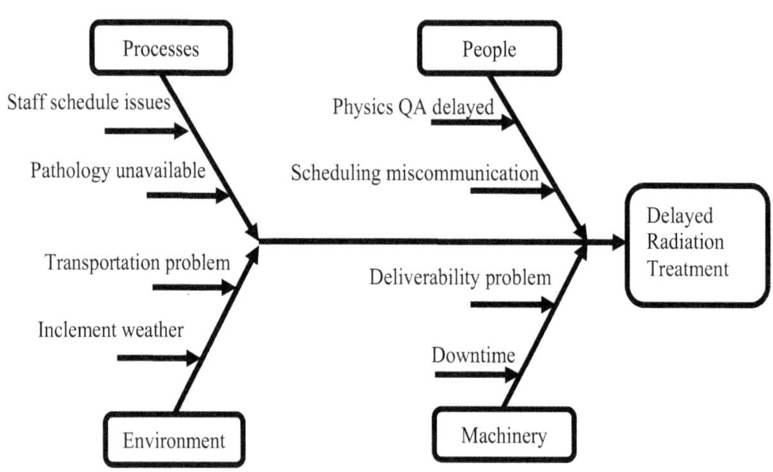

Figure 7–8. Ishikawa (fishbone) diagram. Causes of the problem (delayed radiation treatment) are identified under broad categories (people, processes, machinery, and environment). In a more complete diagram, secondary and, perhaps, tertiary causes would also be indicated using additional branches from the primary causes.

7.2.7 Toyota Production System

The Toyota Production System began in post-World War II Japan and developed over the next several decades. The central developers were Sakichi Toyoda, Kiichiro Toyoda, and Taiichi Ohno. A central theme of the Toyota Production System is the elimination of waste, of which excess inventory has received considerable attention. The other categories of waste in this system are overproduction, waiting, transporting, processing, motion, and correction. The Toyota Production System uses a number explicit techniques, such as:

- *intelligent automation* (jidoka) in which the system stops if something goes wrong,
- *just-in-time production* (related to low inventory) and the associated concept of pull scheduling (versus push-based systems, described below),
- *value stream mapping* (also called "material and information flow mapping"),
- *process mapping* (described in Chapter 4),
- *fail-safing* (poka-yoke), and
- *the kaizen system of continual improvement.*

These concepts are defined and explored in the following paragraphs.

Jidoka, as used in the Toyota Production System, means "automation with a human touch." An important example is the design of fault-identifying capabilities within a device that will cause it to stop until the problem is corrected. A frequently mentioned example is the 1896 development by Sakichi Toyoda of a loom which would automatically stop if the thread broke, thus preventing defective products from being produced (Baudin 2007). Most treatment devices in radiation therapy (such as linear accelerators and high dose rate brachytherapy units) incorporate principles of jidoka.

Pull scheduling is a manufacturing concept used to make production more efficient. Since any unfinished production (called "work in process," WIP) or internal inventory increases the production overhead, it is desirable to minimize work in process. Manufacturing generally involves a sequence requiring some components to be completed for incorporation into a product further along in the manufacturing sequence. For example, a modern aircraft requires passenger seats which, in turn, require components such as arm rests, which are composed of molded plastic, metal framing, and other parts that must be acquired before the component is assembled. One solution is to predict future demand for a component and then begin its production so that it will be ready when needed. This is an aspect of material requirements planning (MRP) that represents a push-based approach to production. In contrast, the trigger for production of a component could instead come from the workers (or customers) who need the component. This is known as pull-based, or "just in time," and characterizes the Kanban approach (Baudin, Woodruff, and Hopp 1990; Spearman and Zazanis 1992). Kanban literally means "signboard," and it was devised by Taiichi Ohno as one aspect of the Toyota Production System. Kanban uses cards, or an electronic equivalent, to signal when stock of a component has been depleted at the point where it is used, triggering production of that component. In a situation where demand is unpredictable, Kanban prevents an excessive inventory from accumulating.

Pull has successfully been employed in medicine, such as for supplies stocking (Nelson-Peterson and Leppa 2007). In contrast, medical situations more suited to a push process are those in which it is vital to have the materials when needed, when prediction is difficult, or where patients are unique. This leads to less efficient processes, according to many managers, who try to make medicine pull. That can cause stress at best and lack of resources at worst.

Within radiation therapy, pull technologies are not generally applicable to patient scheduling, planning, or treatment processes because each step is unique for a patient due to the uniqueness of each patient's individual prescription and image-based anatomy. Pull is also not applicable in radiation therapy because there is no "inventory" of identical components to manage. Nor is pull applicable to routine quality assurance, which does not vary in demands from month to month.

Value stream mapping is a method of identifying and eliminating time waste and improving efficiency in processes (Tapping et al. 2009). The patient planning process in radiation therapy is a suitable process for evaluation. Although it resembles process mapping, value stream mapping is concerned with the duration of value-added activities that benefit the patient (client) versus the duration of non-value-added activities. Non-value-added activities typically include any time required for an outside process to be completed (such as the wait for external imaging or pathology) or internal queuing (such as the wait for an available dosimetrist to do the planning, a physicist to do plan review, or a linear accelerator with which to do the dose measurements). But there is also a gray area where the determination of whether some activities are non-value-added is a matter of perspective. One issue in radiation therapy is that treatment courses are occasionally canceled before they begin for reasons unrelated to the planning interval. Thus staggering the work will reduce the overall work load since some tasks will be cancelled before time has been invested in them. The process of value stream mapping begins with an assessment of the activities and the wait intervals existing in the current implementation of the process. This requires measuring the intervals for perhaps several dozen patients. An example is shown in Figure 7–9. This current-state value stream map shows the central role of the patient care coordinator, not only in regard to awareness of the treatment planning process, but also in regard to patient interaction. The map also highlights the impact that delay in the patient's decision, receipt of the pathology, or additional imaging can play in the wait times during the planning process. The next step is to generate improvements to reduce the non-value-added intervals and to incorporate them into a future state value stream map. The final step is to implement the new value stream map. One radiation therapy department used value stream mapping to find opportunities for process efficiency which included the following changes: same-day physician review of simulation images for adequacy, consolidation of appointments, and alignment of dosimetrist and physician schedules to preempt communication delays (Orenstein 2009).

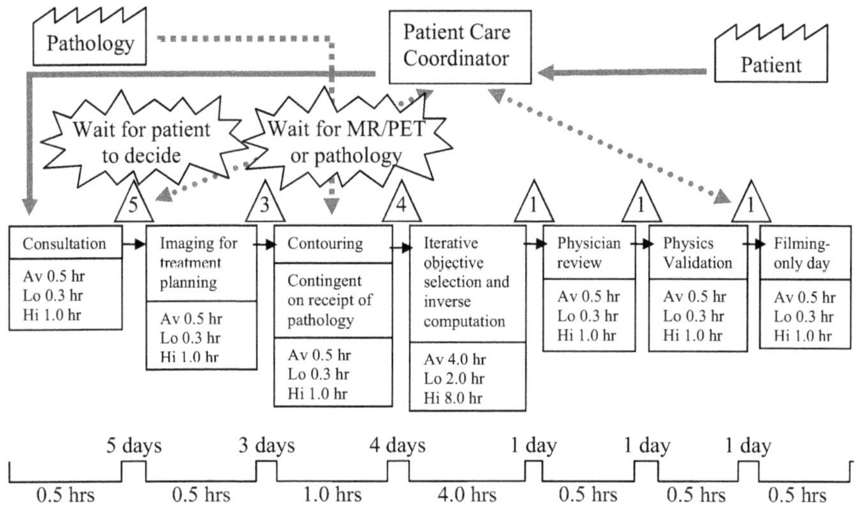

Figure 7–9. Value stream mapping applied to the treatment planning process. Each box represents a step in the treatment planning process. Minimum (Lo), maximum (Hi), and mean (Av) time estimates are given for each activity in the process. These estimates are illustrative and were not derived from actual measurements. The triangles indicate non-value-added wait times in days. The line at the bottom of the figure consists of a series of upper and lower line segments that represent the timeline of value-added activities versus non-value-added activities. The lower line segments represent the time actually creating value for the patient, while the upper line segments can be thought of as waste time, which generally represents queue time before a service can be performed.

Poka-yoke, which means mistake proofing, was developed by Shigeo Shingo while at Toyota (Shingo 1986). Poka-yoke is tied with Shingo's notion of 100 percent inspection. Although Deming considered inspections for quality to be relatively ineffective, Shingo shifted the focus from detection of defects themselves to detection of conditions that could cause defects, thus making it possible to perform corrective actions before the defect occurs (Hinckley 2001).

Kaizen, which is a combination of the Japanese words for "change" and "continuous improvement," is considered one of the two foundational principles of the Toyota Production System, the other being respect for people. The concept was popularized through the publication of *Kaizen: The Key to Japan's Competitive Success* by Masaaki Imai (Imai 1986). The goal of kaizen is the elimination of waste in the form of over-work ("muri"), unevenness ("mura," leading to pull systems), and waste in the process ("muda"). Several aspects of kaizen in the workplace are involvement of all employees at every level, implementation of ideas using the PDCA cycle, and focusing on small incremental improvements (although it is also possible to have a focus week, known as a "kaizen blitz") (Graban and Swartz 2012).

In the 1990s, attempts were made to bring the Toyota Production System into U.S. automotive industry. But these techniques, such as low inventory and quality circles, failed to meet expectations. This led to a reanalysis of the underlying ideas of the Toyota Production System. Steven Spear and H. Kent Bowen have identified four underlying characteristics of the Toyoda Production System: work processes are highly specified, process chain communication (in terms of stock and resources) is simple and direct, the pathway for products and services is simple and direct, and any improvements will be accomplished "through the scientific method, under a teacher's guidance and at the lowest possible organizational level" (Spear and Bowen 1999). To manifest these characteristics, the vision of the Toyota Production System is to mold a company into what has been called "a community of scientists performing continual experiments." When problems arise, the employee is empowered to seek out countermeasures. The word "countermeasure" is meant to emphasize the provisional aspect of the corrective active, in contrast to a "solution" involving wider system changes. Countermeasures include many of the tools and practices of the Toyota Production Model, such as Kanban. Thus, these tools and practices should not be viewed as the core of the Toyota Production Model, but rather as means to achieve the four characteristics listed above.

7.2.8 Total Quality Management

Total Quality Management (TQM) emerged as a term in the 1980s to cover the principles of quality introduced by Deming, Juran, and others. TQM focuses on continuous improvement and sustaining quality with a focus on customer satisfaction under the tacit assumption that the customer knows best what he or she wants. Cua, McKone, and Schroeder have identified nine traits of total quality management, four of which are particular to TQM (cross-functional product design, process management, supplier quality management, and customer involvement). The remaining five traits are human-oriented and strategic-oriented practices that are also common to "just-in-time" and "total productive maintenance" (information and feedback, committed leadership, strategic planning, cross-functional training, and employee involvement) (Cua, McKone, and Schroeder 2001).

In contrast, Williams says TQM consists of three elements: tools, techniques, and training (Williams 1994). The tools center around control charts. The techniques include employee participation in the areas of decision-making processes, the PDCA cycle, brainstorming sessions, and quality circles. The training instills a sense of pride in the equipment with which the trainee works. For example, Williams writes that Japanese workers "become totally responsible for inspection, operation, and repair of their machinery."

In the list of techniques just given, brainstorming deserves special mention. It was developed by Alex Osborn in order to improve the ideas generated in a group setting (Osborn 1957). Since that time, its efficacy has been called into question based on controlled studies of group interactions involving brainstorming (Mullen,

Johnson, and Salas 1991). More recently, a shortcoming of brainstorming has been attributed to production blocking, which occurs naturally as people take turns speaking. Production blocking limits the quantity of ideas expressed, but it does not necessarily reduce the quality of ideas and, in fact, in group settings, participants favor the more feasible ideas (Stroebe, Nijstad, and Rietzschel 2010).

Yong et al. have researched the criticisms of TQM that portray it as faddish, overly formulaic, hazy and ambiguous, and unable to deliver increased productivity (Yong 1999). This has raised the question of whether the underlying problem with TQM is the failure among companies professing to have adopted it to have implemented it completely, a state referred to, in full irony, as *partial TQM*. Yong's paper identifies four causes for the failures of TQM implementation: lack of a long-term strategy, lack of time, lack of resources and infrastructure, and lack of action and consistency.

7.2.9 Lean

The Lean manufacturing movement stems from the 1990 publication of the book *The Machine That Changed the World* (Womack et al. 1990). This book describes the differences between the auto industries of the U.S., Europe, and Japan. It coined the term "Lean" to describe how the Toyota Production System used small batch production instead of mass production as a means to increase efficiency. James P. Womack and Daniel T. Jones subsequently published *Lean Thinking* (Womack and Jones 1996), which described Lean in terms of these characteristics:

- "Specify the value desired by the customer."
- "Identify the value stream for each product providing that value and challenge all of the wasted steps (generally nine out of ten) currently necessary to provide it."
- "Make the product flow continuously through the remaining value-added steps."
- "Introduce pull between all steps where continuous flow is possible."
- "Manage toward perfection so that the number of steps and the amount of time and information needed to serve the customer continually falls."

The promise of Lean is to add more to a department's quality improvement efforts than a one-time process transformation. This is as result of many successive smaller process improvements that will add up. Nonetheless, Lean must be used properly to have the most beneficial effects. After all, it's easy to make a process so Lean and efficient that it becomes unstable and prone to error.

Lean has been used successfully in radiotherapy (Kim et al. 2007). In this work, the authors use Lean techniques to streamline the treatment of patients referred for radiation treatment of bone and brain metastasis. Lean tools—such as value as defined by the customer, value stream mapping, and one-piece flow (a just-in-time concept which strives for zero wait time between process steps)—were used. The result was a much more efficient process for getting patients treated. Before the Lean event, 27 steps were required to get a patient treated in the same day as the referral. The process was reduced to 16 steps, and the percentage of patients being treated on the same day as the referral rose from 43% before to 94% after the Lean intervention. Clearly, the delivery of care can benefit from Lean.

Application of Lean principles is not a sure path to improvement, as evidenced by the following two examples. The policy for U.S. stockpiles of vaccines received criticism in 2006 due to the use of just-in-time practices for the supply chain. The concern was that the just-in-time approach was incompatible with the "just-in-case" mindset needed for prompt government intervention (Wysocki and Lueck 2006). As a second example, in 2004, Park Nicollet Health Services in the Minneapolis–St. Paul was able to realize a $7.5 million benefit which they attributed to Lean production practices, such as the use of "Rapid Process Improvement Workshops" for improving patient care and safety, improving patient access, and reducing waste in inventory and processing (Varkey, Reller, and Resar 2007). However, in early 2009 Park Nicollet laid off seven percent of its workforce. Although the downturn in the economy may have been partly to blame for the layoffs, the frequency of changes and the increase in workload leading to burnout were problems cited by staff (Yee 2009).

Appropriately applied, Lean is focused on creating process stability, predictability, and learning. Much as Six Sigma does (discussed later), Lean can teach and facilitate the spread of problem-solving techniques in a department. That being said, there is no single silver bullet that can address every quality and safety improvement need. One should be aware of the many tools out there and use the appropriate tool for the job.

7.2.10 Motorola's Six Sigma

Six Sigma was developed at Motorola in 1986, taken up by General Electric in 1995, and quickly spread to other companies. Six Sigma is a quality improvement methodology based on a focused, systematic, and statistically based process improvement approach. The goal of Six Sigma is to eliminate or reduce defects (errors) in a product or process by reducing process variability. As defects and variability are reduced, overall performance improves and the needs of the customer or patient are better satisfied (Pyzdek 2009). Six Sigma uses facts, data, statistical analysis, and the understanding of customer needs to design processes that have high reliability and defect rates limited to 3.4 per million defect opportunities. This defect rate is not the frequency of outliers exceeding a 6σ deviation, but rather the frequency for a 4.5σ deviation, where a 1.5σ shift has been applied to a process with a nominal reliability of 6σ. This 1.5σ shift has been variously attributed to account for either long-term drift of a process with 6σ short-term tolerance, or to the detectability of shift in the mean of a process with 6σ tolerance (Bothe 2002). See Figure 7–10 for a graphical depiction of these tolerances. Bringing Six Sigma to health care began in the late 1990s. In 1998, Dr. Mark Chassin published an article in *The Milbank Quarterly* titled "Is Health Care Ready for Six Sigma Quality?" Chassin supported methods like Six Sigma as part of an overall framework for change and patient safety (Chassin 1998).

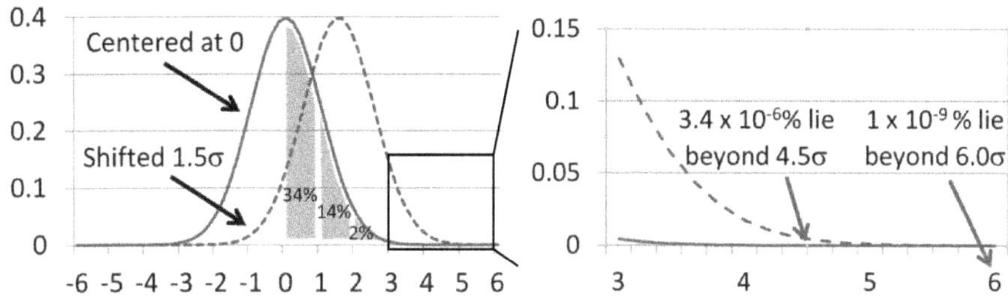

Figure 7–10. Plots of the normal distribution depicting the origin of the Six Sigma name. The right figure is an enlargement of the boxed region in the left figure. The plot on the left shows a normal distribution centered at zero and at 1.5σ, taken to be the amount of variation in precision tolerable to the manufacturer. The area between 0 and 1σ is 34% of the total area under the curve. Similarly, the area under the second standard deviation is 13.6% and under the third deviation is 2.1%. By summing the areas on both sides, we obtain the 68–95–99% rule for points falling in a normal distribution, which says that 68%, 95%, and 99% of the points lie within one, two, or three standard deviations of the mean. The amount exceeding 4.5σ on one side of the graph is 3.4×10^{-6}% and the amount exceeding 6σ is 1×10^{-9}%. It is assumed that a process may vary by 1.5σ, which means that an accuracy of 3.4 outliers per million will be achieved in a process kept to a nominal rate of 6σ for defects.

Six Sigma projects follow the DMAIC process, which is similar to Deming's PDCA cycle. DMAIC stands for Define, Measure, Analyze, Improve, and Control (Brook 2010; George 2004). The process is composed of the following phases:

Define. This phase focuses only on the problem and helps us understand the nature of it (defects, errors, waste, variation, etc.) Problem definition is essential before investing time and resources in starting a project. The problem statement should be: Specific, Measurable, Achievable, Relevant and Time bound (SMART). The project scope, goals, resources, and timeline are identified during this phase. Project charters provides a framework to organize this information.

Measure. This phase aims to confirm the current state of performance (baseline) by developing a clear and meaningful measurement system, as well as defining the data that is needed and how it will be collected. Some of the tools that are used during this phase are: process mapping, check sheets, histograms, process observation, among others.

Analyze. The purpose of this phase is to identify essential elements of an "ideal" product or service, as well as the root causes that influence the problem or defects. Some of the tools used during this phase are: cause and effect diagrams, scatter diagrams, control charts, Pareto diagrams, root cause analysis, failure modes and effects analysis (FMEA), and regression analysis.

Improve. This phase aims to confirm the cause-and-effect relationship of the problem/defect by constant experimentation and to determine the effectiveness of the proposed solutions. During this phase, the best solutions are developed, selected, and implemented with controlled risks. Some of the tools used during this phase are: affinity diagrams, future state maps, PDSA cycles, standardization, and simulations.

Control. This phase aims to ensure that the implemented solutions become a permanent element of the process so that the improvements are maintained after the project is closed. It is important to develop a transition strategy from the project team to the operational team, as well as to define the metrics to monitor and ensure sustainability. Some tools used during the phase are: project closure documentation, control charts, control plan, performance dashboards, and process documentation.

It is important to emphasize that Six Sigma is useful when the project or issue calls for a more rigorous, statistical approach and when the team has the time and resources necessary to successfully complete the project.

Six Sigma has several strengths (Kubiak and Benbow 2009). First, it emphasizes data and facts, focusing on measurable, attainable, and quantifiable results. Second, it integrates human elements, cultural changes, and process elements of improvement. Third, it uses tools and techniques that address processes in a sequential and disciplined fashion. When properly implemented, Six Sigma demonstrates the commitment of upper management, thus improving the overall company/practice morale and team cohesion.

Six Sigma also has several limitations (Kubiak and Benbow 2009). First, it may not work well for certain products or processes that may need higher or lower standards than 3.4 defects per million. Second, some projects might not call for this level of analysis. Alternately, the data necessary may be difficult, costly, or impossible to measure or gather, or may be insufficient. In this case, the value of reducing variation may not actually be worth the time and resources that it requires to implement Six Sigma, especially when time and resources are a significant constraint or when the solution can be achieved using easier, less expensive, and faster methods.

7.3 History of Quality Improvement Inside Health Care

7.3.1 Semmelweis and the Scientific Method

Ignaz Semmelweis served as an assistant obstetrician in the Vienna General Hospital in 1846. One of his responsibilities was to be a clerk of records. He was dismayed that maternal mortality rates in his clinic due to puerperal fever were about 10% while those at an allied clinic for teaching midwives (and elsewhere) were below 4%. His facility also performed autopsies, and after ruling out possible causes, he conjectured that some unknown "cadaverous material" was causing the fever. In 1847, after instituting a hand-washing policy using a solution of chlorinated lime, mortality rates dropped from 18% to about 2% (Semmelweis 1983). Semmelweis's work is distinctive in its early use of the scientific method in medicine to establish a causal connection, even though its underlying germ theory was not confirmed until decades later. Regrettably, the absence of the underlying germ-based explanation of disease spread prevented his findings from being taken seriously. Widespread acceptance had to wait for the germ theory discoveries of Louis Pasteur and others in the 1860s (Semmelweis 1983).

7.3.2 Codman and The Joint Commission

In 1910, surgeon Ernest Codman proposed that hospitals should track every patient to see whether treatment was effective and, if not, why not (Donabedian 1989). He also started the first morbidity and mortality conferences while at Massachusetts General Hospital. Subsequently, he founded his own hospital in order to carry out follow-up assessments. He then published the results in a book titled *A Study in Hospital Efficiency* (Codman 1918). Codman reported that of the 337 patients discharged between 1911 and 1916, there were 123 errors.

Dr. Codman participated in the founding of the American College of Surgeons and its Hospital Standardization Program, which led in 1951 to the Joint Commission on Accreditation of Hospitals, later renamed the Joint Commission on Accreditation of Healthcare Organizations, and now known as The Joint Commission (Clark et al. 2011). The Joint Commission performs hospital accreditation and publishes annual *National Patient Safety Goals* in order to highlight specific recommendations for patient safety. As a part of the accreditation process, the Joint Commission has requirements for sentinel event reporting.

7.3.3 Donabedian's Structure, Process, Outcomes Model

Avedis Donabedian is considered the founder of the study of quality in health care. In particular, he coined the term "outcomes" to refer to patient follow-up assessment through record review, indirect observation, and direct observation (Donabedian 1966). Donabedian took an expansive view of patient outcomes and, in addition to mortality and morbidity, he considered things like how the patient felt and the patient's awareness of the risk factors.

The Donabedian model encompasses three aspects of health care: structure, process, and outcome. The structure concerns attributes of the care delivered. The process refers to compliance with good medical practices. The outcome correlates the service rendered with indicators of the patient's future health and condition. Unlike E. A. Codman's narrow focus on end results, the Donabedian approach takes in a wider consideration of structure, process, and outcome upon which decision-making should be based (Mulley 1989).

7.3.4 National Committee for Quality Assurance

The National Committee for Quality Assurance was established in 1990 and accredits individual physicians, health plans, and medical groups. NCQA's approach to quality improvement is based on the following pattern: measure, analyze, improve, and repeat. The NCQA has at least five different programs to measure aspects of healthcare quality using the following three approaches: process of care, outcome of care, and the experience of patients and their families (National Committee for Quality Assurance 2007).

7.4. Achieving Continuous Quality Improvement

7.4.1 The Institute of Medicine's Six Aims

The Institute of Medicine (IOM) has defined quality as "the degree to which health care services for individuals and populations increase the likelihood of desired outcomes and are consistent with current professional knowledge" (Lohr 1990). This definition is directly based on outcomes and assessment, rather than the process itself.

The Institute of Medicine has so far published eleven books in the Quality Chasm series. Three notable titles are *To Err Is Human: Building a Safer Health System* (Kohn, Corrigan, and Donaldson 2000), *Crossing the Quality Chasm: A New Health System for the 21st Century* (Institute of Medicine 2001), and *Patient Safety: Achieving a New Standard for Care* (Corrigan et al. 2004). The first book is notable in its claim that between 44,000 and 98,000 preventable deaths occur in hospitals each year. It goes on to consider why errors are so prevalent, followed by a discussion of reporting systems, legal protection, and performance standards. The second book recommends six specific areas for improvement in healthcare—that it should be safe, effective, patient-centered, timely, efficient, and equitable. The third book covers safety systems, adverse event/near miss analysis, and standardized reporting.

7.4.2 ISO 9000/9001 Accreditation

ISO 9000 and ISO 9001 (International Organization for Standardization 2008) are sets of standards for quality management systems to ensure products meet customer expectations, satisfy regulatory requirements, and maintain continuous improvement. The standards are centered on the notion of a process approach that is depicted as a cycle of four nodes:

- Management responsibility
- Resource management
- Product realization
    - external input: customer requirements
    - external output: the product
- Measurement, analysis, and improvement (external input: customer satisfaction; external output: continual improvement of the quality management system)

This is supplemented with the PDCA methodology mentioned earlier.

The principles of ISO9001 have been used in approaches to quality in radiation therapy. According to *Towards Safety Radiotherapy* (BIR 2008), the Bleehen report *Quality Assurance in Radiotherapy* (Bleehen 1991) recommended a framework based on the International Standard ISO 9001:2000 (formerly BS5750/ISO 9000) in order to satisfy "the need for procedures to be documented, audited and continually reviewed." In the 1990s there was an attempt in the United Kingdom (Mayles

1999) and by the European Society for Radiotherapy and Oncology (ESTRO) (Leer et al. 1998) to implement ISO 9000 concepts in radiotherapy.

7.4.3 Malcolm Baldrige Award

The Malcolm Baldrige National Quality Improvement Act of 1987 established the Baldrige National Quality Program and its associated award (Baldrige Performance Excellence Program 2011). The purpose of the program was to instill a focus on quality among U.S. companies so as to enhance competitiveness. In 1999, the program was expanded to encompass health care and education. As of 2009, 11 health-care organizations have been named recipients of the award.

For healthcare centers, the evaluation is based on the seven categories of the Health Care Criteria for Performance Excellence framework. This framework encourages the use of models and tools described elsewhere in this chapter. In particular, the Baldrige health care criteria (1) focus on results, (2) are non-prescriptive and adaptable, (3) integrate key health care themes, (4) support a systems perspective to maintaining organization-wide goal alignment, and (5) support goal-based diagnosis (Baldrige 2011).

7.4.4 Guidance from ACR/ASTRO and ACRO

The American College of Radiology (ACR) *Practice Guideline for Radiation Oncology* (Samuels et al. 2009) provides guidance for quality improvement in a radiation therapy environment. This document requires that there will be a continuing quality improvement (CQI) program overseen by a CQI committee, which will maintain meeting minutes and ensure that recognized problems will be addressed. The CQI program should include the following 13 components:

1. Closed chart review based on a sample
2. Physics QI program report review
3. Review of cases with dose deviation exceeding 10% or plan second check deviations exceeding 10%
4. Implementation of new modalities or techniques
5. Review of all charts for which an incident report was filed
6. Review of unplanned treatment interruptions
7. Review of outcome studies
8. Individual physician peer review
9. Patient outcomes
10. Retention of records
11. Patient-related outcome data
12. Patient satisfaction and quality-of-life audits
13. Other general information that helps to assure quality

7.4.5 Other Applications in Radiation Therapy

A number of publications in the past 10 years have reported attempts to apply quality methodologies to radiotherapy. The tools investigated include statistical process control and control charts for:

- general external beam QA (Pawlicki, Whitaker, and Boyer 2005; Pawlicki and Mundt 2007; Pawlicki and Whitaker 2008; Able and Bright 2009),
- monitor unit verification (Pawlicki et al. 2012; Nordström et al. 2012),
- IMRT QA (Basran and Woo 2008; Gérard et al. 2009; Pawlicki et al. 2008), and
- HDR brachytherapy QA (Able, Bright, and Frizzell 2012).

One innovative concept is the distinction between tolerance levels which bound the expected range of a parameter (such as linac dose output per MU for reference conditions) based on past measurements when the process is operating normally, versus action limits which bound the acceptable range of the parameter considered to be safe for the desired outcome (Pawlicki et al. 2012). That is, the former term refers to the internal properties of the parameter (how much variation should be expected) while the second term refers to the external requirement to achieve a desired outcome (how much variation we can allow). The tolerance levels must be within the action limits in order to have reliability. This distinction is useful because the situation in which a parameter crosses a tolerance level suggests that something has changed which could eventually cross an action level, but which can be dealt with preemptively so that this latter circumstance does not occur. For example, the linac beam symmetry may exceed the tolerance level, suggesting that it is drifting out of calibration. After tuning the beam magnets, the new symmetry may not be ideal, but it does establish a new mean set point above and below which new tolerance levels can be assigned. This is a much more sensitive approach than simply relying on action levels for ensuring appropriate calibration.

It is essential to look not only at the present values of the important process parameters (e.g. beam output, flatness, and symmetry), but to also look at parameter trending over time. Although this is a common practice for radiotherapy device QA (Klein et al. 2009), by coupling it with tolerance levels, it is possible to distinguish normal statistical variation from abnormal device changes. The resulting plots are called process behavior charts (also known as control charts), described earlier in this chapter.

7.5 Summary

- Many quality improvement tools are described in the literature with unfamiliar vocabularies and methodologies, making it particularly challenging to create sufficient buy-in from the staff to create a work culture to embrace their use.
- While conventional quality assurance is based on specified tolerance limits (e.g, from TG 142), Statistical Process Control introduces the addi-

tional concept of control limits that represent the natural tolerance limits of a process, typically taken as a 3σ variation from the mean and well within the specified tolerance limits.

- The Lean approach to quality, which grew out of the Toyoda Production System, is focused on the elimination of waste. This is accomplished by reducing the rate of errors, reducing unused inventory, reducing any other sources of inefficiency, and focusing on continual improvement.
- Six-sigma is focused on the goal of zero defects as a quality indicator. Six sigma projects accomplish this through a methodology using the following steps: Define, Measure, Analyze, Improve, and Control (DMAIC).
- Early efforts to improve quality in medicine focused on statistical methods to reduce preventable harm, such as the spread of infection. In the twentieth century, the focus shifted to outcomes-based medicine. More recently, many of the tools and frameworks from the quality movement have been applied within health care.
- In the past two decades, many projects have been undertaken to bring quality improvement techniques from industry into radiation therapy. Most of these projects have been to improve quality assurance by using statistical methods, to refine our understanding of quality by bringing in objective metrics, and to introduce quality management frameworks and techniques so as to improve the efficiency of our departments.

References

Able, C.M., M. Bright, B. Frizzell. (2013). "Quality control of high-dose-rate brachytherapy: treatment delivery analysis using statistical process control." *Int J Radiat Oncol Biol Phys* 85:828–33.

Able, C.M., M. Bright. (2009). "Quality control of external beam treatment delivery: mechanical parameters." *Med Phys* 36:2428.

Baldrige Performance Excellence Program. (2011). Health care criteria for performance excellence. Retrieved from http://www.nist.gov/baldrige/publications/hc_criteria.cfm.

Barbosa, E.P., M.A. Gneri, A. Meneguetti. (2013). "Range control charts revisited: simpler Tippett-like formulae, its practical implementation, and the study of false alarm." *Commun Stat Simul Comput*. 42:247–262.

Basran, P.S., M.K. Woo. (2008). "An analysis of tolerance levels in IMRT quality assurance procedures." *Med Phys* 35:2300–2307.

Baudin, M. *Working with Machines: the Nuts and Bolts of Lean Operations with Jidoka.* New York: Productivity Press, 2007.

Baudin, M., D. Woodruff, W. Hopp. (1990). "CONWIP: a pull alternative to kanban." *Int J of Prodn Res* 28:879–894.

Bleehen, N.M. *Quality Assurance in Radiotherapy.* London: Standing Subcommittee on Cancer of the Standing Medical Advisory Committee, 1991.

Bothe, D.R. (2002). "Statistical reason for the 1.5? shift." *Quality Eng* 14:479.

British Institute of Radiology (BIR). *Towards Safer Radiotherapy.* London: Royal College of Radiologists, 2008.

Chassin, M.R. (1998). "Is health care ready for Six Sigma quality?" *Milbank Q* 76:565–91.

Codman, E.A. *A Study in Hospital Efficiency?: as Demonstrated by the Case Report of the First Five Years of a Private Hospital.* Boston: Thomas Todd Co., 1918.

Cua, K.O., K.E. McKone, R.G. Schroeder. (2001). "Relationships between implementation of TQM, JIT, and TPM and manufacturing performance." *Journal of Operations Management.* 19:675–694.

Deming, W.E. *Out of the Crisis.* Cambridge, MA: Massachusetts Institute of Technology, Center for Advanced Engineering Study, 1986.
Donabedian, A. (1966). "Evaluating the quality of medical care." *Milbank Q* 44:166–203.
Donabedian, A. (1989). "The end results of health care: Ernest Codman's contribution to quality assessment and beyond." *Milbank Q* 67:233.
Eisenberg, J.L., J.S. Clark. *The Joint Commission Survey Coordinator's Handbook*, 13th ed. Danvers, MA: HCPro, 2011.
Gérard, K., J.-P. Grandhaye, V. Marchesi, H. Kafrouni, F. Husson, P. Aletti. (2009). "A comprehensive analysis of the IMRT dose delivery process using statistical process control (SPC)." *Med Phys* 36:1275–1285.
Graban, M. and J.E. Swartz. *Healthcare Kaizen: Engaging Front-Line Staff in Sustainable Continuous Improvements.* Boca Raton, FL: CRC Press, 2012.
Harry, T., M. Whitaker, T. Pawlicki. (2012). "Theoretical analysis of the effects uncertainties have on treatment outcomes." *Med Phys* 39:3760.
Hinckley, C.M. *Make No Mistake: An Outcome-Based Approach to Mistake-Proofing.* Portland, OR: Productivity Press, 2001.
Hindle, T. *Guide to Management Ideas and Gurus.* London: Profile, 2008.
Hutchins, D.C. *Just in Time*, 2nd ed. Aldershot, Hants, England: Gower Technical Press, 1999.
Imai, M. *Kaizen: the Key to Japan's Competitive Success.* New York: McGraw–Hill, 1986.
Institute of Medicine. *Crossing the Quality Chasm: A New Health System for the 21st Century.* Washington, D.C.: National Academy Press, 2001.
International Atomic Energy Agency. *Absorbed Dose Determination in External Beam Radiotherapy: an International Code of Practice for Dosimetry Based on Standards of Absorbed Dose to Water.* Vienna: IAEA, 2000.
International Organization for Standardization. *Quality Management Systems: Requirements.* Geneva: ISO, 2008.
Ishikawa, K. *What is Total Quality Control? The Japanese Way.* Englewood Cliffs, N.J.: Prentice Hall, 1985.
Juran, J.M. (1986). "The quality trilogy." *Qual Prog* 19:19–24.
Juran, J.M. *Juran on Planning for Quality.* New York: Free Press, 1988.
Juran, J.M. *Managerial Breakthrough.* New York: McGraw-Hill, 1964.
Kim, C.S., J.A. Hayman, J.E. Billi, K. Lash, T.S. Lawrence. (2007). "The application of Lean thinking to the care of patients with bone and brain metastasis with radiation therapy." *J Oncol Pract* 3:189–193.
Klein, E.E., J. Hanley, J. Bayouth, F.F. Yin, W. Simon, S. Dresser, C. Serago, F. Aguirre, L. Ma, A. Arjomandy, C. Liu, C. Sandin, T. Holmes. (2009). "Task Group 142 report: quality assurance of medical accelerators." *Med Phys* 36:4197–212.
Kohn, L.T., J. Corrigan and M.S. Donaldson. *To Err is Human Building a Safer Health System.* Washington, D.C.: National Academy Press, 2000.
Kubiak, T.M. and D.W. Benbow. *The Certified Six Sigma Black Belt Handbook*, 2nd ed. Milwaukee, WI: ASQ Quality Press, 2009.
Kutcher, G.J., L. Coia, M. Gillin, W.F. Hanson, S. Leibel, R.J. Morton, J.R. Palta, J.A. Purdy, L.E. Reinstein, G.K. Svensson. (1994). "Comprehensive QA for radiation oncology: report of AAPM Radiation Therapy Committee Task Group 40." *Med Phys* 21:581–618.
Leer, J.-W.H., A. McKenzie, P. Scalliet, D.I. Thwaites, and European Society for Therapeutic Radiology and Oncology. *Practical Guidelines for the Implementation of a Quality System in Radiotherapy: a Project of the ESTRO Quality Assurance Committee.* Brussels: ESTRO, 1998.
Lohr, K.N. *Medicare. A Strategy for Quality Assurance*, vol. 1. Washington, D.C.: National Academy Press, 1990.
Massoud, R., K. Askov, J. Reinke, L.M. Franco, T. Bornstein, E. Knebel, C. MacAulay. (2001). A modern paradigm for improving healthcare quality. Quality Assurance Project of USAID. Retrieved from http://www.hciproject.org/node/899.
Mayles, W.P.M. *Physics Aspects of Quality Control in Radiotherapy.* York, UK: Institute of Physics and Engineering in Medicine, 1999.
Montgomery, D.C. *Introduction to Statistical Quality Control*, 7th ed. New York: Wiley, 2012.
Mullen, B., C. Johnson, E. Salas. (1991). "Productivity loss in brainstorming groups: a meta-analytic integration." *Basic Appl Soc Psych* 12:3–23.

Mulley, A.G.J. (1989). "E.A. Codman and the end results idea: a commentary." *Milbank Q*, 67:257–261.
National Committee for Quality Assurance. *The Essential Guide to Health Care Quality.* Washington, D.C.: National Committee for Quality Assurance, 2007.
Nelson-Peterson, D.L., C.J. Leppa. (2007). "Creating an environment for caring using Lean principles of the Virginia Mason production system." JONA 37:287–294.
Nordström, F., S. Af Wetterstedt, S. Johnsson, C. Ceberg, S.J. Bäck. (2012). "Control chart analysis of data from a multicenter monitor unit verification study." *Radiother Oncol* 102:364–370.
Orenstein, B.W. (2009). "Lean management—Wisconsin's ThedaCare improves its radiation oncology care process." *Radiology Today* 10:14.
Osborn, A.F. *Applied Imagination: Principles and Procedures of Creative Thinking*. New York: Scribner, 1957.
Pawlicki, T., A.J. Mundt. (2007). "Quality in radiation oncology." *Med Phys* 34:1529–1534.
Pawlicki, T., B. Chera, T. Ning, L.B. Marks. (2012). "The systematic application of quality measures and process control in clinical radiation oncology." *Semin Radiat Oncol* 22:70–76.
Pawlicki, T., M. Whitaker, and A.L. Boyer. (2005). "Statistical process control for radiotherapy quality assurance." *Med Phys* 32:2777–2786.
Pawlicki, T., M. Whitaker. (2008). "Variation and control of process behavior." *Int J Radiat Oncol Biol Phys* 71(1 Suppl):S210–214.
Pawlicki, T., S. Yoo, L.E. Court, S.K. McMillan, R.K. Rice, J.D. Russell, J.M. Pacyniak, M.K. Woo, P.S. Basran, J. Shoales, A.L. Boyer. (2008). "Moving from IMRT QA measurements toward independent computer calculations using control charts." *Radiother Oncol* 89:330–337.
Phillips-Donaldson, D. (2004). "Gurus of Quality—100 Years of Juran—as his 100th birthday celebration draws near, Joseph M. Juran shares insights and highlights from his long, productive years in quality." *Quality Progress* 37:25.
Pyzdek, T. *The Six Sigma Handbook: A Complete Guide for Green Belts, Black Belts, and Managers at All Levels,* 3rd ed. New York: McGraw-Hill, 2009.
Ross, P.J. *Taguchi Techniques for Quality Engineering: Loss Function, Orthogonal Experiments, Parameter and Tolerance Design.* New York: McGraw-Hill, 1988.
Samuels, M.A., S.A. Rosenthal, B.J. Youngberg, and A.C. Hartford. *ACR Practice Guideline for Radiation Oncology.* Reston, VA: American College of Radiology, 2009.
Scholtes, P.R., B.L. Joiner, and B.J. Streibel. *The Team Handbook*, 3rd ed. Madison, WI: Joiner, 2003.
Semmelweis, I.F. *The Etiology, Concept, and Prophylaxis Of Childbed Fever*, K.C. Carter, trans. Madison, WI: University of Wisconsin Press, 1983.
Shewhart, W.A. *Economic Control of Quality of Manufactured Product*, 2nd ed. New York: D. Van Nostrand Company, Inc., 1931.
Shingo, S. *Zero Quality Control: Source Inspection and the Poka-Yoke System.* Stamford, CT: Productivity Press, 1986.
Spear, S., H.K. Bowen. (1999). "Decoding the DNA of the Toyota production system." *Harvard Bus Rev* 77:96–106.
Spearman, M.L., M.A. Zazanis. (1992). "Push and pull production systems: issues and comparisons." *Oper Res* 40:521–532.
Stroebe, W., B.A. Nijstad, E.F. Rietzschel. (2010). "Beyond productivity loss in brainstorming groups: the evolution of a question." *Adv Exp Soc Psychol* 43:157.
Svensson, G.K., N.A. Baily, R. Loevinger, and R.J. Morton. *Physical Aspects of Quality Assurance in Radiation Therapy.* New York: American Association of Physicists in Medicine, 1984.
Taguchi, G. and Y. Yokoyama. *Taguchi Methods: Design of Experiments.* Dearborn, MI: ASI Press, Japanese Standards Association, 1993.
Taguchi, G. *Introduction to Quality Engineering: Designing Quality into Products and Processes.* Tokyo: Asian Productivity Organization, 1986.
Taguchi, G. *Taguchi on Robust Technology Development: Bringing Quality Engineering Upstream.* New York: ASME Press, 1993.
Tague, N.R. *The Quality Toolbox*, 2nd ed. Milwaukee, WI: ASQC Quality Press, 2005.
Tapping, D., S. Kozlowski, L. Archbold, and T. Sperl. *Value Stream Management for Lean Healthcare.* Chelsea, MI: MCS Media, 2009.

Taylor, F.W. *The Principles of Scientific Management.* New York: Norton, 1967.

Varkey, P., M. Reller, R. Resar. (2007). "Basics of quality improvement in health care." *Mayo Clin Proc* 6:735–9.

Vlachogiannis, J.G. (2003). "Taguchi's technique: an effective method for improving X-ray medical radiographic screen performance." Proceedings of the Institution of Mechanical Engineers, Part H. *J of Eng in Med* 217:375–384.

Wheeler, D.J. and D.S. Chambers. *Understanding Statistical Process Control*, 2nd ed. Knoxville, TN: SPC Press, 1992.

Williams, R.L. *Essentials of Total Quality Management.* New York: Amacom, 1994.

Womack, J.P. and D.T. Jones. *Lean Thinking: Banish Waste and Create Wealth in Your Corporation.* New York: Simon & Schuster, 1996.

Womack, J.P., D.T. Jones, and D. Roos. *The Machine that Changed the World.* New York: Rawson Associates, 1990.

Wysocki, B.J. and S. Lueck. "Just-in-time inventories make U.S. vulnerable in a pandemic." *The Wall Street Journal*, Jan. 12, 2006.

Yee, C.M. "Hard times curtail Park Nicollet's ambition." *Star Tribune*, May 23, 2009.

Yong, J., A. Wilkinson. (2001). "Rethinking total quality management." *Total Qual Manage* 12:247–258.

Appendix 7.1: Standard Notation for Discussion of Statistical Process Control

In the discussion of SPC the following variables and definitions have been used. For reference, see Montgomery's Introduction to Statistical Process Control (Montgomery 2012).

n = number of measurement in each sample
m = number of preliminary samples
x_i = a measurement
$\{x_1, ..., x_n\}$ = a subgroup of measurements of size n
μ = the intrinsic (unmeasurable) mean of the distribution
$\bar{x} = (x_1 + \cdots + x_n)/n$ = mean of the subgroup
$\bar{\bar{x}} = (\bar{x}_1 + \cdots + \bar{x}_m)/m$ = grand mean of the means of the subgroups
σ_x = the intrinsic (unmeasurable) standard deviation of the distribution
$\sigma_{\bar{x}} = \sigma_x/\sqrt{n}$ = the intrinsic (unmeasurable) standard deviation of the measure of $\bar{x}$
$\hat{\sigma}_x = \bar{R}/d_2 = \bar{s}/c_4$ = a measured estimate for σ_x
$R = \max(x_1, ..., x_n) - \min(x_1, ..., x_n)$ = the range of the subgroup
$\bar{R} = (R_1 + \cdots + R_m)/m$ = the mean range of the m preliminary samples
σ_R = the intrinsic (unmeasurable) standard deviation of the range
$d_2 = \bar{R}/\sigma_x$ = the "mean of the range in the sample" divided by the "standard deviation"
$d_3 = \sigma_R/\sigma_x$ = the standard deviation of the distribution of the relative range
$\hat{\sigma}_R = d_3/d_2\,\bar{R}$ = a measured estimate for σ_R
$c_4 = s/\sigma_x$ = the "sample standard deviation" divided by the "population standard deviation"
$s = \sqrt{((x_1 - \bar{x})^2 + \cdots + (x_n - \bar{x})^2)/(n-1)}.$ = an estimate of the standard deviation $\sigma_{\bar{x}}$ of the subgroup
$\bar{s} = (s_1 + \cdots + s_m)/m$ = the average of the standard deviations for the m preliminary samples

$$UCL(x)_{\bar{R}} = \bar{\bar{x}} + 3\hat{\sigma}_{\bar{x}} = \bar{\bar{x}} + 3\hat{\sigma}_x/\sqrt{n} = \bar{\bar{x}} + 3\bar{R}/\left(d_2\sqrt{n}\right)$$
$$CL(x) = \bar{\bar{x}}$$
$$LCL(x)_{\bar{R}} = \bar{\bar{x}} - 3\hat{\sigma}_{\bar{x}} = \bar{\bar{x}} - 3\hat{\sigma}_x/\sqrt{n} = \bar{\bar{x}} - 3\bar{R}/\left(d_2\sqrt{n}\right)$$

$$UCL(x)_{\bar{s}} = \bar{\bar{x}} + 3\hat{\sigma}_{\bar{x}} = \bar{\bar{x}} + 3\hat{\sigma}_x/\sqrt{n} = \bar{\bar{x}} + 3\bar{s}/\left(c_2\sqrt{n}\right)$$
$$CL(x) = \bar{\bar{x}}$$
$$LCL(x)_{\bar{s}} = \bar{\bar{x}} - 3\hat{\sigma}_{\bar{x}} = \bar{\bar{x}} - 3\hat{\sigma}_x/\sqrt{n} = \bar{\bar{x}} - 3\bar{c}/\left(c_2\sqrt{n}\right)$$

$$UCL(R)_{\bar{R}} = \bar{R} + 3\hat{\sigma}_R = \bar{R} + 3(d_3/d_2)\bar{R}$$
$$CL(R) = \bar{R}$$
$$LCL(R)_{\bar{R}} = \bar{R} - 3\hat{\sigma}_R = \bar{R} - 3(d_3/d_2)\bar{R}$$

$$UCL(s)_{\bar{s}} = \bar{s} + (3/c_4)\bar{s}\sqrt{1 - c_4^2}$$
$$CL(s) = \bar{s}$$
$$LCL(s)_{\bar{s}} = \bar{s} - (3/c_4)\bar{s}\sqrt{1 - c_4^2}$$

$$UCL(x)_{\bar{s}} = \bar{\bar{x}} + \left(3/\left(c_4\sqrt{n}\right)\right)\bar{s}$$
$$CL(x) = \bar{\bar{x}}$$
$$LCL(x)_{\bar{s}} = \bar{\bar{x}} - \left(3/\left(c_4\sqrt{n}\right)\right)\bar{s}$$

Chapter 8

Information Management

**Derek Brown, Ph.D.[1], Eric Ford, PhD.[2], A. J. Mundt, M.D.[3]
Sasa Mutic, Ph.D.[4], and Todd Pawlicki, Ph.D.[5]**

[1]Adjunct Associate Professor, Departments of Oncology and Physics and Astronomy, University of Calgary, Calgary, Alberta,

[2]Associate Professor, Department of Radiation Oncology, University of Washington Medical Center, Seattle, Washington

[3]Professor and Chair, Department of Radiation and Applied Sciences, University of California–San Diego, La Jolla, California

[4]Professor and Co-director, Division of Medical Physics, Washington University School of Medicine, St. Louis, Missouri

[5]Professor and Vice Chair of Medical Physics and Technology, Department of Radiation Medicine and Applied Sciences, University of California–San Diego, La Jolla, California

8.1 Introduction

In this chapter we discuss aspects of in-house training, documentation, information technology, and managing change. These components are an integral part of any quality management program, but they are typically absent from most textbooks on the topic. This chapter is meant to introduce the issues and provide the reader with direction for future study.

8.2 In-house Training

To many physicists and others in the field, the phrase "training" is often synonymous with vendor-supplied device training. This type of training, while important, is often a one-time-only affair, typically at the time of roll-out of a new technology or technique. As such, it does not address the need for ongoing training. The discussion here specifically focuses on a structure for training that is developed and maintained by the department itself to meet its own needs. The term training refers to an educational program designed for the purpose of teaching and certifying competencies in individual trainees. For practical purposes, we will adopt ESTRO's definition of competency—"The ability to adequately perform a professional act in a specific environment by integrating knowledge, skills and attitude."

Dunscombe recently sorted through over one hundred recommendations compiled from seven published documents aimed at improving quality and safety in radiotherapy (Dunscombe 2012). In that analysis, the leading recommendation is "training" (Donaldson 2007, Huq et al. 2008, ICRP 2010, Hendee and Herman 2011, ASTRO 2012). In fact, training is the only recommendation that appears in every one of these documents. There is little question that the people and organizations who are thinking about, and publishing papers on, quality and safety in radiotherapy consider training to be a value added endeavor.

There are two main reasons that training and competency certification show up so frequently in these documents. The first is that training and competency certification are helpful in reducing the likelihood of incidents. The second is that they can be used to harmonize clinical practices and improve efficiency. Deming also recognized the importance of training as a means to improve quality, which will be discussed later in this chapter.

These reasons are best illustrated using a simple, realistic example of a typical training program in a typical clinic anywhere in the world. Jack and Jill, two physicists, work in a typical, large, 6-linac clinic. Jill is a seasoned veteran; she's seen it all. Jack is a new hire; he's board certified, but has little clinical experience. Jack has been hired to fill a hole in the brachytherapy program, and Jill has been assigned to get him up to speed. Jill, being the expert, has no problem with this, and she is happy to take Jack under her wing. Over the next few weeks, Jill tells Jack everything she knows about brachytherapy and even walks him through a few cases. At some point, Jill decides that Jack has seen enough and tells their manager that Jack is now competent to perform brachytherapy duties on his own. Sound familiar?

Many of us have experienced this style of informal training, and there are several obvious problems with this familiar model. Only Jill knows what knowledge Jack possesses with respect to brachytherapy. There was no formal documentation to follow during training, so Jill could easily have missed teaching Jack any number of important points. Jack, not given an opportunity to know what he does not know, is left feeling uncomfortable and unsure. Although there is a dearth of studies demonstrating a direct link between training and competency certification and safety/quality in radiotherapy, it would be difficult to argue against the assertion that the afore-mentioned problems increase the likelihood of incidents.

Furthermore, no documentation or training materials were produced or used, so when Jill retires, her vast reserves of knowledge will be lost to the clinic. When Jill trains the next new hire, she's likely to provide slightly different information. That new hire and Jack will now perform the same task slightly differently and, as clinical staffers, they will be less interchangeable. Now we have issues with efficiency and the harmonization of clinical practices.

Training, documentation, and competency certification can reduce the likelihood of incidents by ensuring that:

- managers are accountable for the adequacy and completeness of the training program,
- the staff is accountable for their individual competencies, and
- all staff receive comprehensive, standardized training.

Furthermore, training, documentation, and competency certification can harmonize clinical practices and improve efficiency by ensuring that all staff are trained in the same way using the same information and knowledge is retained when staff retire or change positions.

8.2.1 Establishing a Training Program

An in-house training program can take many forms, depending on the resources and expertise available. An important first step is to define the key components of the program and to build the program around those components. Below is a list of key components that were found useful during the development of a training program (Brown et al. 2011):

- The program should incorporate a multidisciplinary perspective and should foster an understanding of the procedure from various clinical perspectives.
- The expectations of knowledge and skill development should be clearly defined and thoroughly tested at each level of training and certification.
- The program should provide standardized training, such that each individual moving through the training program receives the same information and develops the same skill set.
- The program should be traceable and searchable. Records of which individuals have completed which levels of training and certification must be maintained and must be easily accessible.

Another important aspect is to define the structure of the training program. Will there be multiple levels? How many? Must all competencies be subject to the same structure, or should there be flexibility? Below is an outline of the structure used in an in-house training and certification program:

- Level 1: An introductory level open to all members of the radiation treatment team, including physics graduate students, residents, and therapy students. Serves to provide a basic overview of the procedure on which people are being trained.
- Level 2: Develops knowledge, builds awareness of the relevant literature, and provides the opportunity for supervised observations of procedures, all with defined objectives. Cross-disciplinary observation is introduced at this level.
- Level 3: Develops practical skills through supervised performance of procedures, increases depth of knowledge through focused literature reviews, and assesses well-defined essential competencies.
- Level 4: Maintenance of certification level that entails performance of a minimum number of procedures per year and a commitment to staying abreast of developments in the literature or by attending relevant continuing education sessions.

8.2.2 Implementation

On-line implementations are preferred owing to their ease of access, automated record keeping, and convenient search functions. Other implementations are certainly possible, but they present challenges, primarily in terms of record keeping. Our training and certification at Tom Baker Cancer Centre—including presentations, quizzes, competency checklists, and readings—is managed using Moodle®, an open-source Learning Management System (Moodle Pty Ltd, Perth, Australia, 2013) and can be accessed on-line at http://www.rtp-learning-centre.ca. The Moodle system hosts electronic presentations and enables automated quiz scoring and record keeping, with advanced filters that facilitate determining which individuals have attained which levels of certification.

Our facility's Moodle website is open to everyone. Users must create an account to access content. Users with specific e-mail domains have access to all content, including quizzes and certification checklists. Users who register using other e-mail domains have access to all content except quizzes and checklists. The infrastructure costs are minimal, as many organizations will have ready access to on-line storage, electronic storage, or paper file storage, depending on the type and form of the training program.

Development costs make up the bulk of resource expenditures for the training and certification program. There are two aspects to training development—creating the environment that will house the training program and creating the content that will provide the training. In our experience, the development of a single specialized procedure training and competency certification track (for example, LDR prostate brachytherapy) requires approximately four weeks full time equivalent (FTE). This includes the time spent by the physics staff, radiation oncologists, and therapists generating and reviewing content. Another 2–4 weeks FTE should be allotted for web development if the developer chooses to house the training and certification program on-line.

Maintenance of the training and certification program is less resource intensive, but is extremely important. An out-of-date program will be detrimental to quality, safety, and efficiency. Maintenance costs are more difficult to estimate. Such costs are primarily dependent on the frequency with which the clinical procedure and any associated hardware and software are changed or updated, as well as the extent to which they are changed or updated. A full upgrade that requires a full change in clinical procedures will require nearly as much work as the original content development took, while a minor upgrade that doesn't change clinical practice will require little effort.

There are three primary prerequisites that are mandatory for the development of a training and certification program:

- Adequate resources must be made available to create a high-quality training and certification program, which includes adequate peer review and testing of the program.
- Adequate resources must be made available to maintain the program so it accurately reflects clinical practice.
- Because it is difficult, if not impossible, to develop training and certification programs for clinical processes that are poorly defined, a training and certification program should be used as the impetus for standardizing and improving process descriptions.

In-house training programs provide benefits in terms of standardization of clinical practices, improvements in quality and safety, increases in efficiency, and retention of knowledge. A recent ESTRO publication also argues that the increasing complexity of radiotherapy now warrants training for specialized procedures in radiotherapy (Potter et al. 2012). It is now extremely difficult for a single physicist to be fully competent in all aspects of radiotherapy. The authors recommend that training in radiotherapy should move to a competency-based model. An in-house training and certification program offers improved management of staff competencies in this complex environment.

One obvious extension of this discussion is the consideration of a national or regional training and competency certification program. Such a program offers the benefit of increased standardization of practice and improved review of training materials and competency evaluations, but it suffers from the loss of clinic- or device-specific details. Regional or national training and certification programs may be best aimed at clinics that are unable to develop and sustain in-house training programs. Another possibility is that generalized regional or national training programs could be complimented by less intensive, but more specific, in-house training and certification programs.

8.3 Documentation

There are at least four distinct sets of documents in a typical radiotherapy department upon which this discussion will focus. These include *policies*, *procedures*, *work instructions*, and *documentation* related to software and hardware used for patient treatments. There is also patient-related documentation, but that is not considered in this context. An effective use of documentation requires a clear definition of what individual types of documents are and how they are created, stored, used, and managed.

8.3.1 Policies

There are two types of policies. The first defines rules for organizational operation. These policies typically apply to employees and define things from vacation to hiring practices, for example. These types of policies are not considered here since they are quite standardized and common among different organizations, their major differences being the ways in which benefits are offered between different organizations. The second type of policies define the values and the guiding principles of an organization. These are mini mission statements and contribute to defining the values and priorities in organizational operation. These policies are subsequently used to guide decision-making in a variety of organizational activities as they embody values of senior management. For example, a policy on QA documentation may state the following: "All commissioning, periodic QA, and incidental QA records will be submitted and archived only in electronic form. Hardcopy records will not be acceptable." A policy like this would be approved by several members of the senior administration (clinical director, administrative director, head of physics, etc.). When it comes to decisions regarding funding of software and hardware that would be used to support management of QA documentation in electronic form, this policy would be used to set the priority on such spending. Also, when it comes to implementation and commissioning of new procedures or equipment, this policy would be a consideration in time or resource estimates needed to complete these projects.

8.3.2 Procedures

The second type of documentation is procedures. A procedure is a set of steps to accomplish something. For example, a procedure can detail step by step how an annual QA report is to be generated for electronic submission. Procedures can generally improve the reliability and quality of operations as they provide standardization for work processes and can reduce variability in daily work. The consistency that is offered by following procedures enables reduction in errors, unnecessary work and activities, and provides a common work platform that can be followed and understood by all workers. Procedures are also valuable tools for training, periodic assessment of current practices, and process improvement. While policies generally do not change frequently, procedures tend to be somewhat dynamic, and reviewing all procedures at least annually may be a useful effort.

8.3.3 Work Instructions

The third type of documentation is work instructions. While policies and procedures usually have a formal approval and documentation process, work instructions are much less formal and are often created by individual employees for their personal use. For example, a physicist can have a set of steps that he or she personally does in creating an electronic QA report which go beyond the instructions contained in a similar procedure. While an organization is generally not concerned with work instructions, erroneous or outdated instructions can lead to errors and variability in processes. Furthermore, it is well understood that policies and procedures are discoverable in legal proceedings and outside audits, and the formality in their creation is there in anticipation of such events. Work instruction may also be discoverable and of concern to an outside inspector. To the extent that these do contribute to patient treatments and organizational operation, it is reasonable to expect that these indeed may be seen by outside entities. While it may be impractical to convert all work instructions to formal procedures, employees must understand that these documents could have impact beyond their personal use, and that care should be used in their creation and use.

8.3.4 Software/Hardware Documentation

The fourth set of documents is records on various software and hardware used in a clinic. These records span from user documentation to commissioning reports and periodic QA. Standardization of these documents and their use can greatly lead to consistency in equipment use and the overall reliability and quality of patient treatments.

The creation and maintenance of organizational documentation can be quite complex and formal, with significant resources required to version documents, approve them, and organize them in logical groups that correlate with organizational goals and processes. An example of a comprehensive documentation system is ISO 9000 (ISO 2005). Such systems tend to be impractical and unrealistic for most clinical radiation oncology practices, so something much less ambitious is often more appropriate. Modern document management systems like Microsoft® SharePoint® (Microsoft Corporation, Redmond, WA), Wikipedia® (Wikimedia Foundation, Inc., San Francisco, CA), or Google Docs™ (Google, Inc., Mountain View, CA) offer simple, effective methods to store and archive all departmental documentation. Version control—ensuring that only the current documents are available for everyone—is quite easy with these systems. Because these systems store documents electronically, archiving is very easy, and documents can be stored well beyond any regulatory time limits. Storing and managing electronic documents also requires a policy for electronic signatures, which can also be easily supported with features of modern document creation software like Adobe® Acrobat® (Adobe Systems Inc., San Jose, CA).

There are no clear standards on the number of policies and procedures that a radiation oncology department should have and maintain. In general, the departmental documentation should be sufficient to support the requirements of various accreditation and regulatory bodies, as well as to enable safe and high quality operations. The exact size and scope of departmental documents will ultimately be proportional to a department's activities, resources, and management approach to quality and safety.

8.4 Information Technology

Information technology (IT) plays a central role in the safety and quality of radiation treatments. Some of the first academic reports on safety in radiation oncology focused on the capability of IT-enabled systems to prevent mistreatments in external beam RT (Fraass et al. 1998, Macklis et al. 1998, Barthelemy-Brichant et al. 1999, Patton et al. 2003, Huang et al. 2005, Yeung et al. 2005, Marks et al. 2007). This was a time when record-and-verify systems were coming into use to replace or augment the manual entry of treatment parameters and guide the use of accessories such as blocks and wedges. While one may typically think about IT-enabled automation as being on the treatment delivery end, its use is much more general. Automatic computer scripts in treatment planning, for example, allow complex segmented breast plans to be generated at the click of a button (Reasons 1997). IT connectivity also largely eliminates the need for manual transfer of data and its associated errors. In a broader sense, modern IT systems provide general clinical information about the patient that can improve safety. To take just one example, at least one oncology information system even includes a pop-up alert to warn about patient allergies.

This benevolent capacity of information technology, however, comes alongside the potential to do great harm. The limitations of IT tools must be recognized. One of the main issues is an inherent over-reliance on automation, a problem that has been much discussed in other industries that are heavily reliant on computerization (Reasons 1997). For example, consider a radiation therapy treatment for head and neck cancer circa 1995. The treatment would be delivered with opposed lateral fields to cord tolerance, followed by matched posterior electron fields and photon fields to full dose. Field shapes would be defined by blocks poured out of Litowitz's metal or equivalent. This old-school approach provided many opportunities for radiation therapists to interact with the treatment unit and patient. The therapist would insert blocks and visualize treatment light fields on the patient.

Fast-forward 10 years, and now that same treatment is delivered with IMRT in a process that is heavily IT-dependent. The treatment is highly automated, and the therapist plays the role of overseeing the operation of a machine that runs on autopilot. The fact that this machine is rarely seen to fail makes it difficult for the therapist to maintain vigilance, or to even know what to look for in terms of possible failure modes.

The risk of such automation is dramatically illustrated in the case of Scott Jerome-Parks who died in 2007 as a result of errors in the IMRT treatment for his head and neck cancer (Bogdanich 2010). There were many causal factors at work in this case, but one factor was clearly the technology of IMRT and its underlying IT tools. A computer error resulted in all the multi-leaf collimator segments being deleted and the full IMRT monitor units being delivered through an open field. In an era of less automation, an equivalent gross error in field shaping would have been obvious upon inspection, but in an era of highly automated IT-enabled field shaping, such errors would often not be obvious to the operator.

8.4.1 Error Pathways

New error pathways are the second unintended consequence of information technology being introduced. One such error pathway that has recently received attention is data transfer between systems. In many clinics, systems from multiple vendors are in place for the various steps in the workflow of care, including planning and imaging, treatment delivery, image-guidance, archiving, and even clinical follow-up or possible future retreatment. Single database solutions do not obviate these problems, and these solutions may even make the transfer process more opaque to clinical staff. The World Health Organization report on risk in radiotherapy (Donaldson 2008) suggests that data transfer problems are involved in 9% of all adverse events. The importance of this is underscored by the AAPM task group created to specifically address this: Task Group 201, Quality Assurance of External Beam Data Transfer. An open question remains as to how to best mitigate data transfer errors. Some groups have created software to perform automatic integrity checks on data transfer (Furhang et al. 2009, Siochi et al. 2009, Moran 2011, Yang and Moore 2012, Yang et al. 2012). This software is not yet commercially available. Though this may be viewed as an IT fix to a problem introduced by IT, it is certainly more appealing than the current solution of manual inspection by a physicist, which is time-consuming and prone to error.

IT-enabled automation comes at another price as well: the possibility of propagating errors systematically throughout the entire course of a patient's treatment, or even across multiple patients. An example of this is the recently publicized misadministration of stereotactic radiosurgery in which the jaws of a linear accelerator were set outside the acceptable limits of the cone system. The result was very high doses outside the intended small radiation field area (Bogdanich and Rebelo 2010). Because the jaws were automatically set in the IT system, the error propagated, and three patients received toxic overdoses. In a by-gone era, the jaws may have been set manually, and this error may not have propagated to more than one patient.

8.4.2 IT Safety and Quality

The above discussion focused on the good and bad of information technology as it is used to accomplish clinical treatments. Another aspect of IT related to safety and quality, though, is the use of IT tools directly in a QM program. This is a rapidly developing area, but a few examples include:

- software to enable patient-specific quality assurance checks, such as second check calculations or patient-specific IMRT QA (Moran et al. 2011), including EPID-based portal dosimetry;

- software to track quality assurance results on daily, monthly, or yearly basis;
- in-room patient position monitoring systems;
- automatic checks of data transfer integrity (Yang and Moore, 2012); and
- RFID-based solutions for patient verification and device tracking.

In addition to these tools, software solutions have been developed by a few radiation oncology groups to enable incident learning in the course of clinical operations (Clark et al. 2012, Bissonnette and Medlam 2010, Mutic et al. 2010). These systems, and similar systems on the hospital level, allow any clinical staff to report incidents and manage follow-up.

In summary, information technology has enabled advances in the quality of radiation therapy that simply would not have been possible without it. That power comes at a cost, however. It creates new ways to make errors—some of which are arguably more severe than in an era of less technology—and it fosters a reliance on technology that can backfire. In the long term, errors need to be designed out of IT systems, perhaps informed by widespread industry-wide incident learning. In any case, a strong quality management system needs to be in place to address the known shortcomings of IT-enabled radiation therapy infrastructure.

8.5 Managing Change

Radiotherapy, like any other technology-based field of medicine, changes rapidly. New hardware and software systems are always being developed, and change is a routine part of the field. Change is also an important part of a robust quality and safety improvement program where issues are identified, remediation proposed, and changes implemented. Many tools and techniques of quality and safety improvement have been described throughout this text. However, it is not sufficient to simply know the tools and techniques to manage change. One must know the strategies of successful change management to have a fully realized quality and safety improvement program. In this section, aspects of managing change are covered.

There are different types of change that range from *standardization*, to *incremental*, to *transformative* (Hacker 2012). Standardization is used to reduce variation and codify existing processes. Standardization is a cornerstone of using Lean and Six Sigma for quality improvement. In most cases, standardization is achieved without encountering significant resistance to change. However, if standardization requires someone to change his or her preferred way of working, then difficulties in the change process can arise.

Incremental change efforts are on a smaller scale than transformative change, but larger than simple standardization efforts. Incremental change is usually targeted at fixing or modifying a narrow problem or procedure. If a department is using an incident learning system, then they are most likely dealing with incremental change on a regular basis.

Another category of change is transformative change. Transformative change is broad and large. It is targeted at modifying the fundamental process structures of systems. In any change situation, there are components or hurdles that are encountered. While some hurdles are unique to a specific environment, many hurdles are universal. Knowing those hurdles in advance, with some simple approaches to mitigate those hurdles, can greatly facilitate an organization's change efforts (Garside 1998).

Change is not accomplished by a managerial declaration, "We must change!" Change is a process. Change happens when the *cost* of change (*C*) is less than the combined cost of impeding factors. The main impeding factors are the *need* to change (*N*), the *end state* of the change (*E*), and the *strategy* to affect the change (*S*). This is neatly described by the following inequality, where *C*, *N*, *E*, and *S* are the components of change. When the cost to individuals or the organization is greater than the components of change (*need*, *end state*, and *strategy*), then change is very difficult to achieve.

8.5.1 Approaches to Change

No single approach to change is best in all situations, but at least two aspects must be in place for change to occur: 1) a technical strategy, i.e., how we'll get this done given our current infrastructure and manpower, and 2) a cultural or organizational strategy. There are some common steps to achieve change that should be considered regardless of the type of change to be achieved (Varkey and Antonio 2010). The readiness for change must be assessed before the change process, which includes assessing the interest in change, the perceived chance of success, and the degree of management support. For larger change efforts (transformational change), other issues come into play, such as aligning the change with the organization's values (whether explicitly written or only implied) and adequate financial resources (time, money, and human capital).

Once this initial step has been completed, then a vision for the change needs to be created, together with a sense of urgency. The leader (person or group requiring change) must articulate a clear, credible vision with standards and enthusiasm. Answers to the questions, "Where are we heading?" and "Why is the project important and critical to do?" should be provided. This is also the time to clarify roles, responsibilities, and expectations. This may seem overly complicated and burdensome for incremental change efforts, but these actions can be very useful in motivating change, even for the smallest change efforts. Actively enabling change is important. The leader must provide necessary resources to those doing the work, rewards for new behaviors, as well as to ensure attention is focused on essential elements and deliverables. A steering team can be very helpful as it can plan, coordinate, and monitor progress toward the end state of the change effort.

Team members can be chosen for their ability to embrace change or, perhaps counter intuitively, because they are resistant to change. Including persons who are widely known to be resistant to change can ultimately create broad buy-in if others see the resistor accepting the change. In any case, members of the team should have appropriate credibility to carry out the change, not just technical knowledge about the problem at hand. Team members should have a systems thinking approach to change, meaning that each team member should understand and appreciate the connectedness of all areas that the change will impact (Garside 1998). Motivation to work on the change project is important, but credibility of the team members is equally important.

The steering team can then create an implementation plan. The plan should include tasks or deliverables at each phase of the plan, and these should be reported to the leader or group requiring change. This should be done at intervals frequent enough so that the leader can quickly make adjustments so as not to derail the change project. Adjustments could include financial, time, or manpower resources, or simply motivational support. The leader of change may be required to demonstrate excitement and communicate early successes, or to be visible and public with commitment to the change. The leader may also be required to identify and mobilize the commitment of key stakeholders, i.e., those that the change will impact the most.

Sometimes these types of efforts are a challenge for those with a medical or technical background, but resources are available to facilitate improvement in those areas (Blanchard and Johnson 1982). For both large and small change projects, it can be helpful to do a pilot test of the change. This can be done in a Plan, Do, Check, Act (PDCA) cycle as described in previous chapters of this text. A PDCA approach can help the steering team identify and fix unforeseen problems with the change before it is widely implemented. Another benefit of PDCA and pilot testing is that it provides a vehicle for promoting early, small wins in the change process—meeting milestones shows that the group is achieving the end state. Lessons learned from the pilot test can be used to effectively disseminate the change to the wider group or organization.

Lastly, the change must be institutionalized and made part of everyday life. This may require baseline data, and the results of the change process may have to be tracked, widely communicated, or regularly audited to ensure that the change has taken hold.

8.5.2 Resistance to Change

There have been volumes written on resistance to change and overcoming that resistance at both the organizational level (Harvard Business Review 2011) and the personal level (Johnson 1998). Resistance to change comes in many forms, such as unclear or unfocused end state (*E*), inappropriate change strategy (*S*), parochial self-interest, office politics, historical inertia ("we've always done it this way"), misunderstanding, or even lack of trust. Effectively dealing with these types of resistance is important to achieving successful change.

It is important for leaders and the steering team to realize that change is typically not a linear process. The steering team and change leader must be flexible with the method being used to achieve change and listen to the key stakeholders. Resistance is a natural reaction to change; it shouldn't be explicitly discouraged, but managed. The steering team and leader must understand the source and concerns of resistance. Resistance might not be explicitly expressed, but it may be manifested in the actions of people that the change is affecting. Passive-aggressive behavior is a change resistance strategy. Resisters should not be viewed as adversaries, but as opportunities to make lasting change. It can be helpful to have empathy for resisters by acknowledging that change is difficult. A constant reiteration of why change is happening will encourage participation in the change.

Another strategy for dealing with resistors is manipulation through the carrot and the stick. Penalties for not adopting change can be effective, but only to a point, and then penalties can have a reverse effect by generating resentment. A stroke (in the field of transactional analysis) is a compliment, which can be negative or positive. As just mentioned, negative strokes should be used with caution, as too many can have a reverse effect. In terms of positive strokes, there are two types: strokes for being and strokes for doing. It is always better to provide strokes for doing rather than for being. In the end, it is important to keep in mind that an appropriate balance between rewards and penalties can facilitate a change effort.

The last, and least effective, strategy for dealing with resistors is coercion, i.e., "the boss rules." Under no circumstances does this approach create lasting change.

8.5.3 Effective Meetings

Comments on managing change would not be complete without input on conducting effective meetings. Most meetings are run poorly with inadequate facilitation, little or no decision-making, no closure, and little or no follow-up. Attendees to meetings need to know what they are coming to, why they are there, and their role in the meeting. First is to have a clear purpose for the meeting (e.g., create a plan, make a decision, measure progress, share information, solve a problem, etc.). A vague purpose also hampers attendee participation because attendees are not sure about what is expected of them. Meetings that have multiple purposes should be delineated in the meeting, i.e., it should be stated when one purpose has been completed and another has begun ("We've updated everyone on X, and now we're going to make a decision about Y").

Sending an agenda with time limits helps attendees stay focused on the task at hand and also prevents the purpose of the meeting from being derailed. A facilitator (not the meeting leader or recorder) can be very helpful to start and end the meeting on time, as well as to summarize the purpose(s) of the meeting once it is achieved. The meeting facilitator should keep the group focused on the issues, guide the flow of the discussion, and remain neutral. The meeting leader has the responsibility for setting the purpose(s) of the meeting, focusing on the outcomes of the meeting, and directing members of the group. A recorder simply captures the ideas, produces the meeting minutes, and distributes them after the meeting.

Occasionally after a meeting, especially standing meetings, it is good practice to evaluate whether or not the meeting achieved its purpose(s). For example, were the objectives clear, was the agenda managed, was the purpose met, and were all the attendees necessary and included in the meeting? It is important to remember that poorly run meetings can be a point of stress for attendees, which may, in turn, make the process of change management that much more difficult.

In summary, change is hard, but strategies exist to facilitate change. Implementing change successfully is not just about identifying changes and applying tools and techniques to make it happen. It also concerns applying the right change management approach that is most relevant to the context. Understanding and applying the simple techniques outlined in this section can help in the task of managing change to improve quality and safety.

8.6 Summary

In this chapter we covered the aspects of in-house training, documentation, information technology, and managing change. A quality management program will contain all these items.

Training

- There is little question that the people and organizations who are thinking about and publishing papers on quality and safety in radiotherapy consider training to be a value added endeavor.
- Training, documentation, and competency certification can reduce the likelihood of incidents by ensuring that managers are accountable for the adequacy and completeness of the training program, that the staff is accountable for their individual competencies, and that all staff members receive comprehensive, standardized training.
- A training program should:
  - incorporate a multidisciplinary perspective and foster an understanding of the procedure from various clinical perspectives;
  - clearly define the expectations of knowledge and skill development;
  - provide standardized training, such that each individual moving through the training program receives the same information and develops the same skill set; and
  - be traceable and searchable.

Documentation

- Policies define the values and the guiding principles of an organization and can facilitate decision-making and time/resource estimates when it comes to implementing and commissioning new procedures or equipment.
- A procedure is a set of steps to accomplish something and can generally improve the reliability and quality of operations as they provide standardization for work processes and can reduce variability in daily work.
- Work instructions are often created by individual employees for their personal use, and while an organization is generally not concerned with work

instructions, erroneous or outdated instructions can lead to errors and variability in processes.

- Standardization of hardware/software documents and their use can lead to greater consistency in equipment use and the improved reliability and quality of patient treatments.

Information Technology

- Information technology (IT) plays a central role in the safety and quality of radiation treatments.
- The limitations of IT tools must be recognized. One of the main issues is an inherent over-reliance on automation. A second unintended consequence of IT is the new error pathways that are potentially introduced.
- IT tools may be incorporated directly into a QM program. A few examples include:
    - software to enable patient-specific quality assurance checks, such as second check calculations or patient-specific IMRT QA, including EPID-based portal dosimetry;
    - software to track quality assurance results on a daily, monthly, or yearly basis;
    - in-room patient position monitoring systems;
    - automatic checks of data transfer integrity; and
    - RFID-based solutions for patient verification and device tracking.

Managing Change

- One must know the strategies of successful change management to have a fully realized quality and safety improvement program.
- In any change situation, there are hurdles that will be encountered. While some hurdles are unique to a specific environment, many hurdles are universal. Knowing those hurdles in advance, with some simple approaches to mitigate them, can greatly facilitate the change efforts.
- No single approach to change is best in all situations, but at least two aspects must be in place for change to occur: 1) have a technical strategy, i.e., how we'll get this done given our current infrastructure and manpower, and 2) have a cultural or organizational strategy.
- Actively enabling change is important. The leader must provide necessary resources to those doing the work. He or she must also provide rewards for new behaviors so as to ensure attention is focused on essential elements and deliverables.
- For both large and small change projects, it can be helpful to do a pilot test of the change. This can be done in a Plan, Do, Check, Act (PDCA) cycle as described in previous chapters of this text.
- Resistance to change comes in many forms, such as unclear or unfocused end state, inappropriate change strategy, parochial self-interest, office pol-

itics, historical inertia ("We've always done it this way"), misunderstanding, or even lack of trust. Effectively dealing with these forms of resistance is important to achieving successful change.

References

American Society for Radiation Oncology (ASTRO). (2013). "Target Safely." Available at: https://www.astro.org/Clinical-Practice/Patient-Safety/Target-Safely/Index.aspx. (Accessed February 13, 2013.)

Barthelemy-Brichant, N., J. Sabatier, W. Dewe, A. Albert, and J.M. Deneufbourg. (1999). "Evaluation of frequency and type of errors detected by a computerized record and verify system during radiation treatment." *Radiother Oncol* 53:149–154.

Bissonnette, J.P. and G. Medlam. (2010). "Trend analysis of radiation therapy incidents over seven years." *Radiother Oncol* 96:139–144.

Blanchard, K. and S. Johnson. *One Minute Manager.* New York: William Morrow and Company, Inc., 1982.

Bogdanich, W. "Radiation Offers New Cures, and Ways to Do Harm." *New York Times*, Jan. 23, 2010.

Bogdanich, W. and K. Rebelo. "A pinpoint beam strays invisibly, harming instead of healing." *New York Times*, Dec. 28, 2010.

Brown, D., P. Dunscombe, Y. Bayliss, and M. Udowicz. (2011). "A multidisciplinary competency certification program for specialized procedures in radiation therapy." *Med Phys* 38:3829.

Clarke, B.G.B., R.J. Brown, J. Ploquin, and P. Dunscombe. (2012, in press). "Patient Safety Improvements in Radiation Treatment through Five Years of Incident Learning." *Pract Radiat Oncol.*

Donaldson, L. *Radiotherapy Risk Profile.* Geneva: World Health Organization Press, 2008.

Dunscombe, P. (2012). "Recommendations for safer radiotherapy: what's the message?" *Front Oncol* 2:129.

Fraass, B.A., K.L. Lash, G.M. Matrone, S.K. Volkman, D.L. McShan, M.L. Kessler, and A.S. Lichter. (1998). "The impact of treatment complexity and computer-control delivery technology on treatment delivery errors." *Int J Radiat Oncol Biol Phys* 42:651–659.

Furhang, E.E., J. Dolan, J.K. Sillanpaa, and L.B. Harrison. (2009). "Automating the initial physics chart checking process." *J Appl Clin Med Phys* 10(1):129–135.

Garside, P. (1998). "Organizational context for quality: Lessons from the fields of organizational development and change management." *Qual Health Care* 7 (Suppl):S8–15.

Hacker, S.K. (2012). "Change ability: individual transformation is the key to organizational improvement." *Quality Progress* 45(8):16–20.

Harvard Business Review. *On Change Management.* Boston: Harvard Business School Publishing Corporation, 2011.

Hendee, W.R. and M.G. Herman. (2011). "Improving patient safety in radiation oncology." *Med Phys* 38:78–82.

Huang, G., G. Medlam, J. Lee, S. Billingsley, J.P. Bissonnette, J. Ringash, G. Kane, and D.C. Hodgson. (2005). "Error in the delivery of radiation therapy: results of a quality assurance review." *Int J Radiat Oncol Biol Phys* 61:1590–1595.

Huq, M.S., B.A. Fraass, P.B. Dunscombe, J.P. Gibbons, G.S. Ibbott, P.M. Medin, A. Mundt, et al. (2008). "A method for evaluating quality assurance needs in radiation therapy." *Int J Radiat Oncol Biol Phys* 71:S170–S173.

International Commission on Radiological Protection (ICRP). *Preventing Accidental Exposures from New External Beam Radiation Therapy Technologies.* Report 112. Amsterdam: Elsevier, 2010.

International Organization for Standardization (ISO). *ISO 9000:2005 Quality Management Principles.* Geneva: ISO, 2005.

Johnson, S. *Who Moved My Cheese?* New York: G.P. Putnam's Sons, 1998.

Macklis, R.M., T. Meier, and M.S. Weinhous. (1998). "Error rates in clinical radiotherapy." *J Clin Oncol* 16:551—556.

Marks, L.B., K.L. Light, J.L. Hubbs, D.L. Georgas, E.L. Jones, M.C. Wright, C.G. Willett, and F.F. Yin. (2007). "The impact of advanced technologies on treatment deviations in radiation treatment delivery." *Int J Radiat Oncol Biol Phys* 69:1579–1586.

Moodle Pty Ltd. Perth, Australia. www.moodle.com.

Moran, J.M., M. Dempsey, A. Eisbruch, B.A. Fraass, J.M. Galvin, G.S. Ibbott, and L.B. Marks. (2011). "Safety considerations for IMRT: Executive summary." *Med Phys* 38(9):5067–72.

Mutic, S., R.S. Brame, S. Oddiraju, P. Parikh, M.A. Westfall, M.L. Hopkins, A.D. Medina, et al. (2010). "Event (error and near-miss) reporting and learning system for process improvement in radiation oncology." *Med Phys* 37:5027–5036.

Patton, G.A., D.K. Gaffney, and J.H. Moeller. (2003). "Facilitation of radiotherapeutic error by computerized record and verify systems." *Int J Radiat Oncol Biol Phys* 56:50–57.

Potter, R., J.G. Eriksen, A.W. Beavis, M. Coffey, C. Verfaillie, J.W. Leer, and V. Valentini. (2012). "Competencies in radiation oncology: a new approach for education and training of professionals for radiotherapy and oncology in Europe." *Radiother Oncol* 103:1–4.

Purdie, T.G., R.E. Dinniwell, D. Letourneau, C. Hill, and M.B. Sharpe. (2011). "Automated planning of tangential breast intensity-modulated radiotherapy using heuristic optimization." *Int J Radiat Oncol Biol Phys* 81:575–583.

Reasons, J.T. *Managing the Risks of Organizational Accidents.* Burlington, VT: Ashgate Publishing, 1997.

Royal College of Radiologists, Society and College of Radiographers, Institute of Physics and Engineering in Medicine, National Patient Safety Agency, British Institute of Radiology. *Towards Safer Radiotherapy.* London: The Royal College of Radiologists, 2008.

Siochi, R.A., E.C. Pennington, T.J. Waldron, and J.E. Bayouth. (2009). "Radiation therapy plan checks in a paperless clinic." *J Appl Clin Med Phys* 10:43–62.

U.S. Department of Veterans Affairs. *Healthcare Failure Mode and Effect Analysis.* Available at http://www.patientsafety.gov/CogAids/HFMEA/index.html# page=page-1.

Varkey, P. and K. Antonio. (2010). "Change Management for Effective Quality Improvement: A Primer." *Am J Med Qual* 25:268–73.

Yang, D. and K.L. Moore. (2012). "Automated radiotherapy treatment plan integrity verification." *Med Phys* 39:1542–1551.

Yang, D., Y. Wu, R.S. Brame, S. Yaddanapudi, D. Rangaraj, H.H. Li, S.M. Goddu, and S. Mutic. (2012). "Technical note: electronic chart checks in a paperless radiation therapy clinic." *Med Phys* 39:4726–32.

Yeung, T.K., K. Bortolotto, S. Cosby, M. Hoar, and E. Lederer. (2005). "Quality assurance in radiotherapy: Evaluation of errors and incidents recorded over a 10 year period." *Radiother Oncol* 74:283–291.

Chapter 9

Radiation Therapy Quality Management Programs

Bruce Thomadsen, Ph.D.[1], Ellen Yorke, Ph.D.[2], Jeffrey Williamson, Ph.D.[3], Jatinder Palta, Ph.D.[4], Saiful Huq, Ph.D.[5], Geoffrey S. Ibbott, Ph.D.[6], and Sasa Mutic, Ph.D.[7]

[1]Professor, Departments of Medical Physics, Industrial and Systems Engineering, Engineering Physics, and Biomedical Engineering, University of Wisconsin, Madison, Wisconsin

[2]Attending Physicist, Medical Physics Department, Mem1orial Sloan Kettering Cancer Center New York, New York

[3]Professor of Radiation Oncology, Department of Radiation Oncology Virginia Commonwealth University, Richmond, Virginia

[4]Professor and Chair of Medical Physics, Virginia Commonwealth University, Chief Physicist, VHA National Radiation Oncology Program, Richmond, Virginia

[5]Professor and Director, Department of Radiation Oncology, Medical Physics Division, University of Pittsburgh Cancer Institute Pittsburgh, Pennsylvania

[6]Professor and Chair, University of Texas MD Anderson Cancer Center Houston, Texas

[7]Professor and Co-director, Department of Radiation Oncology, Washington University School of Medicine, St. Louis, Missouri

9.1 Introduction

This chapter presents three examples to illustrate the application of the concepts discussed in this book. Because of limitations of length, each example will consider particular aspects of radiotherapy. In general, that would be the way to address design of quality management since looking at the whole program in a facility likely would become a project too ponderous to come to completion.

9.2 Quality Management for Radiotherapy (Ellen Yorke)

This book has discussed many quality management tools and many strategies for implementing them in individual clinics. In this section, we review the application of some of these newer methods to the general external beam radiotherapy process, with examples drawn partly from literature and partly from author Ellen Yorke's experiences. The time-honored physics QA activities—commissioning new machines, devices, and planning systems—are very important, but they are not part of this section. We know how important these activities are for patient care and how failures in these areas can have severe implications for many patients (IAEA undated, Bogdanich and Rebelo 2010). Many reports from AAPM and other organizations provide guidance on how to accurately perform such QM. There are also many papers on how to perform this work efficiently, and innovative physicists and vendors supply ideas and devices to aid these efforts.

In the last decade, it has become evident that the physicist's role in assuring safe and high-quality treatments goes beyond the traditional physics QA of periodic machine and chart checks. Many older physicists—who trained when the field was smaller and interactions between physicists, physicians, and therapists were closer—have appreciated this on an intuitive and informal level for a long time. As

Table 9–1. Process tree for emergency (on-call) treatments in my department

Patients with CT simulation	Patients without CT simulation
Consult with attending MD/resident – History and physical	Consult with attending MD/resident – History and physical
Decision to treat/treatment site(s); includes MD review of previous RT	Decision to treat/treatment site(s); includes MD review of previous RT
RT notations made in hospital database	RT notations made in hospital database
Immobilization at simulator (CT scan performed, setup information documented, alignment marks placed)	Immobilization at treatment machine (localization radiographs); setup information documented, alignment marks placed
Prescription by attending MD (disease site_ Rx point, dose/fraction, # fractions, energy, specify beam arrangement)	Prescription by attending MD (disease site_ Rx point, dose/fraction, # fractions, energy, specify beam arrangement)
Approval of field apertures (DRRs or radiographs) by attending MD	Approval of field apertures (radiographs at treatment machine) by attending MD
Separation and other relevant patient-specific measurements at simulator or from CT scan	Separation and other relevant patient-specific measurements at treatment machine
MU calculation (first calculation by therapist, check by resident)	MU calculation (first calculation by therapist, check by resident)
Entry of treatment information into Aria (therapist)	Entry of treatment information into Aria (therapist)
Patient set up at machine	Setup finalized
Radiograph(s) acquired, saved to database, approved (attending or resident)	Radiograph(s) acquired, saved to database, approved (attending or resident)
Treatment delivered	Treatment delivered
On first working day after initial treatment, all aspects of the case are reviewed by a dosimetrist or physicist; attending physician approves radiographs if not already done.	On first working day after initial treatment, all aspects of the case are reviewed by a dosimetrist or physicist; attending physician approves radiographs if not already done.

a green post-doc from a college physics teaching background and training in a small, but lively, academic center, I attended the 1982 AAPM summer school. C. J. Karzmark gave a lecture on the then cutting-edge technology of conventional simulators in which he emphasized the importance of communication between all the members of the therapy team—a lesson that has resonated with me through my whole time in medical physics. Clear lines of communications, clear procedures, clear documentation, understanding, and learning from our mistakes—these ideas are not new. What is new is the widely publicized, structured effort to make these and related concepts as much a part of our professional life as daily output measurements.

9.2.1 Process Mapping

The ASTRO document "Safety is no Accident" gives an overview of these ideas, including suggestions for applying the concepts to one's own workplace (Zietman et al. 2012). For each clinical process of interest, it is useful to lay out the process steps, preferably within a small group that includes representatives of all the disciplines involved, as discussed in Chapter 4 of that document. There are cases where these groups are all physicists (or all therapists). There are other cases where physicists, physicians, therapists, nurses, administrators, and technical staff from other departments (e.g., diagnostic radiology) would need to be involved (Ford et al. 2009, Terezakis et al. 2011, Perks et al. 2012, Denny et al. 2012, Ciocca et al. 2012, Sawant et al. 2010, Novak et al. 2005, Israelski and Muto 2004, Thomadsen et al. 2003). The process tree (or process map) is often presented in the form of a flow chart. Figure 1 in Ford et al. (2009) or Figure 1.1 in Zietman et al. (2012) and reproduced here as Figure 9–1, are process maps for the entire external beam process. Task Group 100 designed a process tree for IMRT which, not surprisingly, also covers the overall external beam process (Huq et al. 2013). The process map does not need to be an elaborate graphic exercise—a list, in typical chronological order, also works. For the chosen format, filling in individual steps within a larger section helps focus on where the process may be weakest and most susceptible to errors. Table 9–1 is a process map for emergency (on-call) treatments as they are delivered at my institution.

9.2.2 Failure Modes and Effects Analysis

Failure Modes and Effects Analysis (FMEA) is a method of prospectively analyzing the weak points of a process (see Section 4.3). Ford et al (2009) published an early application of this technique to the total external beam enterprise, and this method is described in several of chapters. Other FMEAs for the whole radiation process or selected subprocesses include Terezakis et al. (2011), Perks et al. (2012), Denny et al. (2012), Ciocca et al. (2012), Sawant et al. (2010), Novak et al. (2005), Israelski and Muto (2004), Thomadsen et al. (2003), and Huq et al. (2013).

Ford et al. identified 127 failure modes in their first analysis (Ford et al. 2009) and 159 in a later one (Terezakis et al. 2011). Denny et al. identified 108 (2012) and TG 100 identified 216 (Huq et al. 2013). It is likely that no matter how many failure modes are identified, others lurk unidentified in the background.

As in Section 4.3, once the failure modes are identified, they are ranked on a scale of 1 to 10 according to their likelihood of occurrence (O), potential severity (S), and difficulty in detection (D), and the Risk Probability Number (RPN) is calculated as their product. The higher the RPN, the more hazardous the failure mode is and the more important it is to reduce that RPN by making it less likely to occur or easier to detect. Orderly procedures, careful commissioning of devices and software, and adequate staff training and device interlocks are some methods that reduce "O". Periodic physics QA, chart checking, peer review, and staffing levels that are adequate to maintain therapist vigilance are some methods that help convert failures from harmful events into close calls (also called near misses)—failures that do not reach the patient and, thereby, reduce D. We are stuck with "S" as it is.

Table 9–2 is a simple FMEA example—an analysis of the single step of calculating the monitor units for a parallel opposed, isocentric treatment (e.g., AP/PA) with the prescribed dose to be delivered to "midplane." This is the most common emergent or "on-call" treatment in my department. The monitor units (MUs) are calculated by an on-call therapist using an in-house computer program and checked by a medical resident using manual calculation from tabulated beam data. If there is an unresolved discrepancy between the two calculations of more than a few MU

(~2%), a senior, supervisory therapist is contacted to resolve the difference. Mandatory in-service training is given to the therapists and residents before they are allowed to take calls. All steps in the on-call treatments are checked by a dosimetrist on the next working day: this is a QA measure that can catch many other failure modes, not merely those associated with the calculation of the MUs. However, depending on the start day and holiday schedules, a patient could have up to four treatments based on on-call work. For this reason, the maximum allowed dose per

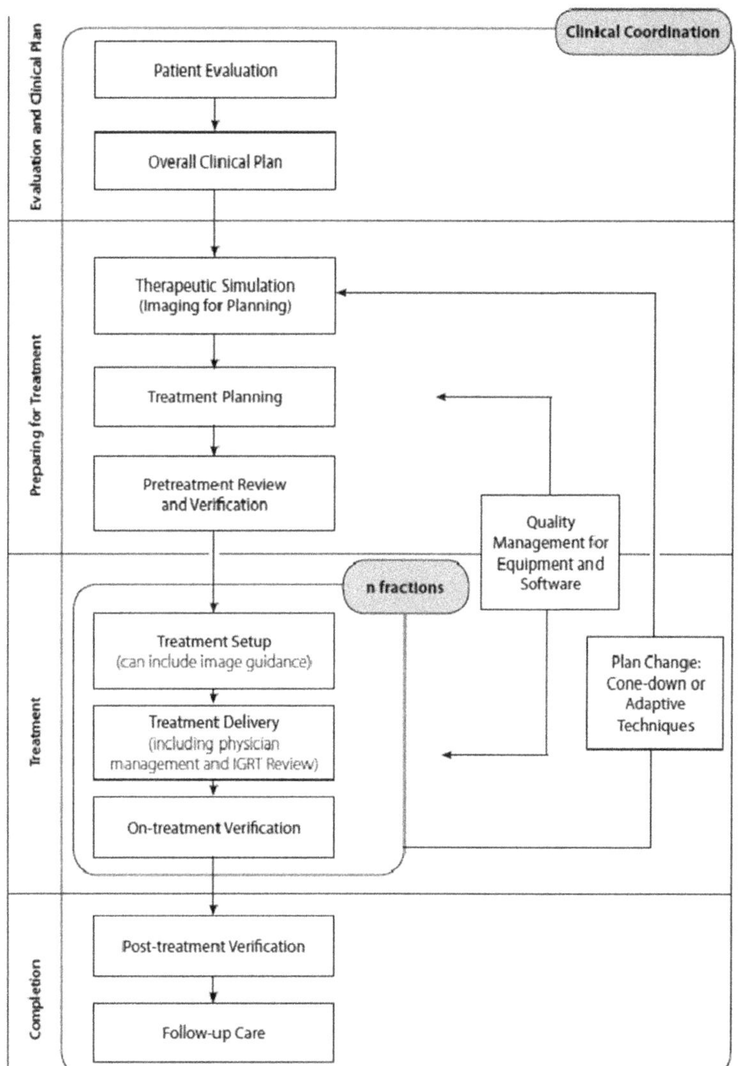

Figure 9–1. The process of care for external-beam radiation therapy. Reproduced with permission from Zietman et al. 2012.

Table 9–2. Example FMEA for on-call monitor unit calculations for parallel-opposed fields treated on-call in author's department. The isocenter and prescription points are always at mid-depth in these cases. This table covers MU calculation risks only. Delivery failure modes (e.g. treat at wrong isocenter) are not included.

Failure	O	S	D1	D2	RPN1	RPN2	Comments
Calculate MU for information from wrong patient	2	8	6	3	96	48	
Calculate MU for wrong beam energy	3	6	4	2	72	36	
Calculate MU for wrong depth (e.g. look up TMR for separation, not sep/2)	4	8	4	2	128	64	
Dose doubling (Calculate MU to deliver total Rx dose from each beam)	3	10	3	1	90	30	
Calculate the equivalent square for the wrong collimator setting	5	3	5	1	75	15	Risk with independent jaws
Error in calculating equivalent square (but collimator setting correct) for Sc	4	2	1	1	8	8	Error can only occur in the manual check calculation
Error in calculating equivalent square for Sp and TMR	4	2	3	1	24	8	Program allows error because the user can input an estimated equivalent square with blocking
Omission of transmission factor for blocks or support devices	5	1	5	2	25	10	
Failures to follow procedure							
MUs used for treatment without second check (failure to follow procedure)	2	9		1		18	These are high potential S
Large discrepancy accepted without senior therapist adjudication	2	9		2		36	because they negate safety mechanisms in place
On-call case not given to dosimetrist before treatment on first working day	2	9		6		108	Might be detected in 'weekly' checks chart rounds

fraction for on-call calculation is set at 4 Gy. On-call treatments are either single-field or parallel opposed; wedges are not used. Finally, several years ago, the department decided that all on-call photon treatments would be isocentric, because of the risk of large errors caused by misuse of the inverse square law in calculations done under severe time pressure.

The information provided for the MU calculation is the prescription (dose/fraction, number of fractions, beam energy), the treatment machine (currently, nine are in use), the collimator settings, a film image or DRR showing the irradiated aperture if blocks or static MLC are to be used, and the patient separation at isocenter and, for very asymmetric fields, at the middle of the irradiated area. Table 9–2 shows the failure modes I identify. You might well find others (and if so, I'd appreciate knowing about them). I used the ranking scale from Ford et al. (2009), which has been publicly available for several years. S assumes the failure continues unmitigated for the entire treatment course. D1 rates detectability on-call (before any treatment is delivered), while D2 rates detectability when the dosimetrist reviews the calculations or afterwards (e.g., on weekly checks or chart rounds). In an effort to relieve stress and allow the on-call personnel to concentrate on the major issues, as of this year they are instructed to ignore transmission factors and to use the equivalent square for the open field to calculate S_P and TMR for all photon fields except whole brain, where they are instructed calculate S_P and the TMR for 16 x16. For very asymmetric fields, they are instructed to look up the TMR for the measured mid-depth at the irradiated field center rather than at isocenter, without worrying about inverse square law corrections. On review, the dosimetrist corrects any resulting minor (usually <2% to 3%) differences in MU caused by these deliberate simplifications. At the physician's discretion, dose adjustments may be made to the remaining treatments. The dosimetrist review is vital in reducing the risk of on-call treatments. This review is done under relatively leisurely conditions by highly experienced personnel and is much more likely to catch major calculation errors and failures in the overall process than the rushed review by a stressed resident or therapist.

The riskiest failures in Table 9–2 are the failures to follow procedure and to make sure that each on-call case is brought to a dosimetrist's attention. To mitigate this potential failure, the department could require that a pause be scheduled in the treatment management system after three to four treatments, because extending the prescription beyond the pause would trigger a complete dosimetrist review. But the honor system has worked to date, so this has not (yet) been done. It is important to emphasize that Table 9–2 covers only MU calculation failure modes. Prescription, simulation, and treatment failures (e.g., treatment at wrong isocenter) are not included. Emergent treatments can have vulnerable processes, and the recently reported actual error and near-misses of one large institution are of interest (D'Souza et al. 2012).

The FMEA approach is useful, but it has several problems, including the following:

1. *There is a definite learning curve for the approach.* This book should help get interested individuals part of the way up the slope.
2. *FMEA requires a substantial commitment by the group doing the analysis.* Members of a clinical department can rarely spend full time on the analysis; they must also carry on with their normal clinical obligations. There must also be an administrative commitment to support group recommendations. Terezakis et al. report that the Johns Hopkins University FMEA was done by a representative committee of 14 people who met over a one-year period (2006–2007) and revisited the analysis in 2009. They estimate that this took "170 hours of staff time" (presumably the time was distributed over all the individuals and the total analysis period). Experience may increase efficiency; Ford et al. (2012) describe a more efficient organizational methodology, building on what was learned in the initial efforts. Denny et al. (2012) describe FMEA for a four-hospital network where the group involved two corporate facilitators and interdisciplinary teams of six to seven individuals from each hospital. The project took approximately eight months to complete, though the total person-hours were not given.

Analysis times are not widely reported and, in general, it is important that a departmental FMEA committee be allowed enough time to address the process of interest in a thorough and comprehensive manner.

3. *The scoring—particularly for O and D—is very variable*. Whether one thinks a failure mode is very rare or only infrequent depends on one's overall experience. Figure 9–2 shows the scores using the Task Group 100 scoring system, which is different from that of Ford et al. (2009), assigned by nine of the Task Group 100 physicists to "failure to detect incorrect linac operation" in the absence of conventional physics QA. The entire task group agreed that this is a high-risk failure, but the RPN varied by more than a factor of two between evaluators.
4. *FMEA's can be hard to track*. Several authors note that the complexity of the radiation oncology process makes it hard to keep track of the FMEA for the entire system (Terezakis et al. 2011, Denny et al. 2012).
5. *Despite dedicated brainstorming efforts, it is unlikely that all possible failure modes will be identified, even if no new technology is introduced to the clinic*. Terezakis et al. (2011) reported that "42% of actual errors reported…were not identified as potential errors in the FMEA." They give several reasons for this. One was that the time spent was insufficient, noting "it is not surprising that all possible events were not conceptualized or anticipated using FMEA in only 170 staff hours." Another is "failure of imagination"—even clinicians with years of experience have not seen everything. Task Group 100 emphasizes that FMEA should be viewed as part of an ongoing and iterative process wherein a department continues to

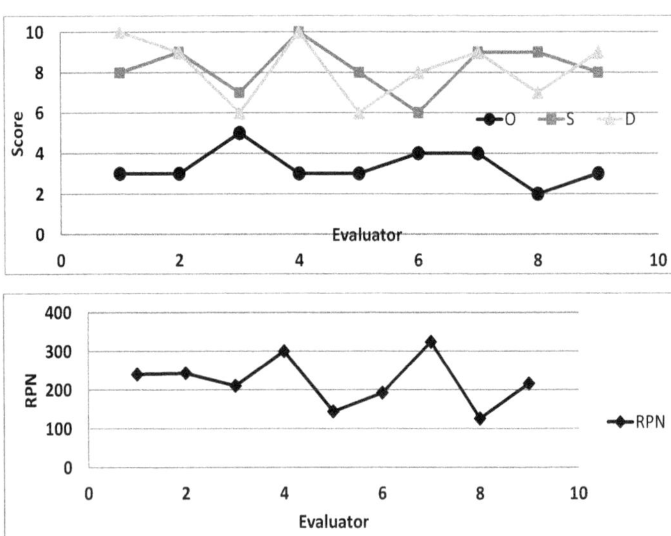

Figure 9–2. The variation in FMEA scores for the nine Task Group–100 evaluators.

evaluate the effectiveness of the mitigating measures it has put into place and strives to correct previously unsuspected failure modes as they are uncovered. Of course, the introduction of new technology—the norm in modern day radiation oncology—is a major perturbation, deserving analysis of its own (Perks et al. 2012, Sawant et al. 2010, Novak et al. 2005).

9.2.3 Incident Reporting

As discussed in Chapter 6, official reporting of medical events is an important aspect of regulatory practice both nationally (Nuclear Regulatory Agency) and within individual states. Regulatory agencies mandate reporting if a patient receives a radiation treatment that disagrees with prescription by a moderate to large extent, independent of whether there is overt clinical harm. The FDA's Center for Devices and Radiological Health (CDRH) maintains a Manufacturers and User Facility Device Experience database (MAUDE) of voluntarily reported adverse events that involve medical devices. Most regulatory databases are open to public search, although it isn't always easy for an interested individual to drill into them for specific information about a scenario or product. There are also databases outside the United States, such as the worldwide Radiation Oncology Safety Information System (ROSIS) (ROSIS and Cunningham et al. 2012) and overall reports such as that of the Health Protection Agency in the United Kingdom (RCR 2008) and the International Atomic Energy Agency (IAEA). Additionally, many individual radiotherapy facilities encourage the reporting of errors (both actual errors and near misses) and maintain a departmental database and use this information to devise error-prevention strategies and to inform and educate the entire treatment team about specific situations and mitigating strategies. A number of institutions have published their experiences with such efforts (Tereszakis et al. 2011, Mutic et al. 2010, Bissonnette and Medlam 2010, Clark et al. 2010, Marks et al. 2007, Macklis et al. 1998, Hunt et al. 2012). In-house incident reporting with an accompanying follow-up study lends itself well to root cause analysis and provides an informed basis for devising mitigating strategies. Also, should an institution wish to do an FMEA, a local database provides information about failure modes observed in the department, giving evidence-based starting points for estimates of their O and D. Recently, consensus recommendations as to how to structure an "incident learning database" has been published under the leadership of the AAPM's Working Group on Prevention of Error (WGPE) (Ford et al. 2012). These recommendations apply to both departmental efforts and a future national database.

Typically (fortunately) the large majority of situations from in-house reports are close calls (also called 'near misses'), minor deviations, or events that affect only a small percent of an individual patient's treatment course and, once discovered, can be corrected later in the treatment course, at least insofar as the total dose distribution is concerned. Unfortunately, as the number of patients receiving hypofractionated treatments increases, radiation therapy becomes less forgiving.

9.2.4 One Individual's Experience with Incident Learning

I find that a combination of departmentally encouraged error reporting (including near misses), structured review of reported cases—including informing an entire group—and continuing creation and evaluation of mitigating strategies appropriate to the local workflow is very helpful. The Memorial Sloan-Kettering Cancer Center (MSKCC) reporting system in past and in current use has been described by Hunt et al. (2012), together with an analysis of the reported errors that made it through to patient treatment over the past decade. In understanding the workflow, it is important to recognize that two independent departments are involved in radiation therapy at MSKCC. The physicians and therapists are in the Radiation Oncology Department, while the physicists and dosimetrists report to the Medical Physics Department. Currently and historically, computerized treatment planning is done on an in-house system. The current treatment management system (TMS) is Varian's Aria (Varian Medical Systems, Palo Alto, CA). There are two Philips (Andover, MA) and one General Electric (GE Healthcare, Yizhuang, China) CT simulators. All the treatment machines are Varian, though of different generations, ranging from an approximately 15-year-old 600C to two TrueBeams. For patients who have a computerized treatment plan (graphic dose distributions, dose-volume histograms), treatment information is electronically downloaded from the planning system to Aria. But field information is manually entered for cases (such as emergent cases) that are not computer planned. Electronic prescription (Aria Patient Manager) has been used for about a year. At the New York City campus, the External Beam Treatment Planning section, consisting of approximately 30 physicists and dosimetrists, is responsible for planning, dosimetric evaluation of previous treatments, MU calculation, and transferring and checking all treatment-related information into the TMS. This section also commissions new planning system modules, physically checks complex patient setups, consults at simulation regarding immobilization and motion management, performs *in vivo* dosimetry, and performs the weekly review of patient charts. The attending physicians are responsible for creating the prescription (initial and final treatment directives), delineating target volumes, delineating or reviewing the delineation of normal organ volumes, providing information to the planners about previous treatments, and approving the final treatment plan. The therapists have primary responsibility for simulation, treatment setup, and treatment delivery. There is also a group of physicists responsible for machine QA and beam data commissioning. So far, they have not been involved in any event reports in my department, though, of course, in the future they could be involved in treatment "failures." In 2012, the main campus was treating approximately 210–220 patients/day on nine machines.

The therapists, physicians, and planners are encouraged to report events and close calls through a hospital-wide program called RISQ (Reporting to Improve Safety and Quality) which uses a commercial system RL6:Risk (RL Solutions, Toronto Canada). The Treatment Planning section has responded enthusiastically to this project. Based on RISQ reports, a joint committee of the two departments, as

Table 9–3. Reported errors in author's department from1/2012 to 7/20/12. The 'actuals' were caught at a treatment; none led to a dose error exceeding 10% over the treatment course.

Failure	Close Call	Actual	Due to Multiple Systems
Minor prescription deviations (e.g. electronic chart format not followed)	38		
Incorrect data in Aria (electronic and manual entry)	27		8
Other planning issues (wrong volumes, beams, constraints, plan quality)	25		2
Incorrect isocenter in setup field information	19	1	3
Delays/inefficiency	19		
Major prescription deviations	10		
Incorrect use of Transmission Factors in MU calculation	7	1	
Incorrect other documentation	7		
Insufficient data in Aria (e.g. compensator mount or alert for bolus)	6		2
Other incorrect MU calculation	5	2	
Incorrect re-creation of or information about prior treatments	4		
Erroneous patient measurements	3		
Software bugs In-house	3		
Prescription changed without notification	2		
Cerrobend blocks	2		
Incorrect DRRs	1		1
Incorrect information given to treatment machine (setup instruction, transcription)	1	1	1
Software bugs _ Commercial systems	1	2	
Wrong patient		1	
New technology problems		1	
Prescribed treatment schedule not followed		1	
Total in each column category	**180**	**10**	**17**

well as the Treatment Planning section on its own, implements mitigating measures. At staff meetings each month, the Treatment Planning Section Chief reviews the relevant RISQ reports for the planners. This is an important consciousness-raising exercise that makes the group keenly aware of the sorts of things that can happen; it certainly increases my vigilance. Table 9–3 summarizes the RISQ reports discussed at the staff meetings between January and August of 2012. Of note, most of the minor prescription deviations were related to physicians' learning curves relative to our newly introduced electronic prescription sheet format; these failures were usually caught by the planner or the plan checker.

New error-mitigation procedures adopted in response to RISQ reports are reviewed in staff meetings and documented on a shared Treatment Planning Section folder on the department server. Measures adopted in responses to the failures in the shaded rows of the table are listed below:

1. Wrong target volumes: For most patients, the target volumes are delineated on the contouring module of the Pinnacle treatment planning system (Philips Healthcare, Andover, MA). The physician notifies the planner when the contours are finished, and the planner transfers the image set and contours to the planning system. Immediately after transfer, the planner is now required to 'lock' the dataset so that it cannot be inadvertently changed without planner awareness and physician approval.
2. Delays and inefficiencies are long-standing problems in many departments, including ours. Delays are not merely annoying—they often lead to rushed planning where errors can occur or plan quality can deteriorate. A new workflow has just been implemented that we believe will assure adequate time for complex planning. The planning 'clock' now starts after receipt of the physician's contours, rather than right after simulation.
3. Manual patient separation (thickness) measurements are routinely performed at the simulation for total body irradiation patients (TBI). Because of the high potential severity of incorrect MU calculations based on erroneous measurements in TBI treatments, mid-pelvic and neck separations are now independently remeasured at the first treatment.
4. It is our continuing practice that all in-house software bugs are promptly reported to the Medical Physics Department's Computer Service for analysis and correction.

Several mitigating measures adopted before the time period of Table 9–3 include:

1. Using the 'compensator mount' feature in ARIA as a 'hard' reminder to the therapists to place bolus. The TrueBeam bolus indicator is different, leading to two close calls in Table 9–3.
2. Implementing 'smart' checklists, with features that are specific to each common disease site and to each treatment machine. The planner fills out the checklist for each patient. The expectation is that this serves as a guided time-out procedure before the plan goes on to the next stage (physics plan check) and thus reduces the incidence of some common errors.
3. To prevent disastrous 'wrong isocenter' errors for hypofractionated treatments when two anatomical sites (two isocenters) are planned from the same simulation session, the two sites are treated sequentially, unless there is a very urgent medical reason for concurrent treatment. In either case, the two hypofractionated plans are placed in different ARIA courses. The benefits of this approach were recently revealed in the context of simple (e.g., parallel opposed, conventionally fractionated) treatments such as two AP/PA pairs with one centered at the level of T8 and the other at the level of L2. In this setting, different courses are not used. Rather, the two beam

sets are entered into the TMS as two separate plans within the same course. However, in one of the 'actual' failures noted in Table 9–3, auto-sequencing brought up a Site 2 field immediately following treatment of the Site 1 PA field, even though these fields were in different plans. Although the therapist noticed the error after only a few MUs, this was an actual failure. Its mitigation (beyond telling therapists not to use auto-sequencing for such cases) has not been finalized. This was not a failure that I, at least, would have foreseen in an FMEA prior to its actual occurrence.

The group culture is particularly important (see Chapter 2). No blame is associated with reported errors, and staff meetings and reviews are handled as a shared learning experience. The planners are encouraged to question the accuracy of the input to treatment plans whenever it is unclear, to seek advice and raise questions with senior planners about any problems they encounter, and to notify the group (and the computer service) about 'bugs' or odd software behavior.

9.2.5 Summary

While this kind of reactive incident-learning approach can keep a group very much on its toes, it has a downside in that most of the corrective measures are directed at a specific failure and are developed only after that failure mode surfaces. This sometimes gives the user (me) the sense of swatting gnats—and too many gnats (minor close calls) can distract one from a hornet. While FMEA is properly criticized for the difficulty in keeping track of various hypothetical failures, a purely reactive approach makes it hard to keep track of failures already encountered in one context that pop up later in another context.

Perhaps it is possible to find an optimal combination of the prospective (FMEA-style) and the reactive approaches. As a simple example, the largest number of close calls in Table 9–3 involve minor problems with prescriptions. These errors cannot be ignored because related errors may potentially have major consequences in other cases. For example, suppose an MD has not formally approved a prescription, and this is noticed and reported to the RISQ database with regard to a disease site where a standard prescription applies to most patients. Or suppose that the beam energy has not been specified on the prescription form, although the physician has approved a graphic treatment plan that uses a particular energy. In either of these cases, one could imagine an odd situation where the usual prescription does not apply or the physician might re-think the original planned beam energy. A "smart" prescription sheet would reject an incomplete prescription, just as many "smart" Internet order forms do not allow a purchase to go through until all necessary fields are filled out.

A great deal of work remains in assuring the safety and quality of patient treatment. Hopefully, we will individually, and as a profession, be able to build on the foundations laid by this summer school.

9.3 Failure Modes and Effects Analysis (FMEA) for Accelerated Partial Breast Irradiation Delivered via High-Dose-Rate Intracavitary Brachytherapy (Jeffrey F. Williamson, Bruce R. Thomadsen, Geoffrey S. Ibbott, and Sasa Mutic)

9.3.1 Introduction

As noted in the introductory chapters of this volume, the main focus of medical physics quality management efforts is monitoring and controlling the technical performance of planning and delivery devices by means of prescriptive quality control/assurance (QC/QA) protocols that specify fixed arrays of tests, outcome tolerances, and frequencies designed to comprehensively test devices such as linear accelerators and remote afterloaders. However, there is growing recognition that errors in the associated clinical processes in which the devices are embedded pose more significant threats to the safety and quality of radiation therapy than failure of the devices themselves (see Section 1.3). Extending the prescriptive quality management (QM) approach to modern radiotherapy processes is daunting due to the high degree of variability in clinic-to-clinic implementations of technologically complex modalities such as intensity-modulated radiation therapy (IMRT) or image-guided high dose-rate (HDR) brachytherapy (Huq et al. 2008, Huq et al. 2013, Williamson et al. 2008). Such a wide variability in process implementation and risk associated with even common generic process steps requires a much higher degree of customization of QM strategies that can only be achieved by those with intimate knowledge of the processes themselves. To solve the problem, Task Group 100 ("application of risk analysis methods to radiation therapy quality management") sought to introduce the radiation oncology community to a suite of generic industrial engineering tools that could be used to analyze any clinical process, leading to the identification of the highest risk steps and the development of risk-informed quality management programs that reflect work practices in individual clinics. TG 100 illustrated this method by performing an analysis and constructing a QM program for a generic IMRT process. This chapter presents the application of the TG–100 methodology to a generic image-guided HDR brachytherapy process.

The TG–100 method consists of four basic steps:

1. Process mapping.
2. Performing a failure modes and effects analysis (FMEA) to identify the highest risk steps and potential failure modes of the process.
3. Performing a fault tree analysis (FTA) to better understand how errors propagate from one step to another to result in a deviation from quality.
4. Design and placement of QC and QA tests to reduce propagation of errors.

FMEA is a formal process that the treatment team can use to identify and describe failure modes characteristic of their process, and then semi-quantitatively rank them in terms of the risk they pose to patient safety and treatment efficacy. As reviewed in more detail by TG 100, FMEA has been used to significantly enhance quality in other areas of medicine—including critical care medicine (Duweet et al.

2005), chemotherapy administration (Sheridan et al. 2006), and anesthesia (Pate-Cornell et al. 1997)—and to guide implementation of advanced intravascular pump technology (Wetterneck et al. 2006).

9.3.2 Materials and Methods

9.3.2.1 The Clinical Scenario: Accelerated Partial Breast Irradiation Via an Implanted Multi-Channel Inflatable Intracavitary Applicator

To make this analysis clinically realistic, a relatively narrow patient population is defined who are candidates not only for breast conservation therapy (BCT), but accelerated partial breast irradiation (APBI) using an inflatable "balloon" intracavitary applicator. Candidate patients undergo lumpectomy and are then referred to radiation oncology to receive treatment to the tissue within 1 cm of the resection-cavity boundaries. Eligibility criteria recommended by ASTRO include age ≥60, unicentric and unifocal T1 lesion less than 2 cm diameter with clear margins (≥2 mm), negative sentinel lymph node biopsies or axillary lymph node dissection, low risk histology, and no neoadjuvant therapy. Relative contraindications are large lesion size, positive margins, unfavorable histology, and positive regional lymph nodes. Technical contraindications include skin-to-balloon distance <5 mm, poor conformance as indicated by trapped volume of air exceeding 10% of the clinical target volume (CTV, often called 'PTV_EVAL' in the literature), and deviations from symmetry exceeding 2 mm (Smith et al. 2009).

We assume that a multiple-channel inflatable applicator (see Figure 9–3) is used such as the Contura applicator that comes in two sizes: 4–5 cm and 4.5–6 cm diameter when inflated. We assume that the applicator is inserted by the radiation oncologist in a separate procedure several weeks after lumpectomy using the "scar entry technique" (SET). In this technique, the lumpectomy cavity is visualized using intraoperative ultrasound, and a single 1 cm incision is made, usually along the original lumpectomy scar, to communicate with the cavity (Zannis et al. 2005).

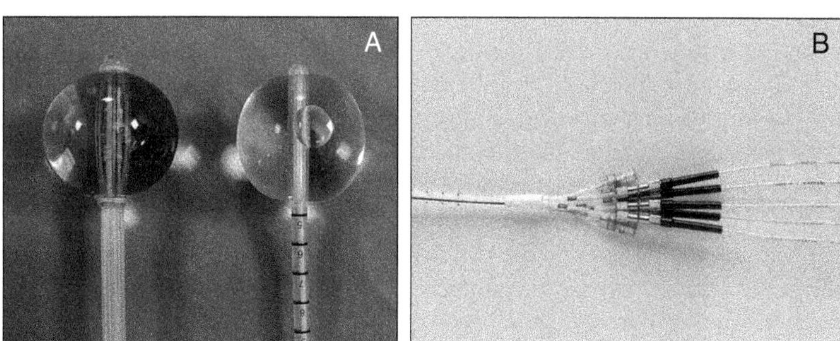

Figure 9–3. A) Photographs of inflated Contura multiple catheter (left) and Mammosite single-catheter (right) applicators. B) Illustrates connecting the five Contura applicator catheters and the Varian VariSource HDR remote afterloader. For this study, a Contura applicator system was assumed.

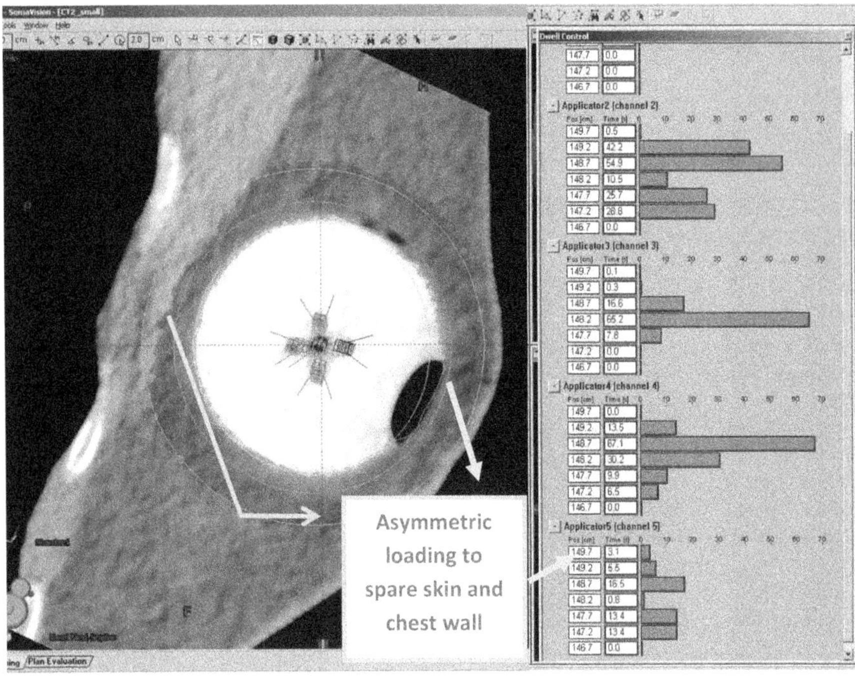

Figure 9–4. Illustration of how the peripheral channels of the Contura applicator are differentially loaded to diminish the isodose surface diameter along the anterior-posterior axis (sparing skin bridge and chest wall) relative to the medial-lateral diameter. The figure also illustrates how the CTV is edited to omit the pectoralis muscle and subcutaneous tissue.

Seroma drainage is usually seen, confirming cavity entrance. The uninflated applicator is then inserted under ultrasound guidance to ensure symmetric positioning within the cavity. The applicator is then inflated with a 1:10 contrast/saline solution and again imaged with ultrasound to confirm good conformance with the cavity walls and to ensure that the "skin bridge" thickness, usually at least 5 mm, is sufficient. Following this process, the patient is taken to the CT to acquire the image set to be used for treatment planning. At this time, the catheter size and orientation will be documented. In addition, the Contura internal catheters will be sounded, with or without the transfer tubes attached, depending on the department's protocol, and the measurements recorded so that the treatment length (distance from indexer reference plane to first programmable dwell position) can be determined. Finally, the radiation oncologist reviews the images in order to assess the adequacy of the placement. The criteria are: good conformance with the cavity walls, applicator symmetry, opposition of the side channels with skin bridge, and acceptable (>5 mm) skin bridge thickness.

We assume a relatively modern department with partial integration of HDR brachytherapy planning and delivery into the departmental electronic medical record (EMR or electronic chart) and computer-controlled radiation therapy (CCRT) systems. CT image sets are stored in the EMR system, which establishes the links between imaging data, plans, patient identifiers, and other demographic data. These datasets are electronically imported from the EMR into the HDR brachytherapy planning system. Other than the skin contour, which is created by the CT, we assume that the target organ at risk (OAR) and avoidance structures are created locally from contouring performed in the planning system and, along with any treatment plans, are electronically stored in the EMR database. All brachytherapy files stored in the EMR database are indexed with respect to patient-identifying information and other demographic data. Any HDR plan from current or past patients can be downloaded electronically from EMR into the HDR remote afterloader (RAL) console computer, which will correctly display the patient's name and medical record number. However, unlike external beam treatments, we assume that the RAL is not integrated into the EMR scheduling system, so that HDR plans are not associated with individual fractions or proposed/completed courses of treatment. Individual treatment fractions are indexed with patient ID and plan only in the RAL console computer once the plan has been imported from the EMR.

The model clinical process assumes that the planner—either a dosimetrist or a physicist rather than the physician—contours the applicator and creates the CTV and avoidance structures according to a rigid protocol illustrated by Figure 9–4. The GTV is assumed to be the surface of the balloon applicator itself. The CTV consists of an isotropic expansion (typically 10 mm) of the GTV minus the applicator, a 5 mm expansion of the skin contour, and the pectoralis muscle. In the literature, CTV is often referred to as "PTV_EVAL." Other avoidance structures needed to constrain the dose distribution may be formed as needed. It is assumed that the attending radiation oncologist needs to approve (and presumably check) the contours prior to commencing treatment planning. Standard planning goals are assumed: at least 90% of the CTV is to receive 90% of the prescribed dose with maximum skin dose ≤125%; rib dose ≤145%; and upper limits on V150 and V200 of 50 and 10 cm^3, respectively (Arthur et al. 2011). Inverse planning maybe used, along with graphical tools for shaping isodose surfaces. As with the TG–100 FMEA for IMRT, we assume that the process has no QA other than the MD approvals referenced in the process tree (see Figure 9–5).

9.3.2.2 FMEA and FTA Methodology

The four coauthors of this section, each of whom is an experienced brachytherapy physics practitioner, participated in this study. In addition, Douglas Arthur, MD, provided invaluable guidance in structuring this process and giving a deeper understanding of the potential clinical failure modes. Originally, this activity was conducted as part of the TG–100 deliberations. However, it was never completed and was abandoned by TG 100 in 2006.

The first step was to perform process mapping which, in the case of this exercise, amounted to defining the process. With feedback from all authors, the process tree was initially developed by Thomadsen and then modified by Williamson. It

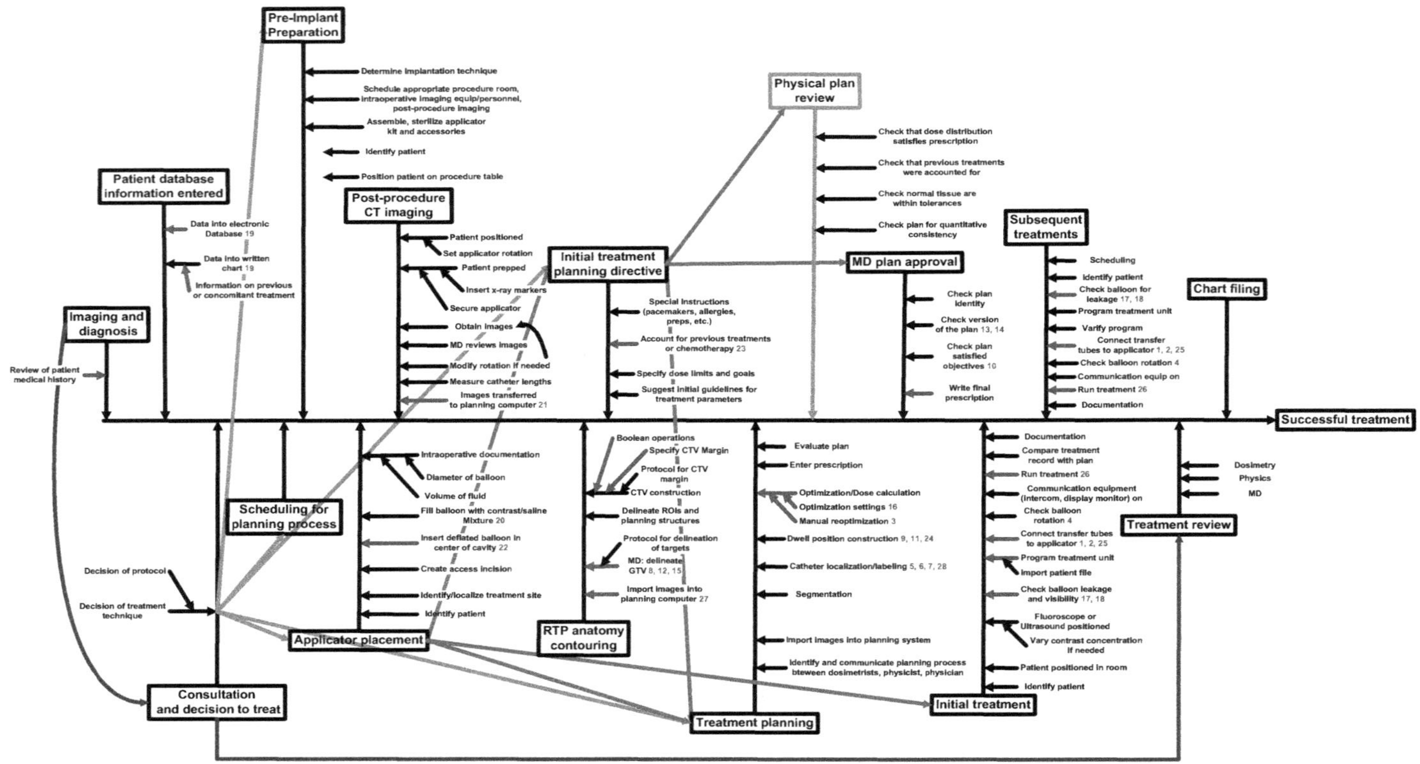

Figure 9–5. Process map of the generic HDR intracavitary APBI process. The lines indicate important causal connections mediated by information flow. Colored lines (see CD) denote flow of general clinical and diagnostic information, translation of clinical intent and decisions into technical treatment variables, and flow of technical data acquired from step to step. No physical plan review was assumed.

Table 9–4. Failure mode subprocess (bough) and the process step (branch) identification

Item Number	Major Processes	Step	Potential Failure Modes	Potential Causes of Failure	JFW Comments and Descriptive Scenario	Potential Effects of Failure	AVE O	AVE S	AVE D	AVE RPN
1	Imaging and diagnosis	RO reviews EMR prior to RO consult	Med Onc or surgeon consultation misinterprets or misrepresents primary clinical findings (imaging studies, path reports, etc), incorrectly stages patient, and recommends BCT and APBI for patient that is not appropriate candidate.	RO bases Tx recommendation on secondary MD report rather than reviewing primary clinical findings and discovering the upstream error.	Upstream physician error potentially discoverable by Rad Onc since primary clinical data is available. We should recommend that the RO performs its duties diligently.	Wrong/very wrong dose distribution	5.00	8.25	5.50	269.3
2	Imaging and diagnosis	RO reviews EMR prior to RO consult	Path or biomarker reports are incorrect due to mislabeling of surgical specimen or biomarker report. Hence, patient is understaged and inappropriately offered BCS by Med Onc and surgeon.	RO recommendation for APBI is fully consistent with prior EMR	An error not easily discoverable by Rad Onc based on the worse case	Very wrong dose	4.25	8.75	8.25	309.5
3	Patient database information	Entry of patient data in RO EMR or written chart	Incorrect patient ID data	Documentation error	Wrong patient ID leading misfiling of demographic and clinical data from hospital DB; identification of wrong patient	Very wrong dose	3.00	8.75	2.75	70.0
4	Patient database information	Entry of patient data in RO EMR or written chart	Correct patient ID data but clinical findings/images from wrong patient loaded into RO EMR	Omission in entry, incomplete patient history	Incorrect clinical findings leads to faulty decision to treat or downstream peer-review correction	Very wrong dose	5.00	7.75	3.75	154.5

Table 9–4, cont. Failure mode subprocess (bough) and the process step (branch) identification

Item Number	Major Processes	Step	Potential Failure Modes	Potential Causes of Failure	JFW Comments and Descriptive Scenario	Potential Effects of Failure	AVE O	AVE S	AVE D	AVE RPN
5	Consultation and decision to treat	Decision of treatment technique and protocol	Clinically inappropriate patient selected for APBI	Misinterpreting of clinical findings incomplete H&P	Even though upstream clinical data are correct, error by RO assessing indications and contraindications to APBI, e.g., SLN+ with surg untreated axilla. RO misrepresents or neglects key finding and offers inappropriate treatment plan to patient	Very wrong dose	4.25	7.75	7.75	252.8
6	Consultation and decision to treat or imaging/ diagnosis	Decision of treatment technique and protocol or imaging/ diagnosis	Patient with radiographically too large or closed seroma cavity selected	RO error in interpreting imaging studies, inappropriate imaging used, or poor imaging quality	JFW: New failure mode	Very wrong dose if not detected; more likely major inconvenience or infection from unnecessary invasive procedure	4.75	6.25	4.75	140.3
7	Consultation and decision to treat	Decision of treatment technique and protocol	Clinically inappropriate patient selected for APBI	Upstream error in documenting clinical findings. Various scenarios with differing levels of discoverability	Because of upstream imaging and diagnosis errors, RO offers inappropriate treatment plan to the patient	Very wrong dose	4.00	7.50	5.50	180.0
8	Pre-implant preparation	Patient identification	Wrong patient identified	Human error, lack of communication; different treating and consulting Ros	Wrong patient setup in procedure room for applicator placement	Very wrong dose	3.50	7.00	5.00	114.0

Table 9–4, cont. Failure mode subprocess (bough) and the process step (branch) identification

Item Number	Major Processes	Step	Potential Failure Modes	Potential Causes of Failure	JFW Comments and Descriptive Scenario	Potential Effects of Failure	AVE O	AVE S	AVE D	AVE RPN
9	Pre-implant preparation/ applicator placement	Applicator selection	Defective applicator selected	No pretreatment/ intraoperative QA	New JFW FM: leaking applicator, asymmetric/ misshaped applicator, blocked applicator blocked, broken interface	wrong dose distribution, patient inconvenience	4.75	5.50	6.00	154.8
10	Applicator placement	Identify/ localize tylectomy cavity	Incorrect position or shape	Personnel inadequately trained	Technical contraindication missed: selected patient has partially closed seroma cavity that is too inflexible to conform to balloon surface or cavity too close to ribs or to skin.	Wrong dose distribution or volume; inconvenience and possibly unnecessary infection due to canceling procedure	5.50	6.00	5.00	153.0
11	Applicator placement	Identify/ localize tylectomy cavity	Incorrect position or shape	Personnel inadequately trained or inexperienced	JFW Scenario: poor quality or misinterpreted US causes MD to create scar entry technique (SET) channel that misses closed seroma cavity	Very wrong dose, location or volume	4.25	8.25	5.25	201.0
12	Applicator placement	Create access incision	Incision not near treatment site	Inattention to detail, personnel inadequately trained	Access incision for SET channel made in inappropriate location	Minor injury; risk of infection from unnecessary incision. If this happens, it would not be detected until after the fact.	4.25	4.00	5.00	84.0

represents a composite of the clinical processes implemented at the University of Wisconsin and Virginia Commonwealth University.

Process Tree. Figure 9–5 shows the process tree that encompasses the major steps of the APBI treatment process described above. The central horizontal trunk of the tree, running from left to right, denotes increasing time, starting with staging and workup of the patient. It continues with entry into the Radiation Oncology EMR and ends with the completion of treatment. The main boughs, representing the major subprocesses, are arranged in approximately chronological order. Smaller "branches" emerging from each bough detail the steps required to execute the subprocess represented by the bough. In principle, each branch could be further composed into twigs and leaves, providing ever finer descriptive detail. In general, neither the coauthors nor the TG–100 membership felt it necessary to subdivide beyond the "branch" level of detail.

Some of the subprocesses may not be completely obvious. "Imaging and Diagnosis" refers collectively to all staging, medical history, disease history, and treatment information accumulated in the hospital EMR prior to the patient seeking consultation in radiation oncology (RO). This would include initial exams, mammographic and ultrasound imaging, biopsy, histopathology, and biomarkers. In addition, prior surgical treatment and pathological staging are included, along with consultation notes laying out the proposed multi-disciplinary treatment strategy. The process steps needed to develop these database records are not considered to be part of the RO process and, hence, are not enumerated. The only RO process step associated with this subprocess is review of the hospital EMR by the radiation oncologist prior to the consultation. The "patient database information entered" bough denotes the referral of the patient to radiation oncology and the creation of an RO-specific EMR which links to the hospital EMR and imaging and diagnosis information therein. "Consultation and Decision to Treat" denotes the RO consultation process, the acquisition of any additional imaging or biomarker information, and the radiation oncologist's formulation of an overall treatment plan, choice of RO modality, and integration of radiotherapy into the overall management scheme. Scalable and editable versions of the process tree—in both the native Visio and derived PDF formats—are located in the electronic appendix associated with this chapter in this book's CD.

FMEA. The FMEA begins by enumerating the important potential failure modes (FMs) associated with each step, i.e., possible unintended subprocess step outcomes that, if not detected and corrected, could cause treatment to fail. The FMs were developed by the coauthors through group discussions held as part of TG 100 during 2005–2006. Based on feedback emerging from this group exercise, Thomadsen completed the development of the FMs and briefly sketched out scenarios to make them concrete. More recently, in 2010 and 2012, Williamson extensively revised these FMs and their descriptions with feedback from Thomadsen and Dr. Arthur. As illustrated by Table 9–4, each failure mode consists of identifying the subprocess (bough) and the process step (branch). A qualitative description of the error in the "Potential Failure Modes" column defines the FM. Additional detail and clarification, when needed, is given in the "Comment." Note that while most

Table 9–5. Descriptions of the O, S, and D values used in the TG–100 FMEA

Score	Occurrence (O)		Severity (S)		Detectability (D)
	Qualitative	Frequency, %	Qualitative	Categorization	Estimated probability of failure going undetected, %
1	Failure unlikely	0.01	No effect		0.01
2		0.02	Inconvenience	Inconvenience	0.2
3	Relatively few failures	0.05			0.5
4		0.1	Minor dosimetric error	Suboptimal plan or treatment	1.0
5		<0.2	Limited toxicity or tumor underdose	Wrong dose, dose distribution, location or volume	2.0
6	Occasional failures	<0.5			5.0
7		<1	Potentially serious toxicity or tumor underdose		10
8	Repeated failures	<2			15
9		<5	Possible very serious toxicity or tumor underdose	Very wrong dose, dose distribution, location or volume	20
10	Failures inevitable	>5	Catastrophic		>20

process steps had one or zero (not included in table), process steps often had more than one FM. When these multiple FMs were thought to present significantly different risks to the patient, they were described in separate rows. Other columns list potential underlying causes and a qualitative description of the potential effects of the process failure on the patient and staff.

The central FMEA deliverable is a numerical estimate of relative risk of harm that the FM presents to the patient. In FMEA, risk is a quantity that is proportional to the product of the probability that the error propagates through the process without detection and the severity of its effects (Stamatis 1995, see also Section 4.3). Specifically, risk is relative to a "Risk Priority Number" or RPN defined as:

$$\begin{aligned} Risk = &\underbrace{\left\{\begin{matrix}\text{Likelihood of}\\ \text{occurrence}\end{matrix}\right\}}_{O}\times\underbrace{\left\{\begin{matrix}\text{severity of}\\ \text{consequences}\end{matrix}\right\}}_{S}\times\underbrace{\left\{\begin{matrix}\text{Likelihood error}\\ \text{is NOT detected}\end{matrix}\right\}}_{D} \\ &\text{Risk Priority Number} = RPN = O\times S\times D \end{aligned} \tag{9.1}$$

The variables O, S, and D, denote probability of occurrence, severity, and probability of not being detected by downstream steps before propagating into outcome, respectively. Each quantity takes a value ranging from 1 (very good) to 10 (very bad). Hence, the highest achievable RPN = 1000 would describe an inevitable but undetectable failure that has catastrophic consequences. An RPN of 1 denotes an error that poses no risk. Table 9–5 outlines the criteria O, S, and D scores developed by TG 100 for radiation therapy process risk analysis. It is evident that O and D are roughly proportional to the log likelihood. The severity descriptor "very wrong" denotes a dose delivery that is highly likely to cause a severe grade 3 or 4 complication in an individual patient or is likely to cause NTCP to change by at least 0.1 to 0.2. A less severe event involving delivery of a "wrong" dose distribution, dose, or volume means a more moderate error that could raise NTCP or lower TCP by 0.05 to 0.10 and is likely to be detectable only in a large population-based outcome study.

The four study coauthors independently rated each FM using the full 1–10 range of the O, S, and D scales without knowledge of each others' score or prior rating experiences based upon earlier versions of the FM table. Figure 9–6 shows a scatter diagram of the raters' responses, compared with the mean value, for all potential failure modes. Obviously, there were wide variations in the values assigned based on each rater's experience and understanding of the potential failure mode. Such variations are normal for an FMEA and emphasize the importance of having multiple raters and using a mean value. Following rating and consolidation of scores into a single Excel spreadsheet, the raters were given an opportunity to review their scores in relation to others, comment on adequacy of the FM descriptions, and to alter their ratings. Of the four coauthors, two took advantage of this mechanism.

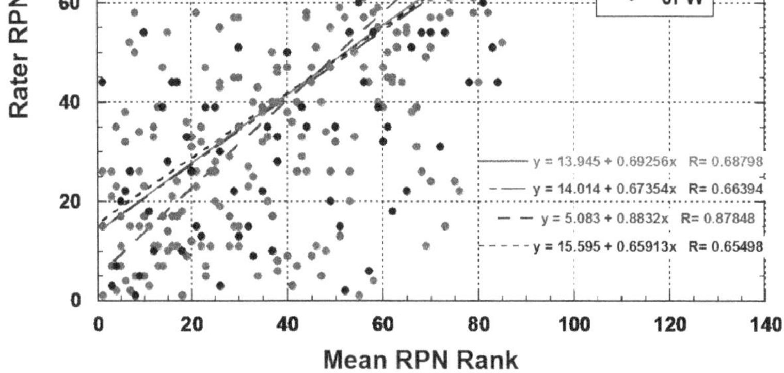

Figure 9–6. Plot of individual raters' FM ranks with respect to RPN vs. the rank based upon the RPN average of the four raters. The equations indicate parameters of a linear fit of individual rater RPN rank as a function of average rater RPN. (Plot points are in color in the book's CD.)

The most useful FMEA result is the rank ordering of FMs according to descending average RPN or S. To this end, the quantities $Q_{i,j}$ where $i = 1,\cdots,N = 95$ denotes the ith FM; $j = 1,\cdots,4$ denotes the jth rater, and $Q \in \{O,S,D,RPN\}$ denotes the quantity under consideration. The average over raters is denoted by $\bar{Q}_i = \frac{1}{M}\sum_{j=1}^{M} Q_{i,j}$ and the mean of rater means is $\bar{\bar{Q}} = \frac{1}{N}\sum_{i=1}^{N}\bar{Q}_i$. Inter-rater concordance was assessed by correlation of each rater, j, with respect to the average over raters where

$$r\left(Q_j,\bar{Q}\right) = \frac{\sum_{i=1}^{N}\left(Q_{i,j}-\bar{Q}_j\right)\left(\bar{Q}_i-\bar{\bar{Q}}\right)}{\sqrt{\sum_{i=1}^{N}\left(Q_{i,j}-\bar{Q}_j\right)^2}\times\sqrt{\sum_{i=1}^{N}\left(\bar{Q}_i-\bar{\bar{Q}}\right)^2}} \tag{9.2}$$

is the Pearson correlation coefficient (PCC). In addition, for $Q = RPN$, the correlation of rank ordering of FMs by each rater relative to rank order given by the average RPN, $\overline{RPN_i}$, was evaluated by substituting rank order for each value $Q_{i,j}$ in the above formula. This yielded the Spearman Rank Correlation.

Fault Tree Analysis Construction. A first approximation to the brachytherapy process fault tree (a fragment of which is shown in Figure 9–7) was derived more or less mechanically from the FMEA. On the left hand side of Figure 9–7, a

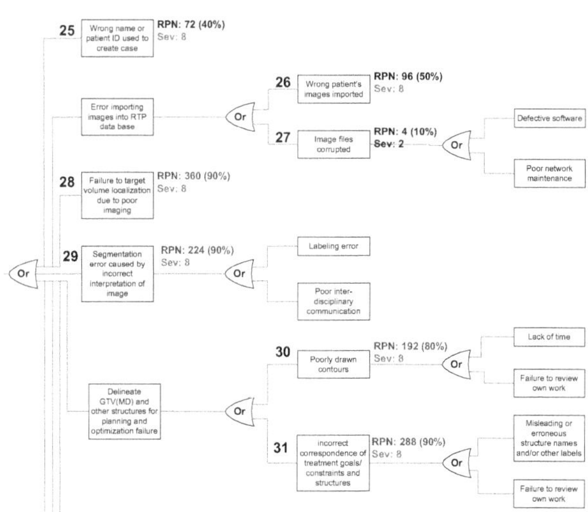

Figure 9–7. A small portion of the fault tree mechanically derived from the FMEA. See the book's CD for a complete, color version of this large figure.

Table 9–6. FMEA inter-rater reproducibility

Rater	Pearson Correlation Coefficient				Spearman Rank Correlation
	O	S	D	RPN	RPN
SM	0.6552	0.8116	0.6239	0.6673	0.5794
GSI	0.5079	0.8441	0.5305	0.6015	0.6064
BRT	0.7781	0.8849	0.8329	0.8847	0.7863
JFW	0.5138	0.8075	0.7944	0.6163	0.5100

single node representing deviation from acceptable treatment outcome is shown. In this case, wrong or very wrong dose or dose distribution represents the single form of failure considered by the FTA. This includes delivering the dose distribution to a patient who either does not need it or cannot benefit from it because ABPI dose distribution provides inadequate target coverage. Propagating to the right is another layer of nodes, each of which represents failure in one of the major subprocesses (process tree boughs), e.g., "errors in transferring images" or "RTP anatomy" failure. This lower layer of faults is connected to the top-level failure by means of lines of causal propagation terminated as inputs to an OR gate, as any one of these error conditions, if unchecked, will cause a "wrong dose" error. The next layer of nodes corresponds to failures in any one of the constituent steps (process tree branches), which are also interconnected to the subprocess fault nodes by OR gates. The 4th level nodes are the individual FMs, as described by the FMEA. The final layer of nodes consists of one node for each of the mechanisms or proximate causes listed in the "Potential causes of failure" column, which propagate into the FM node via an OR gate. This rather generic or formulaic FTA does not take into account the propagation of error from one branch of the tree into others. For example, a dwell position localization error does not become a treatment failure unless it is undetected by the subsequent treatment planning subprocess. Causal links between FMs in one branch or bough of the process and FMs elsewhere in the tree can only be identified by clinical process participants who are familiar with the mechanisms of failure. Hence, additional analysis and review of the process are required to identify such links. Two examples of such enhanced FTAs are shown in figures 9–8 and 9–9. For example, an "Image and diagnosis error," e.g., a mistake by the surgical pathologist in assessing specimen margin status, can cause the radiation oncologist to inappropriately offer APBI to the patient.

9.3.3 Results

9.3.3.1 FMEA

The complete FMEA with individual rater scores for all 95 FMs is located in the electronic appendix in order of FM number, descending average RPN, and descending S. Table 9–6 shows the inter-rater reliability in terms of the SCC, which measures the correlation of each rater's scores with the mean scores. The average RPN values have PCC values ranging from 0.60 to 0.88 while Spearman Rank Correlations were slightly lower, ranging from 0.51 to 0.79. These values are comparable to inter-rater agreement achieved in the TG–100 IMRT FMEA and similar to values reported in other areas of medicine (Duwe et al. 2005, Marwick et al. 2007). The best inter-rater consistency was achieved in severity scores, S, and the poorest in probability of occurrence, which is not unexpected, since almost no HDR brachytherapy-specific data on event occurrence exists.

Tables 9–7 and 9–8 (appearing only on the CD included with this book due to their size) show the highest risk (in terms of RPN number) and severity (in terms of S number) FMs in rank order. A number of observations can be made.

1. S and RPN rankings are poorly correlated. Only three of the eight FMs with $S \geq 9$ have RPNs exceeding 250, which corresponds to the 29th percentile. Many high-S FMs were considered relatively low risk by the raters because the probabilities of occurrence and evading detection were quite low. For example, the signature HDR brachytherapy catastrophe—detachment of the active source in the patient—has the highest S (=10) of all FMs, but it is highly unlikely to occur and is highly detectable. Redundant detection methods have been built into the hardware (cable length and radiation source detectors in the VariSource afterloader) since the 1991 Indiana, PA incident. In addition, universally accepted user safety precautions have been adopted, e.g., area radiation monitors in all treatment rooms and surveys with hand-held monitors. For the purposes of QMP design, the S rankings were ignored.
2. The highest-risk FM is #76, with an RPN value of 374, channel-to-applicator mismatch during setup of the patient, due to attaching incorrect transfer tubes to the applicator catheters. This is due to one rater who gave this mode an RPN of 900, without which it would have had a mean RPN of 199. Two related modes (51 and 49) had mean RPNs near 280, while an equivalent error pathway (FM #25: error in recording applicator rotation during CT imaging) had an RPN of 181. Potentially, if the central catheter loading were to be erroneously loaded into the peripheral catheter opposing the skin bridge, a significantly higher dose skin dose than planned could be delivered. Another high-risk FM is systematic reduction of balloon radius due to leakage (FM #92, RPN = 370) due to not imaging the patient on each day of treatment. A similar error (FM #84, RPN = 308) involves planning the treatment for an applicator contour that does not match the diameter of that implanted in the patient.
3. Perhaps the most striking finding is that of the top 20 modes, nine of them (FMs 76, 52, 26, 46, 75, 50, 53, 51, and 49) involve catheter mismatch and treatment length errors, i.e., positioning the radioactive source at the wrong location. These errors arise from a variety of causes, including wrong choice of transfer tube or systematic coupling interface length correction; inaccurate sounding measurements; inaccurate identification of

distal-most dwell position; and channel-applicator number mismatches from a variety of causes. Thomadsen's analysis of 44 HDR brachytherapy misadministrations supports this concern: fully 50% of the reported events involved positioning the source at the wrong location (Thomadsen et al. 2003).

4. Three RMs related to physician decision-making had high RPNs. Two FMs (rank 7 and 18) involved entailed suboptimal "decision to treat" errors by the radiation oncologist due to undiscovered "upstream" errors in secondary referring physician consultation notes or primary surgical,

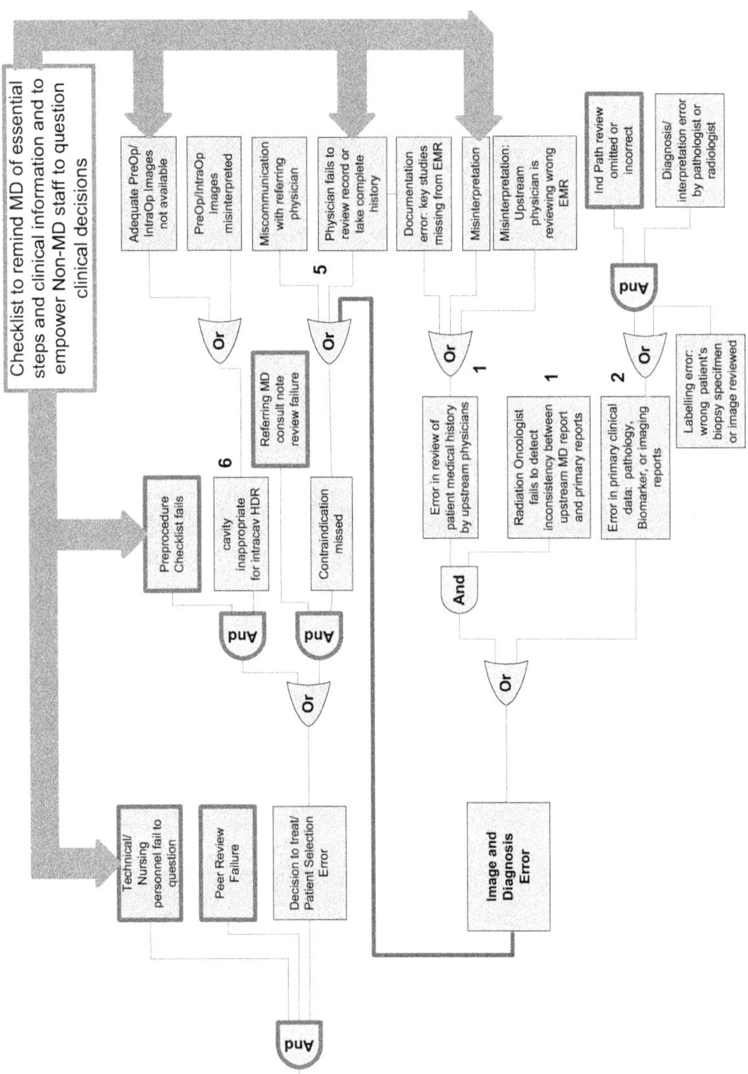

Figure 9–8. Fault tree for "decision to treat" errors. The thick-outlined (red on the CD) boxes and AND gates denote additional quality assurance steps incorporated into the process. The black numbers denote the failure mode number. See text for description of heavy arrows and box.

pathology, or in-laboratory data recording. Offering APBI to a clinically inappropriate patient (FM #5) given an error-free EMR also poses a high risk (26th highest RPN of 253).

5. Other high-risk FMs include error in constructing CTV (FM # 40, RPN of 264), setting incorrect inverse planning goals (FM # 54, RPN of 293), and incorrect balloon radius (FM # 84, RPN of 308).
6. Several device QA and commissioning failures have high rankings: wrong source strength (FM #59, RPN of 263), faulty planning data (FM #56,

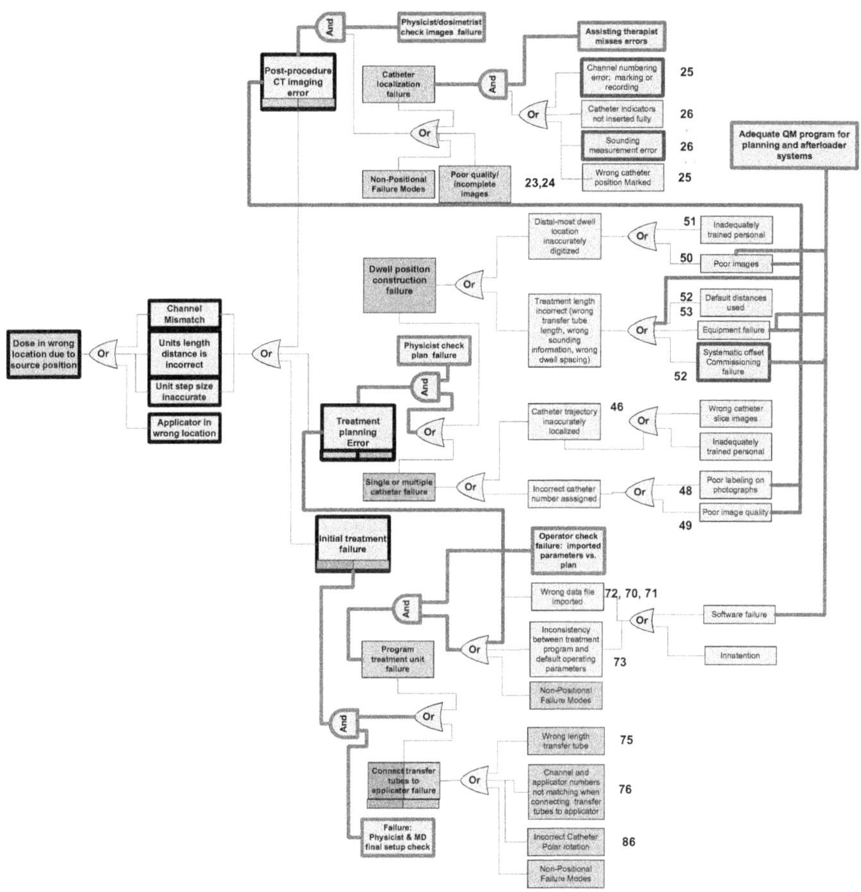

Figure 9–9. Fault tree for source positioning error failures, illustrating the complex cross-linkages between post-procedure CT imaging, treatment planning, and initial/subsequent treatment. The red pathways (see color figure on CD) and actions are additional QA/QC steps as discussed in the text. The black numbers identify the failure mode number.

RPN of 260), and incorrect decay correction or source strength programmed into afterloader (FM #82, RPN of 226).

9.3.3.2 FTA Results

Detailed fault-tree analyses are shown for "decision to treat" and "dose in wrong position" failures in Figure 9–8 and Figure 9–9 respectively. Based just on these two subprocess FTAs, a number of observations can be made. Without the conscious addition of QC and QA tests, the basic clinical process has few inherent barriers to detect and block errors occurring at early steps from propagating into later steps. Unchecked failure propagation is indicated by "OR" gates, whereas an "AND" gate requires all inputs to be positive for failure in order to propagate the error to the next stage. Only one inherent barrier (radiation oncologist review of hospital EMR to detect inconsistencies and errors prior to patient consultation) was identified in these two processes. Second, many of the FMs are linked across subprocesses (denoted by heavy blue lines in figures 9–8 and 9–9) in somewhat surprising and implementation-specific ways that can only be uncovered by detailed analysis of the specific clinical process. For example, errors in "imaging and diagnosis"—i.e., errors in staging, diagnosing, and treating the patient or documenting patient history prior to radiation oncology consultation—do not directly cause brachytherapy delivery errors. These errors are propagated (denoted heavy blue line in Figure 9–8) into the "decision to treat" subprocess where they may or may not be detected by the radiation oncologist. Errors in patient selection due to erroneous inputs need to be distinguished from "primary" patient selection errors that are due to errors in radiation oncologist judgment, lapses in taking history and reviewing records, or ill-advised accommodation to referring physician or patient preferences. Patient selection errors are FMs in which a patient, by virtue of disease-specific factors, is inappropriately chosen to receive APBI or perhaps any form of breast conservation therapy, ignoring such well-established contraindications as multi-centric or -focal disease, near or positive margin, or large/unknown burden of regional disease. A more intracavitary brachytherapy-specific class of FMs is selecting patients who are poor technical candidates for the procedure, e.g., resection cavity that is too close to skin surface, too large, or is no longer open. This error pathway poses lower risk to the patient primarily because it is relatively easily discoverable, manifesting itself during the insertion process as implantable or during post-procedure imaging by poor conformance or unacceptably narrow skin bridge. However, the consequences of such errors are not negligible: an unnecessary invasive procedure, delaying selection of more appropriate therapy, and pressuring the radiation oncologist to accept an implant of marginal quality.

The propagation of source-positioning FMs (Figure 9–9) is even more complex. These errors encompass three major subprocesses: post-procedure CT imaging (PPCT), treatment planning (TP), and initial/subsequent treatment (IST). Incorrect positioning of the source can occur in at least three ways:

1. **PPCT subprocess**. Making errors in "catheter sounding" during PPCT, i.e., physically measuring a distance in each catheter that enables the treatment length programming parameter to be calculated for a reference point

in each catheter that can be clearly visualized on the planning CT. In addition to treatment length errors, channel mismatch errors can arise from failing to control or specify catheter rotation or by misnumbering the catheters with respect to anatomy.

2. **TP subprocess**. One FM is failure to detect propagated PPCT localization errors. In addition, new FMs can arise from errors initiated in the TP subprocess. These include failing to accurately identify the treatment length reference point, to specify active dwell locations relative to this point, or to correctly calculate treatment length from the PPCT sounding data. In addition, the catheter trajectories may be inaccurately segmented, leading the optimization program to produce dwell weight distributions that are suboptimal with respect to the physical catheter geometry. In addition to PPCT channel numbering errors, the planner can create additional opportunities for channel mismatch by mislabeling the radiographically segmented catheter trajectories.
3. **IST subprocess**. In addition to failure to detect PPCT or TP localization errors (denoted heavy blue lines in Figure 9–9), there are many opportunities to cause source-positioning errors during the treatment process. One of the most insidious systematic errors is to use transfer tubes that are too long, causing the dwell positions to be placed proximal to their intended locations. Some afterloading systems offer 1.5 and 1.0 m long transfer tubes. For the VariSource system, programmed treatment lengths differ by 14 mm depending upon whether "quick connect" or "standard" indexer coupling interfaces are used. In addition, the wrong treatment parameter file could be opened on the afterloader console computer, since it is not fully integrated into the EMR/CCRT system. Finally, the catheter rotation could be incorrect or the transfer tubes not connected in the correct correspondence.

9.3.4 Derivation of Quality Management Interventions

A general description of QM interventions and ranking according to general efficacy is given in the TG–100 report (Huq et al. 2013, see also Chapter 5). In addition, the report provides general guidance on how to use the FTA and FMEA to shape the QM program, including on how to position QM interventions in the FTA. As described in Chapter 5, in descending order of effectiveness, the major classes of interventions are:

- forcing functions (interlocks, computerized verification, and electronic data transfer);
- protocols, including checklists, standard labels, and written procedures;
- independent verification and redundant checks;
- policies, including device QA/QC; and
- education.

Reviewing the "potential causes of failure" column and the FTA for upstream errors that can propagate into more than one downstream FM, draws attention to antecedent conditions that can give rise to multiple FMs. Addressing these as early as possible in the process improves efficiency and robustness. However, there is no cookbook-like, formulaic approach for constructing the QM program. In the following paragraphs, we illustrate the process involved by developing proposed QM interventions for the highest-risk FMs for the hypothetical HDR intracavitary ABPI process.

Rank	RPN	Step#	Process	Step
#1	374	76	Initial treatment	Connect transfer tubes to applicator
	FM: Channel and applicator numbers not matched			
	Related FMs: **52** (RPN 349; rank 3): Systematic error in treatment length computation **26** (RPN 347; rank 4): Errors in catheter sounding measurements **46** (RPN 326; rank 5): Inaccurate catheter localization **75** (RPN 310; rank 7): Incorrect length transfer tube selected **50** (RPN 288; rank 13): Distal-most dwell position inaccurately localized: wrong offset **53** (RPN 284; rank 15): Random error in treatment length computation **51** (RPN 281; rank 16): Distal-most dwell position inaccurately localized: poor image quality **47** (RPN 286; rank 14): Catheter trajectory localization error **86** (RPN 302; rank 10): Incorrect balloon rotation: initial treatment **49** (RPN 278; rank 17): Multi-catheter localization error from poor image quality			

FM #76 is representative of an entire class of potential failures that dominate the high-risk FM list: placing the radioactive source at the wrong location in the catheter due to channel mismatches, inaccurate dwell position digitization, or treatment length parameter errors. An FTA diagramming the entire family of errors is shown in Figure 9–9. The first step in planning QM interventions is to review the FTA and "potential causes of failure" to determine if there exist common progenitor conditions that give rise to multiple FM pathways.

The FTA shows the most common potential causes are "inattention" and "lack of training" in situations where error-prone measurements or data transcription must be performed, or where the operator must choose between several alternatives, e.g., which length transfer tubes to use or which offsets to incorporate into the treatment length calculation. These errors cannot be mitigated by forcing. Hence, we must select interventions from the next level: checklists, clearly written protocols, and independent checks. These interventions require clear and concisely written technical procedures and forms for capturing all key data. The procedure document should clarify the clinical workflow and describe who does what at each step of the process. For each step, the procedure defines the actions to be taken and clarifies what correct execution is. Without such a well-defined procedure, training cannot occur and deviations cannot be consistently identified. In the case of treatment-length estimation and channel numbering, the procedure should specify for Contura applicators:

- How rotational orientation of the catheter is specified, the channel numbers labeled, and numbering and orientation documented.
- As recommended by TG 59, a data capture form or detailed written protocol should unambiguously specify the choice of transfer tubes and indexer coupling interface (Kubo et al. 1998). In addition, the sounding measurements and tools used should be graphically identified along with the radiographically visible reference point, the length parameter formula, and the values of any offsets or corrections.

To ensure that sounding measurements are consistently error-free, QC in the form of step-by-step verification by a second person should be employed. This is most efficiently accomplished by having two people perform sounding measurements: one to perform measurements and a second person to observe and record the data (Kubo et al. 1998) as illustrated in Figure 9–9. In view of the high risk posed by source-positioning errors, having an additional QA check by the physicist—at the time of planning, check channel numbering, and dwell localization on plan, as well as treatment length—seems warranted. Each of the many FMs that cannot be eliminated by forcing must be checked prior to delivering each fraction. The patient setup check should be documented via a checklist, including, for example, verifying applicator orientation (FM #86) and catheter-channel correspondence (FM #76).

Finally, "forcing" should be used where possible. "Manual forcing" can be achieved by laying out in advance only the correct transfer tubes, coupling interfaces, data forms, and sounding tools, minimizing the likelihood that the wrong accessories are used. In the hypothetical APBI process considered in this chapter, the afterloading system is only partially integrated into the CCRT/EMS system. Fully integrating the RAL into the scheduling system would reduce the likelihood that the wrong treatment plan is imported into the control console computer, as well as ensuring that the correct number of fractions is recorded and delivered. Since this integration has yet to be achieved in our model process, careful manual verification of each downloaded treatment parameter against the printed treatment is essential.

Another common progenitor cause is "commissioning and device QA" failure. These FMs can result in poor image quality, systematic errors in treatment length computation, inconsistencies in default settings (e.g., step length) in the afterloading console and planning software, and issues in accurately importing treatment plan parameters from the CCRT/EMR system. All of these progenitor causes can be mitigated, if not eliminated, by implementation of a comprehensive device QA/QC protocol as defined by TG 56 for afterloading devices and other task groups for cone-beam and simulation CT (Nath et al. 1997, Klein et al. 2009, Mutic et al. 2003).

Rank	RPN	Step #	Process	Step
#2	370	92	Subsequent treatments	Documentation of patient changes
	FM: Patient implant geometry changes			
	Related FMs: **85** (RPN 318; rank 9): Incorrect balloon radius due to leaking **3** (RPN 70; rank 87): Wrong identifying/demographic data in RO EMR **13** (RPN 126; rank 70): Wrong size applicator placed into cavity **14** (RPN 105; rank 75): Mispositioned applicator **18** (RPN 80; rank 81): Wrong volume of fluid in applicator			

In each of these cases, suboptimal applicator positioning (poor conformance, subsequent leaking of originally adequately positioned applicator, etc.) was unnoticed and passed on to the next step. Failures in the original "Applicator Placement" subprocess pose less risk than in case of treatments because PPCT is likely to reveal any significant problems. To eliminate the possibility of treatment geometry changes invalidating the approved treatment plan, the applicator should be imaged prior to administering each fraction. A minimum requirement would be to use fluoroscopy or planar radiography to confirm full balloon inflation. Some form of 3D imaging (CT or ultrasound) is preferred, since more subtle deviations from conformance and orientation can be detected. The related FMs can be addressed by interactive use of 3D imaging (usually ultrasound) during initial placement. Intraoperative imaging would likely improve placement quality, as well as eliminate delays due to adjusting applicator position during or following PPCT.

Rank	RPN	Step #	Process	Step
8	310	2	Imaging and diagnosis	RO reviews EMR prior to RO consult
	FM: Path or biomarker report is incorrect due to due to mislabeling of specimen/report or due to sampling/interpretation error by pathologist. Hence patient is understaged and inappropriately offered BCS by medical oncologist and surgeon			
	Related FMs: **1** (RPN 269; rank 18): Inappropriate treatment offered due to error in referring MD consultation **4** (RPN 154, rank 58): Inappropriate treatment offered due to loading another patient's records into RO EMI **5** (RPN 253; rank 26): Inappropriate treatment offered due to radiation oncologist assessment error **45** (RPN 244; rank 33): Previous thoracic RT missed **22** (RPN 253; rank 15): Drug allergies, pacemaker neglected			

All of these FMs occur in the physician-dominated realm of determining what form of radiotherapy, if any, is appropriate to offer the patient following a tylectomy or lumpectomy. The FTA corresponding to these errors (except for FM# 45) is shown in Figure 9–8. Three of these modes (1, 2, and 5) have RPNs greater than 250.

In FM #2, an upstream error in histopathologic diagnosis or biomarker assessment occurs, resulting in understaging, so that BCT or APBI are inappropriately recommended by the patient's surgeon, medical oncologist, and radiation oncologist. In the normal course of radiation oncology practice, this type of error would be very difficult to detect, since all data in the upstream EMR are consistent. The literature indicates that in at least some practice settings, such errors occur at disturbingly high frequencies (Raab and Grzybicki 2010). For example, by using a standardized cytologic–histologic correlation criterion to flag suspect cases, Raab et al. (2005) found histopathologic diagnosis errors ranging from 1.8% to 19%, with errors of about 13% in the breast cancer group. Of these errors, 2/3 were sampling errors, 1/3 were misinterpretations, and 40% were associated with some level of patient harm (2% severe harm and 45% moderate harm). Another large study found an incidence of specimen labeling errors of 0.25%, of which 73% involved labeling slides with the wrong patient name and 27% with the wrong site (Layfield and Anderson 2010). In a study of 340 breast cancer patients diagnosed from 1997–2001, pathologic second opinions resulted in some change in pathologic diagnosis or prognostic factors in 80% of the cases (Staradub et al. 2002). Major changes that altered surgical therapy occurred 7.8% of the time. For example, of the cases with assessable margins, second opinions altered margin status 7.5% of the time. The Susan G. Koman for the Cure Foundation found this situation sufficiently alarming that they called for national action to improve robustness of histopathologic breast cancer diagnosis (Perkins et al. 2006). Their consultants commented that some biomarkers commonly used for treatment selection, e.g., hormonal responsiveness of the tumor, have error rates as high as 20%.

Since the literature seems to indicate that interpretation and sampling errors, rather than labeling errors, dominate the overall error rate, independent review of the histology slides appears to be an effective QC intervention. Re-review of histopathology slides is often practiced by academic institutions, at least for pathologic diagnoses from outside surgical pathologists. At Thomadsen's institution, the radiation oncologist is required to review the pathology slides for all cases. At the very least, radiation oncologists should review with their surgical pathologists the QM procedures employed in their labs to ensure that error rates are monitored and are acceptably low, satisfying themselves that the most robust state-of-the-art methods are used for biomarker assessment.

A related class of upstream EMR errors (FM #1) are clinical misjudgments by the surgical and medical oncologist, e.g., the surgical oncologist recommends BCT without further re-excision based upon the mistaken belief that specimen margins were negative. This error can be avoided through careful review of the surgical pathology records by the radiation oncologist rather than simply basing the radiation therapy treatment decision on the surgeon's review of prior medical records.

The last FM we consider (FM #5, RPN = 253), is independent error by the radiation oncologist: recommending HDR intracavitary APBI to a patient for whom the choice is suboptimal or contraindicated even though the upstream EMR is error-free. This can take two forms: neglecting clinical or technical contraindications, e.g., offering intracavitary brachytherapy to a patient with a partially or fully closed seroma cavity due to image misinterpretation or lack of an appropriate imaging study. As shown in Figure 9–8, perhaps the most effective method of detecting such an error is presenting the case at peer review conference, such as new case review or chart rounds. Since cases are not typically reviewed until the treatment plan is available (and in the case of APBI, treatment delivery initiated and, perhaps, substantially complete), many "decision to treat" errors may not be detected by standard weekly or biweekly new patient conferences until it is too late. Since APBI has relatively straightforward indications and contraindications, another option is to prepare a checklist (indicated by the orange box and arrows in Figure 9–8) which would remind residents and attending physicians of key clinical history findings that require documentation (e.g., history of prior radiation therapy, connective tissue disorders, lumpectomy with clear margins, negative sentinel lymph node biopsies, etc.) as well as common indications and contraindications (Smith et al. 2009). Obvious "decision to treat" errors could be caught early in the process—without constraining the radiation oncologist's flexibility to deviate from the policy when in the patient's best interests—by making the consultation note and key staging information available in the EMR, making the checklist available to the technical and nursing staff, and empowering staff to question obvious deviations from the standard clinical policy.

Interestingly, a physician-dominated subprocess that does not give rise to high-risk FM scenarios is applicator placement. This is due mostly to the fact that applicator placement errors are highly discoverable, most of them appearing as technical failures on PPCT. Their relatively low ranking might also suggest that from the physicist perspective, correct applicator placement is an easily learned surgical skill with relatively little variability in outcome.

Rank	RPN	Step #	Process	Step
6	318	85	Initial treatment	Run treatment
	FM: Incorrect balloon radius			
	Related FMs:			

Deviations from actual and assumed balloon radius could easily cause 20% or greater dose delivery errors due to an undetected leaking balloon, inaccurate balloon contouring, or errors in systematic planning system image importation, e.g., wrong voxel size or scaling factor. The leaking applicator scenario is addressed by requiring ultrasound (U.S.) or fluoroscopic imaging of each fraction (FM 92 above). The main protection against balloon contouring and other planning errors would be QC and QA checks of the treatment plan, which are not assumed by the high RPN of 308. However, to reduce RPN to below 100, detectability as well as frequency must be reduced. This would require using fluoroscopy or US to quantitatively measure balloon diameter for the first fraction or requiring a quick CT/simulation exam before first treatment. Although not mentioned in the FMEA, other undetected planning errors (wrong CTV thickness or fraction size) can be addressed by requiring the HDR operator to review key treatment plan parameters.

Rank	RPN	Step #	Process	Step
11	299	63	Plan approval	Run treatment
	FM: Wrong plan			
	Related FMs: **64** (RPN 259; rank 23): Wrong plan approved and exported into EMR **65** (RPN 199): Wrong final WD only patient **66** (RPN 258; rank 24): Wrong WD multiple patients and fractionation schemes **62** (RPN 210): Plan from different patient is approved and loaded into EMR under current pat ID **57** (RPN 162): Planner fails to recognize plan fails to satisfy treatment goals			

In FMs #63 and 64, the planner creates a plan that fails to realize the planning goals previously approved by the radiation oncologist. By hypothesis, import of images from EMR into planning system should automatically embed the patient's name and other identifying information into the plan. However, several plans under the same patient ID (as different courses in BrachyVision) can be created. Several error mechanisms suggest themselves: (a) a physician could approve a valid plan but the planner exports an invalid practice plan into the EMR (FM #64); (b) the physician reviews an invalid plan, but fails to note that it violates previously specified criteria (FM #63) or (c) following approval of a valid plan, a trainee planner alters the plan, rendering it suboptimal since approval does not "lock down" the plan (another FM #64 variant). Fortunately, the planning criteria are relatively simple, uniform, and easily realizable so long as the skin bridge thickness meets minimal (>5 mm) criteria. One approach is to specify a strict plan documentation protocol. At Williamson's institution, the planner adheres to a procedure that rigidly specifies the applicator orientation with respect to isodose planes, isodose curves and colors, views to be plotted, and dose-volume histograms to be plotted. The complete plan is printed in hardcopy form in advance of physician approval. The graphical information is designed to demonstrate at a glance that the planning criteria are met. The physician signs off on the hardcopy, which creates a complete, permanent record of the approved plan against which the less stable electronic versions can be compared. During each subsequent treatment, the treatment parameters loaded into the afterloader console are compared against the hardcopy plan. An alternative approach would be full integration of the planning, treatment delivery, treatment scheduling, and electronic chart. In this case, the plan is approved on-line by the physician, which automatically locks down the treatment parameters so only a specific patient and set of parameters can be used in that time slot.

FM #62 involves loading a plan from a different patient into the EMR under the current patient's identifying information. In theory, this shouldn't happen since the EMR allows only information with the same header information to be stored under a patient's ID. A more likely scenario is that once the patient's plan is exported into the afterloader console computer from the EMR, it is possible to load any plan stored therein into the afterloader control due to lack of integration of scheduling and treatment components of the EMR. Once again, the hardcopy vs. electronic plan check can detect both types of failure.

FMs #65 and 66 assume that the physician completes the final written directive following plan approval, which becomes the official expression of clinical intent for regulatory purposes and might also be integrated into the manual daily charting system. It is possible that the physician could choose a fractionation scheme that is different from the initial planning directive. A vigilant planner could spot this inconsistency and replan the treatment according to the final directive, which is, in fact, the wrong directive. The coauthors recommend avoiding this problem by having a clear written procedure that defines the initial directive as the definitive expression of clinical intent. Any deviation between the final and initial directives would then be brought to the attention of the physician's attention as a possible error by the physicist reviewing the plan or during the final check preceding each treatment.

Rank	RPN	Step #	Process	Step
12	293	54	Treatment planning	Optimization settings
	FM: Optimization method, dose-point locations, prescribed dose, and other treatment goals specified incorrectly			
	Related FMs: **55** (RPN 247, rank 30): Random entry error in setting optimization parameters **59** (RPN 263; rank 20): Wrong source strength **56** (RPN 260; rank 21): Dose calculation error **60** (RPN 260, rank 22): Prescribed dose specified to wrong structure **58** (RPN 230; rank 36): Planner uses graphical tools to shape prescription isodose failing to note that other planning goals are violated			

All of these FMs involve treatment-planning errors other than source positioning failures that have already been dealt with. The first line of defense against source strength (FM #59) and dose-calculation errors (#56) is a comprehensive device commissioning and QA protocol that includes the planning system, thereby ensuring that half-life, source-strength units, and dosimetry programs are correctly specified and compatible with the clinical process. To guard against failure to update source strength in planning system following source replacement, a forcing strategy employed by BrachyVision and VariSource can be recommended, wherein the afterloader console automatically rescales dwell times to reflect its programmed source strength value independent of the planning system value. If this strategy is not supported, the planner should manually confirm the planning system air-kerma strength. Independent plan-specific treatment time verification should also detect errors in dosimetry parameters, source strength, and prescription to the wrong structure. The strategies outlined above to mitigate plan approval FMs, i.e., rigidly prescribed standards of plan documentation and evaluation, should make these deviations from planning goals easily detectable, enabling the various plan optimization failures to be intercepted. Prescribing to the wrong structure is a risk for most planning systems, since contours are allowed to have arbitrary names. One approach to minimizing such errors is to adopt a fixed naming and color nomenclature, e.g., "PTV_EVAL" for what this chapter calls "CTV" (Smith et al. 2009). Training contours and preliminary contours used to form CTV should be named differently. Placing dimension lines denoting the CTV and applicator surface diameters on the transverse-plane isodose plot makes it easy for downstream plan reviewers to verify CTV dimensions.

Rank	RPN	Step #	Process	Step
19	264	40	RTP Anatomy	CTV construction
	FM: CTV radial thickness incorrect or avoidance structure incorrectly formed			
	Related FMs: **39** (RPN 254, rank 25): Planner fails to construct CTV and prescribes treatment to balloon surface **35** (RPN 246; rank 32): Inaccurate contouring of applicator surface **36** (RPN 224; rank 38): Inaccurate delineation of pectoralis or skin thickness **41** (RPN 221, rank 40): Error in Boolean combination **Other related FM:** 33, 36, 38			

All of these FM should be intercepted by requiring the radiation oncologist to approve planner contours and structures either prior to planning or during plan approval subprocess. The contouring is sufficiently well defined and straightforward that the physicist (who is presumably more available than the physician) should be able to review the planner's contours promptly. Sticking to the rigidly prescribed structure names, colors, and dimension lines would also help. Finally, a manual calculation check should be sensitive to FM #40 errors (Santanam et al. 2012).

9.3.5 Conclusion

The main benefit of FMEA and FTA is that they compel the physicist to abandon one-size-fits-all prescriptive protocols that are adapted without much consideration for the specific clinical implementation of the APBI process. FMEA is essentially a formalized process that forces the physicist to work collaboratively with staff and physicians to diagram the entire clinical process and conceptualize the possible ways in which the process can fail. These team members work "in the trenches" in their areas of expertise and therefore have a more real-world, comprehensive feel for what can go wrong given the frailties of the specific software, hardware, and infrastructure resources provided to implement the process. FMEA provides the team with a common language for describing the impact of failures, semi-quantitative formal risk assessment, and a root-cause taxonomy that can be helpful in crafting effective interventions. FMEA naturally incorporates mitigation of both high-frequency, quality-eroding events and low-frequency, high-impact catastrophic events as endpoints in the process. In the end, the QM interventions developed are more global than is customary for physics-centric, device-performance-driven QA. For example, the example here propels several physician-dominated steps involved in basic clinical decision-making to the top of the high-risk list of concerns. While most of the QM interventions are quite obvious, their construction is driven by knowledge of the specific vulnerabilities characteristic of the specific hardware, software, and logistic arrangements characteristic of one's own clinic.

The authors want to emphasize that the main benefit to FMEA are not the risk scores themselves, but providing the organized process that defines what is being analyzed—a common language and platform for participants in the process to efficiently and meaningfully assess potential errors and their risks to the patient. The most important products are not the RPNs, but rather a better understanding of the process, its weaknesses, and options for improving it. All involved staff—including therapists, dosimetrists, physicians, and nurses—voice how they see things, facilitating communication that leads to a more comprehensive and less biased understanding of the analyzed processes. There are other ways to accomplish what FMEA does, but FMEA provides an efficient method and rules that everyone understands and can efficiently follow. Finally, the collaborative process provides a forum for establishing group consensus and buy-in on QM interventions chosen to improve process robustness.

9.3.6 Acknowledgments

The authors would like to thank Dr. Douglas Arthur for his invaluable insights into ranking possible failure modes in the physician-dominated process steps of breast brachytherapy. Dr. Dorin Todor and Ms. Lynn Gilbert, CMD, provided helpful guidance in identifying strengths and vulnerabilities of the approach developed at VCU, upon which the senior author based the hypothetical process analyzed in this chapter. Silas Bernardoni provided help on the fault trees. Finally, the authors would like to thank Dr. Nitai Mukhopadhyay, Department of Biostatistics at Virginia Commonwealth University, for his help in evaluating inter-rater reliability. The mention of specific commercial products and vendors is incidental and is intended neither as endorsement nor criticism of these products.

9.4 Application of Risk Analysis Methods to IMRT Quality Management (Jatinder R. Palta, M. Saiful Huq, and Bruce Thomadsen)

9.4.1. Introduction

Prescriptive approaches to technical quality management, such as those promulgated by the AAPM and other professional organizations, will continue to play a role in the future. However, with the adoption of prospective quality management techniques proposed by TG 100 (Huq et al. 2013)—such as process mapping,

FMEA, and fault tree analysis—we can envision a future in which such technical quality management documents are informed by a more rigorous, although still partly subjective, analysis of the technology involved and of the causes and consequences of suboptimal performance. Familiarity with the prospective error management techniques discussed by other speakers in this summer school will facilitate the transition to quality control protocols that are weighted toward those tests most likely to be most effective in preserving the safety of the patient and enhancing clinical outcomes. It is imperative that we explore such a paradigm shift at this time, when patient expectations are rising while available resources are falling. This section is a concise review of the example in the TG–100 report. The figures and tables come from that report, and credit should go to the authors of that report for the concepts and materials used. Those authors include M. Saiful Huq, Benedick A. Fraass, Peter B. Dunscombe, John P. Gibbons, Jr., Geoffrey S. Ibbott, Arno J. Mundt, Sasa Mutic. Jatinder R. Palta, Frank Rath, Bruce R. Thomadsen, Jeffrey F. Williamson, and Ellen D. Yorke.

Until recently, the emphasis in radiotherapy quality management, particularly by the medical physics community, has been on the technical performance of radiotherapy equipment. In recent years, however, there has been increasing recognition that a major source of quality and safety impairment arises from weakness or variability in radiotherapy processes. Whereas there are a limited number of linear accelerator designs, for example, there are at least as many designs of processes as there are radiotherapy clinics. The high degree of commonality between linac designs lends itself to the development of more or less generic machine quality control protocols, which, therefore, can be prescriptive. The wide variability in processes, however, requires a much higher degree of customization that has to be carried out by those with intimate knowledge of the processes themselves.

This chapter presents an example quality management program for the planning and delivery of intensity-modulated radiation therapy (IMRT). It is based on the process mapping, FMEA, and FTA of a generic, but clinically realistic, IMRT process. The prioritization of risk in this IMRT process is based on the consensus of 10 senior physicists and one physician, and it is limited to the system vulnerabilities identified by this group. The recommendations include clinical process changes, documentation and training requirements, and culture changes, as well as traditional device- and process-oriented quality assurance and quality control checks. These recommendations *are not* to be viewed as prescriptive practice guidelines or universally applicable recommendations from TG 40, TG 142, and similar prescriptive guidance documents (Kutcher et al. 1994, Klein et al. 2009). These recommendations, and the risk analyses upon which they are based, are intended to illustrate to the reader how to develop risk analyses for their own clinical processes, and how to use the results for designing and formulating their own QM and decision-making processes for IMRT and other advanced-technology treatment procedures.

It is notable that the analytical FMEA approach led TG 100 to propose many recommendations that are consistent with the recommendations and checklists of the ASTRO white paper, "Safety considerations for IMRT" (Moran et al. 2011). As well, several FMEAs for large-scale radiation therapy processes have been pub-

Table 9–9. Identified steps and failure modes for example FMEA of IMRT

Process Number	Process Description	Number of steps in process	Number of failure modes
1	Patient database information entered	1	3
2	Immobilization and positioning	4	7
3	CT simulation	10	14
4	Other pretreatment imaging	6	7
5	Transfer images and other DICOM data	3	8
6	Initial treatment planning directive (from MD)	9	9
7	RTP anatomy contouring	15	31
8	Treatment planning	14	53
9	Plan approval	2	11
10	Plan preparation	11	30
11	Initial treatment (Day 1)	7	20
12	Subsequent treatments (Day N)	9	23

lished. Among these, two are single-institution analyses (Ford et al. 2009, Perks et al. 2012), one is for a hospital network (Denny et al. 2012), and one is for electron beam IORT (Ciocca et al. 2012). In common with the TG–100 analysis, these emphasize that FMEA is a valuable tool, although the analysis requires dedication and a multidisciplinary team approach. The TG–100 report also proposes that FMEA will be helpful in identifying high-risk features as new technology is introduced, and several FMEAs dealing with specific radiation therapy equipment have been published (Sawant et al. 2010, Novak et al. 2005, Israelski and Muto 2004).

9.4.2. Risk Analysis of a Generic IMRT Clinical Process

9.4.2.1 IMRT Process Mapping

The risk analysis for the IMRT process is based on 12 subprocesses listed in Table 9–9. Figure 9–10 shows the main branches of the IMRT process tree that fall within the purview of therapeutic medical physicists. Clinics undertaking FMEA for IMRT are encouraged to examine their own practices, although the example described in this chapter may be general enough to include the workflow in many clinics.

Because an important goal of the FMEA is to develop the most effective QM program without assuming the use of customary QM procedures, several familiar subprocesses that are purely quality management steps are omitted from the present FMEA. Omitted steps include pretreatment chart checks, routine linac and IMRT QA, physician review, and weekly chart checks, since including these steps would bias the results. The expectation is that those QA steps that were truly necessary would find their way back into the QM program as a result of the FMEA/FTA.

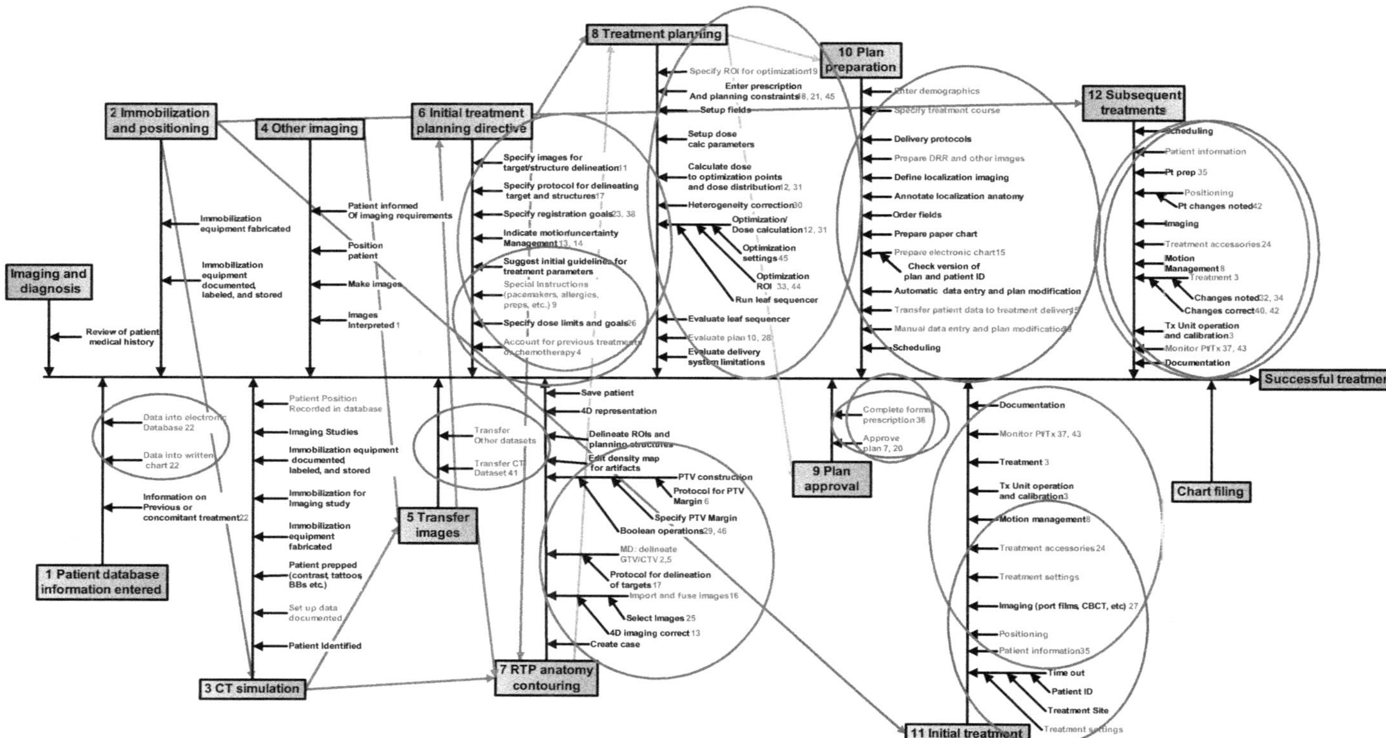

Figure 9–10. Process map for IMRT in the absence of any quality management. The black arrows show the normal flow of the process, proceeding from left to right on the largest scale and from outward to inward within a given step. Color is important to understanding this map. See the book's CD for a color version and a detailed description.

9.4.2.2 IMRT Failure Modes and Effects Analysis

To create the FMEA, the TG–100 members reached a consensus on the steps within each subprocess and identified as many failure modes for each step as they could imagine. Based on its collective experience, the TG listed possible causes for each failure mode and described clinical situations where they felt the failure could occur. During the later analysis, additional failure modes were discovered that were not initially included in the FMEA. Therefore, it is recommended that failure modes discovered later should simply be added to the analysis or addressed at that time. The FMEA serves as a tool toward improved safety and quality and is not an end in itself.

A total of 216 failure modes (FMs) were eventually included in the analysis. The distribution of these FMs between the different boughs of the process tree shown in Figure 9–10 is presented in Table 9–9. The FMs were then organized in order of decreasing average Risk Probability Number (RPN). The TG has described a consensus nomenclature for severity and scales for occurrence, severity, and lack of detectability (O, S, and D), which were used in the analysis. As previously described, the FMEA was performed assuming no deliberate QA or QC measures, such as those recommended by TG 40 for the entire radiation therapy process or TG 142 for linac QA. Thus. estimates of O and D were based entirely on checks that are inherent in routine clinical processes downstream. Despite the lack of specific QA and QC checks, there are opportunities to detect failures such as faulty immobilization, which causes problems with patient setup at "Day 1" or "Day N" treatment, leading to a medium value for lack of detectability D. On the other hand, without conventional linac QA, incorrect dose calibration would be very difficult to detect during an individual patient's treatment, leading to a high D value for this FM. Although TG–100 members assigned O and D *as if no QA were in place (*allowing failures to be detected only through normal procedural steps further downstream), the evaluators' individual experiences and biases undoubtedly influenced their O and D values in the hypothetical absence of traditional QA.

Method of Analysis. A second method identified the most and least hazardous steps according to the highest (or lowest) average values of RPN. Average RPN values ranged from 19 to 388. Process steps with the highest-ranked FMs (i.e., the top 20%) were marked on the process tree with their ranking numbers in red (Figure 9–10). This visually highlights particularly hazardous branches and boughs of the process map. Steps which occur before or after the highest-ranked 20% failure modes were also shown in Figure 9–10 to show where the high-risk steps lie in the overall process. Process steps with a ranking close to the highest-ranked 20% failure modes were also marked in Figure 9–10 because many steps at the 20th percentile level had almost equivalent RPN scores, and addressing them was considered of equal importance. Additionally, the steps where failure was judged to result in a high severity score (average S $\geq$ 8) were given special attention, even if their overall RPN was not high. The rationale is that prevention of these failures should have high priority without regard to the TG members' estimates of their likelihood of occurrence or detection. These steps were marked on the process tree in green.

To gain further insight, each evaluator's O, S, D, and RPN values were plotted individually for selected high- and low-risk steps, and the correlation between the scores assigned by two different evaluators, and between an evaluator's scores and the median values, were examined. This analysis illustrated that individual evaluators were in qualitative agreement as to the most and least hazardous steps, despite frequent quantitative differences in their individual RPN values. Such statistical methods would not be necessary in a single-institution FMEA of its own process, although there would likely be some individual scoring differences requiring averaging, discussion, or negotiation (Ford et al. 2009, Perks et al. 2012, Denny et al. 2012). Ford et al (2009) note that the rank order of the failures is more important than the absolute scale of the RPN values, so it would be of interest to know if risk rankings of single-institution FMEAs of the IMRT process are very different from those of TG 100 and, if so, to understand why.

The next step in the process was to use the entire FMEA together with Fault Tree Analysis (FTA) to develop a risk-based QM program. In clinical use, the resulting program would be adopted for a trial period and then reevaluated using the department's error-reporting mechanisms. For some steps, a successful program would lead to a decrease in the RPN values of previously high-ranked failure modes due to reduced O or D values. Persistent, high-risk, or high-S failure modes, and newly realized failure modes, would prompt new QM efforts, continuing in an ongoing effort.

Results. Table 9–10 shows the 10 most hazardous steps evaluated according to the highest average RPN values. Standard-of-practice physics QA (independent checks of treatment plans and both physical and electronic charts, linac QA, rigorous commissioning of treatment planning systems, etc.) would substantially lower the RPN value for only a few of these failure modes, primarily due to reduction in lack of detectability (D). Human factors, not device performance failures, were the most commonly cited causes of the highest risk failures. This is consistent with observations in other studies of patient safety in radiation therapy (Thomadsen et al. 2003, Clark et al. 2013). Factors noted included inadequate training, intra- and inter-departmental miscommunication, lack of consistent procedural guidelines, and loss of attention by people performing a task. A somewhat less frequently cited cause was time pressure. Physicians are deeply involved in seven of these highly hazardous steps, emphasizing the necessity of a fully interdisciplinary approach to radiation therapy QM.

The distribution over the process tree of the failure modes in the 20% most hazardous categories shown in Figure 9–10 suggests the following observations:

1. Several GTV and CTV delineation steps that can lead to a geographic miss emerged as high-risk failure modes. Target structures are usually defined by a physician. QM measures, such as physician peer-review, are required to keep such failures from propagating through the process to affect patient treatment. PTV delineation was judged less hazardous, perhaps because the PTV is derived from the GTV and CTV and because the PTV

is, in fact, a QM measure aimed at reducing the impact of setup and localization errors.

2. The "initial treatment-planning directive" is the physician's instructions to the planner regarding the planning goals and constraints. Most "initial treatment-planning directive" steps carry high risk, high severity, or both. Although physicists can take limited QM measures on their own—such as establishing standardized procedures and ensuring adequate dosimetry staff training—physician peer review may be the most direct and effective QA measure for this subprocess.
3. Treatment planning provides many opportunities for failure, especially since nearly all treatment planning failures become systematic treatment failures if not found. The FMEA supports the traditional concentration of physics QM efforts in the area of treatment planning and indicates the necessity of these efforts. Many of the riskiest steps are concerned with specifying anatomical regions of interest. Dose calculation, image transfer, and conversion of contoured regions of interest to 3D structures for plan optimization and evaluation were identified as hazardous steps. Conventional QM strategies, including planning-system commissioning and routine planning QA procedures, can reduce the risk of many of these FMs.
4. Failures at many steps in the plan approval and plan preparation subprocesses may have serious consequences because of their high-S scores (10 of 11 failures in Plan Approval and 13 of 30 failures in Plan Preparation have average $S \geq 8$).
5. Treatment delivery steps are critical. Most of the individual treatment delivery steps do not have the very highest RPNs, since there are many opportunities to detect treatment-related errors before they affect a complete treatment course, unlike treatment preparation errors, which typically introduce systematic errors affecting the entire treatment course. However, treatment delivery does include a number of failures that are within the top 20% of hazardous categories. This is another check on the FMEA, as it confirms that it is reasonable to subject treatment delivery processes to strong QM measures. Linac hardware failure in the absence of normal device QA practices was rated in the upper 10% of risk.
6. The high RPNs associated with motion management may reflect the fact that respiratory and other motion management methods are relatively new and unfamiliar to many institutions, that routine planning and delivery processes have not been fully modified to address these issues, and that the sensitivity of overall dose-delivery accuracy to execution errors and uncertainties of motion-management has not yet been determined.
7. Several high-risk failures in the treatment delivery steps are related to failure to act or to wrong actions or actions carried out at the wrong times. As is often the case when something in a complex process happens incorrectly, the reaction to that problem can cause new and potentially worse errors downstream.

Table 9–10. The 10 highest average RPN steps and their corresponding potential failure modes, causes of failure, and effects of failure from the TG 100 FMEA

Rank (Process Tree Step #)	Sub-process # Description	Step Description	Potential Failure Modes	Potential Causes of Failure	Potential Effects of Failure	AVG O	AVG S	AVG D	AVG RPN
1 (#31)	4 – Other pretreatment imaging for CTV localization	6. Images correctly interpreted (e.g. windowing for FDG PET)	Incorrect interpretation of tumor or normal tissue.	Inadequate training (user not familiar with modality) Lack of communication (inter-disciplinary)	Wrong volume	6.5	7.44	8.0	388
2 (#58)	7 – RTP anatomy	Delineate GTV/CTV (MD) and other structures for planning and optimization	1. >3*sigma error contouring errors: wrong organ, wrong site, wrong expansions	Lack of standardized procedures Hardware failure (defective materials/tools/equipment) Inadequate design specification Inadequate programming Human failure (Inadequate assessment of operational capabilities)	Very wrong dose distributions Very wrong volumes	5.29	8.43	7.86	366
3 (#204)	12 – Day N treatment	Treatment delivered	LINAC hardware failures/wrong dose per MU; MLC leaf motions inaccurate, flatness/symmetry, energy – all the things that standard physical QA is meant to prevent.	Poor design (hardware) Inadequate maintenance Software failure Lack of standardized procedures (weak physics QA process) Human failure (incorrectly used procedure/practice) Standard linac performance QM failure (not further considered here) Inadequate training	Wrong dose Wrong dose distribution Wrong location Wrong volume	5.44	8.22	7.22	354

Table 9–10, cont. The 10 highest average RPN steps and their corresponding potential failure modes, causes of failure, and effects of failure from the TG 100 FMEA

Rank (Process Tree Step #)	Sub-Process # Description	Step Description	Potential Failure Modes	Potential Causes of Failure	Potential Effects of Failure	AVG O	AVG S	AVG D	AVG RPN
4 (#48)	6 – Initial treatment planning directive (from MD)	Retreatment, previous treatment, brachy, etc.	Wrong summary of other treatments. Other treatments not documented.	Lack of staff (rushed process, lack of time, fatigue) Human failure (inattention) Lack of communication Human failure (reconstructing previous treatment) Human failure (wrong info obtained) Information not available	Wrong dose	5.33	8.56	7.33	333
5 (#59)	7 – RTP anatomy	Delineate GTV/CTV (MD) and other structures for planning and optimization	2. Excessive delineation errors resulting in <3* sigma segmentation errors	Lack of standardized procedures Availability of defective materials/tools/equipment Human failure (materials/tools/equipment used incorrectly) Human failure (inadequate assessment of materials/tools/equipment for task) Inadequate design specification Inadequate programming Inadequate training Human failure (inadequate assessment of operational capabilities) Human failure (inattention) Human failure (failure to review work) Lack of staff (rushed process, lack of time, fatigue)	Wrong dose distribution Wrong volumes	5.86	6.57	8.00	326
6 (#65)	7 – RTP anatomy	PTV construction	3. Margin width protocol for PTV construction is inconsistent with actual distribution of patient setup errors	Lack of standardized procedures Lack of communication Inadequate training Human failure (inattention) Human failure (failure to review work) Lack of staff (rushed process, lack of time, fatigue)	Wrong dose distribution Wrong volumes or suboptimal plan	7.29	5.43	7.86	316

Table 9–10, cont. The 10 highest average RPN steps and their corresponding potential failure modes, causes of failure, and effects of failure from the TG 100 FMEA

Rank (Process Tree Step #)	Sub-Process # Description	Step Description	Potential Failure Modes	Potential Causes of Failure	Potential Effects of Failure	AVG O	AVG S	AVG D	AVG RPN
7 (#136)	9 – Plan approval	1. Plan OK to go to treatment	3. Bad plan approved	Lack of communication Human failure (inattention) Lack of standardized procedures Human failure (incorrectly used procedure/practice) Inadequate training	Very wrong dose Very wrong dose distribution Very wrong volume	4.89	8.00	7.89	313
8 (#200)	12 – Day N treatment	Set treatment parameters 2	Special motion management methods (e.g., gating, breath-hold) not applied or incorrectly applied	Poor design (software) Poor design (hardware) Inadequate training Human failure (operator not observe counterintuitive patterns on screen)	Wrong dose Wrong dose distribution Wrong location Wrong volume	6.22	6.67	7.11	310
9 (#46)	6 – Initial treatment planning directive (from MD)	Specify special instructions, viz. pacemaker, allergies, voiding, bowel prep, etc.	Special instructions not given Wrong special instruction (e.g., allergy, pacemaker)	Lack of standardized procedures (documentation) Lack of staff (rushed process, lack of time, fatigue) Human failure (inattention) Lack of communication Human failure (wrong or inadequate info obtained)	Non-radiation related injury	5.25	8.75	6.50	306
10 (#126)	8 – Treatment planning	13. Evaluate plan (DVH, isodose, dose tables, etc.)	1. Inadequate evaluation	Human failure (not enough time/effort spent) Inadequate training Poor evaluation strategy Human failure (incorrect final prescription)	Wrong dose Wrong dose distribution	5.63	7.00	7.13	303

9.4.2.3 IMRT Fault Tree Analysis

General Features of the Fault Tree Derived from the FMEA. FTA is a tool that allows one to visualize potential locations for effective and efficient QM measures, since the propagation of FMs through the process is more visually illustrated in the FTA than in the FMEA. The TG–100 members developed a complete fault tree for the entire IMRT process. Table 9–11 lists the most common progenitor causes for the failure modes graphically portrayed by the FTA. The dominant category is Human Failure. In the IMRT FMEA, TG–100 members suggested various forms and underlying causes of human failure; interestingly, poor employee performance was rarely cited as a major cause. Human errors have many causes, e.g., loss of attention, biased expectations, distractions due to multiple demands, bad judgment in the face of a deviation from the normal process, fatigue, or overwork. Human failure rates can be most efficiently reduced by "forcing functions." A forcing function is defined as something that prevents a failure-causing behavior from continuing until the problem has been corrected. Linac interlocks are familiar examples, and they often prevent errors from being made in the first place (NPSF). Unfortunately, forcing functions often require highly technical solutions and are works-in-progress rather than immediate solutions. Close collaboration between clinicians and vendors is important in this regard. Human failure rates can also be reduced by strategies that address underlying causes by providing, for example, properly lighted and distraction-free environments, good ergometric design of computer and device graphic user interfaces (GUIs), and efficient and orderly flow of information. These strategies, along with good supervision and training in the establishment of a safety culture, can reduce, but never completely eliminate, human failures. Published studies reporting radiation therapy computational and transcription task error rates suggest failure rates of the order of 0.5%–1% (Dunscombe et al. 2007, D'Souza et al. 2012, Margalit et al. 2011, Arnold et al. 2010, Bissonnette and Medlam 2010, Huang et al 2005, Mutic et al. 2010, Marks et al. 2007, Macklis et al. 1998, Calandrino et al. 1997). This likely represents the best that can be achieved under optimal conditions.

The next two most common categories, Lack of Standardized Procedures and Inadequate Training, along with the Lack of Communications and Information Problems, all reflect latent organizational flaws. These problems cannot be addressed efficiently by adding more QA or QC checks, but instead require the redesign, or at least improved documentation, of the current process. Establishing standard procedures and protocols, assuring personnel are trained appropriately (with exams for verification), and designing clearly understood lines of communication and information flow create an environment that reduces the likelihood of occurrence of many potential paths for failures. It is also very important for department managers to provide a work environment that is free of clutter, interruptions, and distractions.

Most of the causes grouped under Lack of Staff result from administrative decisions. As an example scenario, if an experienced dosimetrist is not available to plan IMRT, the physicist often generates the treatment plan for IMRT and performs

Table 9–11. Most common classifications for the possible causes for the failure shown in the IMRT Fault Tree Analysis (Appendix C of the TG-100 report)

Category	Occasions
Human failures	230
Lack of standardized procedures	99
Inadequate training	97
Inadequate communication	67
Hardware/software failure	58
Hardware	9
Software	44
Hardware or software	5
Lack of staff	37
Inadequate design specifications	32
Inadequate commissioning	18
Use of defective materials/tool/equipment	12

the pretreatment plan check. This is a dangerous situation, since an error is more readily detected by an independent check than by the person who made the error in the first place. Inadequate staffing levels can also produce a rushed environment or fatigue, leading to user errors. In general, such problems cannot be addressed merely by process design, documentation initiatives, or by adding QA checks. Instead, they can only be solved through administrative decisions that are informed by current staffing studies (Abt 2003, Klein 2010, ACR 2012).

The large number of potential failures attributed to Hardware and Software Failures and to Design Failures illustrates how highly dependent radiation therapy quality is on robust and accurate equipment performance. Preventing device failures from propagating into events requires: 1) careful specification of device performance characteristics during purchase, including reliability and safety features; 2) comprehensive commissioning in the context of the process to be used to assure proper operation of equipment; 3) training of personnel on how to recognize and respond to machine failures; and 4) appropriate periodic equipment QA that monitors equipment operation.

Comprehensive commissioning (see Chapter 4) identifies inadequacies in both equipment *and procedures* before beginning patient treatment. The commissioning not only checks the operation of equipment and provides the information necessary for its use, but it also establishes the limits of reliable operation for equipment and systems. Commissioning of procedures entails coordination between all involved personnel as tested by walking them through trial runs. Time spent during commissioning can save time and increase reliability during routine operation. Commissioning provides a detailed and real-world understanding of a device's features, providing the basis for rationally integrating it into the department's clinical practice. It is very important that hospital administrators and department leaders allow adequate time and personnel resources for commissioning tasks. This is an area where "haste makes waste."

Simple Example of FTA Guidance in QM Design. As seen in the discussion above, many failure causes cannot be addressed through conventional, device-performance QA or physics chart checks, but rather require system redesign, administrative changes, or a broader type of commissioning. For example, Figure 9–11 shows just the potion of the fault tree dealing with the high-risk Radiation Treatment Planning (RTP) Anatomy failure mode error delineating the GTV, CTV, or other structures. The annotated fault tree in Figure 9–12 indicates the general type of QM methodology required to mitigate the principal progenitor causes. The red diagonal line indicates a cause best addressed by more complete training, establishing clear communication modalities (including forms and checklists) and establishing protocols, policies, procedures, and expected outcomes. Those boxes with red arrows reflect causes that could be eliminated by providing appropriate resources for the facility (administrative decisions), and those with green arrows by comprehensive commissioning. Of the 156 causes that may give rise to an undetected RTP anatomy failure, 133 can be at least partially addressed by the measures described above. The remaining 23 causes, mostly user errors, may be addressed by peer review of the contoured structures at the end of the subprocess.

Suggested Use of FTA. While much of the analysis and many of the resultant QM measures could be derived from the FMEA alone, the FTA graphically illustrates the propagation of errors from one process step to another, helping to identify what structural changes to make to the process and the optimal placement of QC and QA interventions.

To achieve maximum efficiency, it is desirable to consolidate proposed QM steps, if possible. Searching the Fault Tree for multiple occurrences of a progenitor cause can provide economy in establishing QC. For example, a generalized review to encompass both the "Initial treatment planning directive" and "RTP anatomy contouring" could provide efficiency. Positioning a check prior to planning will avoid wasted planning effort based on erroneous imaging datasets, incorrect contours, or unrealistic treatment goals. Another example is the progenitor cause of "Defective Equipment" which can be addressed by a department-wide preventive maintenance program that covers all clinical hardware and software. If a failure at a single step feeds to different downstream failures, preventing the common failure reduces the likelihood of the several resultant failures. One can also examine the Fault Tree and process map to look for junctions where QA activities could cover multiple potential failures.

9.4.3. Risk-informed Design of IMRT QM Programs

The next step in the overall plan to improve quality and safety is to use the FTA and FMEA risk- and process-oriented information to design a QM program for 216 FMs described by TG 100 in a generic IMRT process. Based on that analysis, the following elements of an IMRT QM program are crucial.

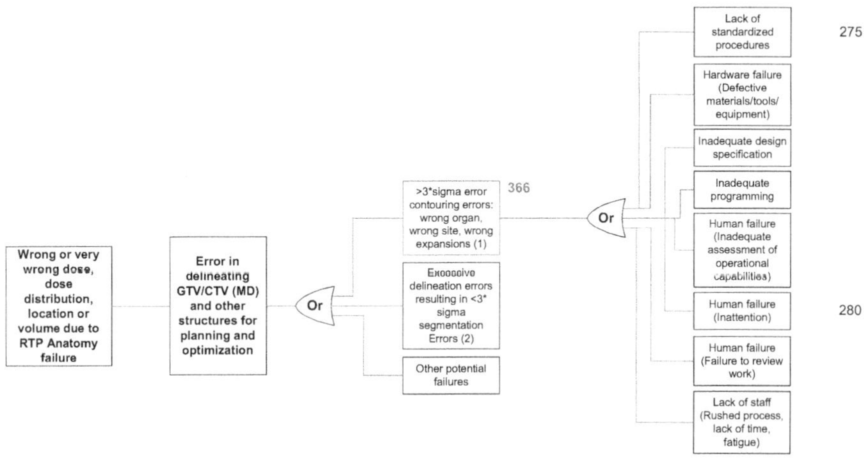

Figure 9–11. A portion of the fault tree for the step RTP anatomy failure involving the failure mode of >3 sigma contouring errors; this failure is in the box with the RPN #366 shown at its upper right corner. The numbers on the far right are line numbers from the full FTA. The numbers on the far right are the line numbers from the full FTA found in the TG–100 report, Appendix C (Huq et al. 2013).

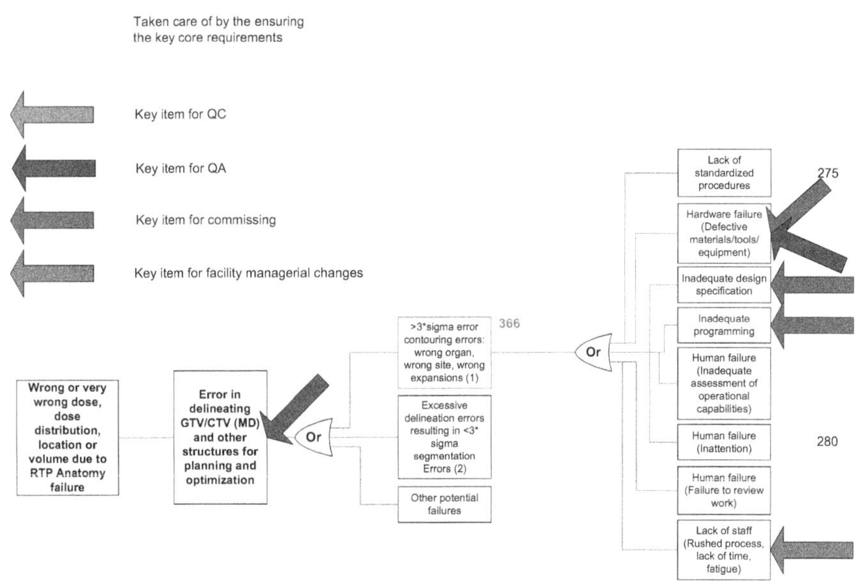

Figure 9–12. The fault tree shown above in Figure 9–11 with the inclusion of quality management. See the book's CD for a color version of this graphic.

9.4.3.1 QM of Pretreatment Imaging

The highest-ranking hazard involves incorrect interpretation of a pretreatment diagnostic imaging (e.g., PET, MR) study for defining the GTV, CTV, or a dose-limiting normal tissue. Progenitor causes listed in the FTA include inadequate training of the study reader or poor interdisciplinary communications. For example, suppose that the radiologist reports that a patient's PET-FDG study reveals a positive para-aortic lymph node, but the radiation oncologist incorrectly identifies this lymph node with a region of high-intensity signal caused by a benign inflammatory condition and, therefore, places the GTV adjacent to the wrong vertebral body.

Consideration of this failure mode, as with many top-ranked failure modes, illustrates a very important fact: it is not possible to establish a complete QM program involving solely medical physicists. Effective QM requires a team approach with members of each specialty participating and making recommendations, particularly for potential failures relating to their areas of expertise. *Full application of the risk-based approach requires the involvement of everyone who participates in the radiation oncology process.* This particular failure mode illustrates a number of general characteristics of physician-driven actions and decisions:

1. They are often critical drivers of down-stream planning activities, which, if performed inaccurately, have a high potential for causing systematic error.
2. Physics, technical, and nursing support staff often lack expertise in physician-driven processes—in this case, image interpretation or even access to the imaging studies—and therefore are in no position to intercept or detect such errors.
3. In the traditional physician-dominated control-and-command model, support staff had little institutional support or standing to challenge physician decisions. As more departments revamp their practices to emphasize a safety culture, this situation is likely to improve (Chera et al. 2012). If a medical physicist or dosimetrist recognizes an occurrence of this failure mode and the departmental culture allows, he or she should, of course, bring it to the attention of the radiation oncologist. However, the main remedy for this failure mode lies with the physician community.

There are at least three broad avenues for intercepting image-interpretation errors.

1. **Peer Review**. Directly addressing this failure mode requires physician-based checks somewhere in the process. The FTA demonstrates that a QA check of the target volumes defined by the prescribing physician, before significant planning effort takes place, is likely the most efficient way to prevent this failure mode. Meaningful peer review of target delineation requires display of imaging studies on which target volume delineation is based, along with the treatment plan and the CT images from simulation. The considerable effort required in order to organize and implement such

peer-review is warranted by the extremely high RPN value of this failure mode, since these errors are likely to be virtually undetectable otherwise.

2. **Adequate Physician Training for Interpretation of Diagnostic Imaging Studies**. Table 9–11 identifies inadequate training as the third most likely progenitor cause of all failures identified. However, the person interpreting the images may think that she or he has the appropriate knowledge. Assuring training in clinical procedures is often problematic, so an institutional policy requiring experts in a given imaging modality to advise new staff on all readings of such images may be beneficial. Important components that can help reduce the risk of this failure mode include physician education through training courses offered by ASTRO and other professional or educational organizations and intra-departmental peer review based on evaluation of actual cases.
3. **Improved Interdepartmental Communication**. A potential cause of this failure mode is failure of the radiology report or reading to address the radiation oncologist's need for quantitative tumor localization (in addition to diagnosis and staging). By communicating these needs to the radiologist, the radiation oncologist can mitigate errors and enhance the value of these imaging procedures to the RT process. In addition, good communication with the radiologist is a low-cost and efficient avenue for the radiation oncologist to become educated in the interpretation of more specialized functional or molecular images.

9.4.3.2 QM of Target and Critical Structure Delineation

The second-ranked failure mode is very large contouring errors (in excess of three times the expected inter-operator delineation error) which will be used to illustrate the evaluation and analysis which can be used to create QM procedures based on review of the fault tree. Figure 9–11 is a section of the IMRT fault tree for the step "Delineate GTV/CTV (MD) and other structures" for which failures can lead to planning or optimization failure. This figure shows only the dominant intermediate and basic events that cause ">3σ error contouring errors: wrong organ, wrong site, wrong expansions" events.

As with the first-ranked FM, radiation oncologists are the only personnel with the specialized knowledge needed to define the contouring protocols for most radiotherapy targets and other critical structures, and should take the lead in developing QM for this step. However, as the FTA shows in Figure 9–11, many failure modes contribute to this type of failure, which means there are a number of different approaches to avoiding or mitigating this error.

In Figure 9–12, the QM steps at the most effective locations in the fault tree of Figure 9–11 are indicated by lines or arrows (the key is in the upper left portion of the figure). These QM steps are described below.

1. **Peer Review**. As in the rank 1 failure mode, a peer review of all structure delineations is the most effective method for intercepting such failures. There are several ways to implement such reviews:

- At facilities with resident training programs, having the attending physician review and edit contours drawn by the resident physician focuses the attending physician's attention on how well the contours match the "standard of care" definitions.
- The larger radiation oncology community could provide assistance to radiation oncologists in small or solo practices through some form of Internet-based peer-review system.
- The Advanced Technology Consortium (ATC) has developed such a peer-review system for formal protocols used by clinical trial groups (e.g., Radiation Therapy Oncology Group (RTOG). ASTRO has developed educational programs, such as the special contouring sessions which are organized during ASTRO meetings. Departments should also encourage the training of dosimetrists and physicists involved in treatment planning, both through in-house efforts (senior dosimetrists training juniors and physicians training physics personnel) and participation in workshops given by professional organizations. This expands the pool of individuals equipped to detect and prevent large contouring errors.
- The medical physicist and dosimetrist should make sure that radiation oncologists are trained in the correct use of contouring software, and may, in the future, assist in implementing new technologies, such as automatic segmentation programs, that may reduce the probabilities for errors. These programs, while in their infancy, already show promise and will likely become better. While the use of such programs may reduce the likelihood of contouring errors and improve the detection of errant contours, they may also open pathways for new failure modes. However, given the large number of failure modes and potential causes of this failure, relying on a single check may leave the process at high risk unless upstream causes of failure are also addressed. Furthermore, peer review may not always be a cost-effective or sufficiently robust approach when contouring errors are relatively frequent. Therefore, the following QM steps are also recommended.

2. **Standardized Procedures**. The lack of uniform procedures and training can dramatically increase inter-observer segmentation variability, i.e., delineation error, beyond the level inherent to the imaging modality. For example, an early study of prostate boundary delineation error on CT revealed extremely large physician-to-physician variability (10% to 20% standard deviations in prostate volume) (Lee et al. 2002). However, when consensus among observers is reached on fundamental issues—such as "are we contouring just the prostate or margins for extra-capsular extension?" or "How do we identify prostatic apex and other boundaries not visible on CT?"—and when observers have an opportunity to be corrected on training cases, much smaller (2% to 4%) variations are observed (Lee et al. 2002). The EORTC/RTOG guidelines for contouring electively treated

lymph-node CTVs in head and neck cancer is an example of a published guideline that can be used as the basis of an institutional consensus-derived segmentation process and associated training (Gregoire et al. 2003). Written departmental guidance on segmenting anatomic and target structures should be developed as part of the Site-Specific Protocol. Such guidelines can also be used as the basis for empowering physics and dosimetry staff to intercept large contouring errors.

3. **Elimination of Hardware Failure/Inadequate Design/Inadequate Programming**. These potential causes are best detected and compensated for during commissioning of planning or other contouring software or, in the cases of a transient hardware/software failure, through periodic QA and preventive maintenance of the system. Adequate commissioning not only assures that the equipment operates as described in the manufacturer's specifications, but it determines how the equipment functions over the range of expected use, particularly outside the normal and intended range. Commissioning must determine the limits of reliable operation and the types of errors that occur with misuse. Commissioning also provides an opportunity to compensate for software design deficiencies, through changes in the clinical process. For example, if the planning system's manual segmentation software is so slow as to challenge physician patience and willingness to review work, using different software (e.g., CT-simulator virtual simulation software) to contour or assign certain segmentation tasks to dosimetrists might be appropriate.
4. **Prevention of Human Failures (inattention, incorrect operational assessment, failure to review own work)**. Minimizing the probability of human failures compromising a patient's treatment (i.e., random execution errors that occur despite training and well-defined procedures) often requires redundant QC or QA checks that operate on the inputs or outputs of the process by adding parallel activities. For example, the "Failure to review own work" could be ameliorated by independent review of the contours at input through the use of an automated anatomic contouring program. In Figure 9–12, an auto-contouring program could be used to check the physician's contour by flagging a large discrepancy as an "AND" gate in parallel with the dosimetrist's contour review or with MD peer review, thus intercepting downstream propagation of human ($>3\sigma$) contouring errors (Kaus et al. 2007). Trying to prevent human failures from compromising clinical care, however, requires considerable resources, and human creativity often finds new ways to fail in ways that were unthought-of at the time the QC was put in place. Peer review of contours is optimal, but a knowledgeable dosimetrist or physicist can often check the consistency of contours and flag many types of potential problems (e.g., contour overlap or accuracy of normal structures). Physicists and dosimetrists should be encouraged to ask questions about structure sets that differ from those they have seen in similar cases.

5. **Avoiding Rushed Process/Inadequate Facilities**. A rushed process may result from poor organization on the operator's part or from managerial decisions leading to inadequate staffing or lead-time before treatment. Minimizing the possibility of such failures requires a commitment on the part of management and medical staff to provide adequate time and resources for the facility to achieve its mission. Given such commitment, all personnel have the responsibility to complete their tasks in a timely fashion.

9.4.3.3 QM of the Treatment Delivery System

Radiological and geometric delivery errors associated with treatment machine failures comprise the third highest risk failure mode (remember that the FMEA assessed risk assuming no specific QA procedures were performed). Since any hardware delivery errors are essentially undetectable in this scenario, a very high RPN number results, underscoring the importance of periodic QA to reduce the risk of machine hardware failures. Most current machine QA guidance (e.g., the recommendations of TG 142) is loosely based on the goal that the total cumulative dose-distribution delivery uncertainty should not exceed 5% or 5 mm when all contributing geometric and dosimetric tolerances are summed in quadrature (Klein et al. 2009). However, most such recommendations are based on TG member consensus and not formal error propagation analyses or endpoint-specific rationales. Of note, reports (e.g., TG 40) do not always specify whether the stated uncertainty refers to 1 or 2 standard deviations (k=1 or k=2). Given the wide variety of techniques currently applied for patient treatment, a single set of QC tolerances and test frequencies may be neither necessary nor sufficient to protect the patient from "wrong dose" or "wrong location" errors or to evaluate risks appropriately. Two brief examples illustrate the issues:

1. Suppose daily on-line IGRT is used for all patients on a given machine. For this situation, the need for traditional (within ±2 mm) localization ODI, light field, and cross hairs may be decreased, and current TG–142 recommendations for QA for these parameters may be too strong, inefficiently using QA resources.
2. On the other hand, current TG–142 recommendations that MLC leaf-positioning error be assessed monthly may be too lax for treatments delivered in few fractions because MLC errors could be missed for this entire treatment course.

TG 100 proposes that ultimately QM for linac failure modes be designed so that the maximum number of fractions where a linac performance deficit *could* remain undetected never results in a patient treatment course exceeding the allowed cumulative positional and dose-delivery tolerances. Here we describe an approach to the determination of test frequencies and tolerances for linac QA, using dose output (Gy/MU, in reference geometry) as an example, followed by very brief comments on other parameters. The approach is discussed in more detail in Appendix E of the report of TG 100.

9.4.3.4 Example Method for Determination of Tolerance and Frequencies for QA Tests of Linac Output

1. **Define the QM goal.** The overall dose-delivery or positional accuracy for the target must be consistent with the department's vision of acceptable quality or the accepted standard of care. In the following example, the goal is "no patient's total dose-delivery uncertainty should exceed 5%" (consistent with TG 40 and TG 142).
2. **Determine the sensitivity of the QM goals to the performance parameter.** In the case of Dose/MU sensitivity, some dose errors are linearly related to the error in a given parameter, such as errors due to miscalibration of the Dose/MU control. Such a relationship is said to have a sensitivity of 1. However, other errors require non-trivial sensitivity analyses (e.g., output constancy and linearity as functions of dose rate, gantry angle, and MU/segment, etc.), since these parameters can affect total dose delivery accuracy, but their dosimetric impact depends on the distribution of gantry angles, the functional relationship between output and gantry angle, MU/segment, and dose rates characteristic of typical plans (Dunscombe et al. 2007, Pawlicki et al. 2005).
3. **Determine the maximum error in the linac performance endpoint for which the machine remains operable.** Linac interlocks that prohibit operation when parameters go out of tolerance are important. Recent failures of symmetry interlocks however, highlight the difficulty in relying on such systems. Most accelerator interlocks can be rendered useless if someone, such as a service person or physicist, adjust the baseline they use to determine operational limits. In this example we will consider two situations: a typical modern scenario where machine interlocks are triggered by dose output errors exceeding 5% and an extreme situation where transient and persistent dose output errors up to 40% are possible without triggering machine interlocks. Although the interlocks of some accelerator models do not allow such a large error, a 40% error has been reported following service when a pot was misadjusted and the interlocks reset to new values. Therefore, this large but not impossible value is chosen for demonstration purposes.
4. **Determine the monitoring frequency needed to achieve the uncertainty goal.** In a treatment of N fractions, a patient can receive up to n fractions with dose output error per fraction of q%, without exceeding an A% dose accuracy limit if all the other treatments are perfect and

$$n \leq AN/q \tag{3}$$

Figure 9–13 is a plot of the number of allowed fractions with output error vs. the error in (n vs. q) for 35, 10, and 5 fraction courses (N=35, 10, and 5) for A=5% and 1.6%. The choice of A=5% is a loose tolerance, since dose output is not the only source of error in a treatment; the tighter tolerance of 1.6% is a more realistic goal which acknowledges other sources of dosimetric uncertainty. Radiobiological effects were not considered in this example. Within this simplified example, if the shortest course treated on a particular machine is 35 fractions, output checks every four days suffice to meet the goal of dose accuracy of at worst 5% if weak machine interlocks permit up to a 40% output error. Shorter treatment courses demand more frequent output measurements; for a typical palliative course of 10 treatments daily checks are needed and, strictly speaking, for fewer than eight treatments, even daily checks are insufficient to assure that no patient experiences a dose output error exceeding 5%. Of course, if machine interlocks prevent treatments with dose output error of 5%—and if one believes these interlocks are infallible—output checks would be unnecessary. However, if output dose error must be smaller (e.g., 1.6% to allow for other sources of treatment uncertainty), our simplified model predicts that daily or more frequent output checks are needed for short treatment courses (e.g., less than five fractions) even if machine interlocks prevent delivery of treatments with output errors exceeding 5%.

The fact that clinical physicists do not typically perform output checks more than once a day, even on machines that deliver single-fraction treat-

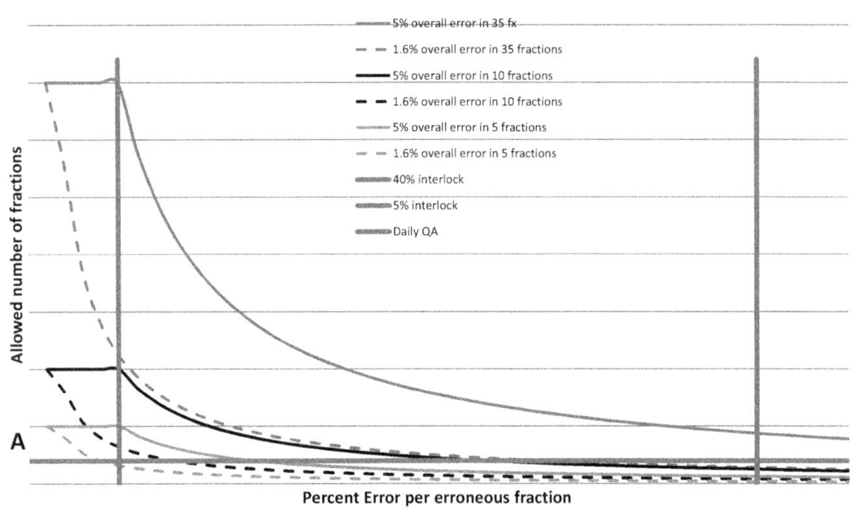

Figure 9–13. The number of fractions that can be delivered with a given error plotted as a function of the percentage error in dose per erroneous fraction, for total allowed dose errors of 5% and 1.6% in treatment courses of 35, 10, and 5 fractions. The tall, vertical lines indicate the two interlocks discussed in the text: a weak interlock where output errors up to 40% can be delivered and a modern interlock which cuts off delivery if the output error exceeds 5%. Line A is at 2 fractions. For situations that fall below this line, the simple model calls for daily or even more frequent output checks. (See the CD for a color version of this figure.)

ments, is based on statistical reasoning that is beyond the scope of the TG–100 report.

5. **Establishing action levels and thresholds**. The above analysis is a simplified model of how one can protect the average patient against "outliers" that embody worst-case scenarios, regardless of how unlikely these scenarios might be. However, QA should also seek to minimize overall *mean* uncertainty of dose delivery by selection of an action level (e.g., the error above which the parameter is readjusted) based on a probability distribution of machine variability. For linac parameters that exhibit significant random variability, but below the fixed threshold levels, process control charts and other statistical techniques should be used to distinguish underlying trends from day-to-day statistical fluctuations (Pawlicki et al. 2003, Kapanen et al. 2006).

9.4.3.5 Comments on Other Dosimetric and Geometric Performance Endpoints

1. **Energy and Beam Flatness/Symmetry**. Selection of a sampling frequency requires both sensitivity analysis and an assessment of how much deviation from flatness or energy an operable linac could exhibit. For many machines, it is unlikely that energy or symmetry errors >10% could occur without concomitant failure of machine output or a sustained effort to retune the machine to operate at the wrong energy. A typical interlock limit for symmetry is 4%. Because the symmetry is the difference between the dose on one side of the field and that on the opposite, a symmetry value of 4% implies that the dose on either side is about 2% off from baseline, or that the symmetry has a sensitivity of 0.5 in dose. Monthly monitoring often means that checks fall within each month, not that checks are performed with no more than 30 days between. This permits a patient who starts treatment soon after one monthly check to receive a full treatment of, for example 35 fractions, with the unit operating with the 4% symmetry error. However, the dose would only be in error by 2%. One potentially serious failure mode is matching a shift in beam energy with a well-intended (but misguided) effort to retune the linac so it operates without triggering interlocks. Such actions, which have happened at several facilities, can lead to significant dose errors. A similar failure can occur if steering is accidentally changed, creating a badly asymmetric beam with the ratio of the currents from each side of the dose monitor chamber set as a new baseline. Thus, any intervention involving beam retuning or steering should trigger independent checks of beam characteristics before returning the linac to clinical use. Simple, non-specific tests can be very useful for checking such failure modes; verifying constancy of a large field, shallow depth beam profile is a highly sensitive check of all beam characteristics sensitive to beam energy, including depth dose (Constantinou and Sternick 1984).

2. **MLC and Jaw Calibration and Operation**. Geometric miss of the target or overexposure of a normal tissue due to MLC problems is potentially a more significant clinical error than a shift in machine output. The common practice of relying on time-consuming measurements for patient pattern-specific MLC verification does not appropriately mitigate all risks of dose delivery errors due to machine performance. If such measurements are made, they should be combined with periodic MLC QA tests designed to span the range of clinical practice comprehensively and performed at a rationally designed frequency.

   Establishing a risk-based, generic MLC QA program requires knowledge of QM goals in relation to the wide variety of possible MLC failure modes, such as random positioning errors (leaf-specific), systematic shifts (for the entire leaf carriage), calibration errors (for example, a 1 mm leaf-gap error results in 5% dose delivery errors for dynamic MLC (dMLC) fields with a 2 cm average gap size), component wear, prescribed intensity variations which drive the MLC to or beyond its mechanical limits or capabilities, and problems compensating for gravity or gantry angle effects (LoSasso 2003, LoSasso et al. 2001). Though random leaf errors have small effects, systematic leaf gap calibration or carriage positioning errors (affecting an entire leaf bank) can influence delivery accuracy significantly (Rangel and Dunscombe 2009, Bayouth 2008, Liu et al. 2008). It has been shown that 1 mm systematic errors can give rise to dose errors of 5% or more for both dynamic MLC and static MLC deliveries (Mu et al. 2008). Using the methodology that led to Figure 9–13, it is possible to determine how often one must perform tests to maintain the 1 mm MLC positioning tolerance. Note that the details of appropriate MLC QA tests vary by manufacturer and system design (LoSasso 2003, LoSasso et al. 2001, Rangel and Dunscombe 2009, Bayouth 2008, Liu et al. 2008).

3. **Other Parameters**. Other machine operating parameters, e.g., radiation vs. mechanical isocenter coincidence and excursion, can be analyzed in a similar fashion. The TG–100 report encourages the use of non-specific, yet highly sensitive, tests of changes in machine geometry integrity, e.g., split field tests rather than more time consuming direct measurements of isocenter coincidence (Macklis et al. 1998).

Table 9–12 suggests frequencies for various QA tests compatible with the IMRT analysis. Though several parameters likely require daily testing, it should be straightforward to include those tests in the morning QA procedure. It is crucial to develop the capability to actively monitor treatment delivery in addition to performing periodic QA. Automated checks of dosimetry, MLC motion, patient setup, patient motion, and other issues during each fraction would help maintain accurate delivery, though this capability is lacking in most equipment (an area that deserves significant developmental effort). Several academic institutions have developed in-house monitoring software, so such capabilities are technologically very feasible and should be made a high-priority item for commercial development by linac and

Table 9–12. Example of accelerator QA checks suggested by the TG–100 analysis for the example IMRT process (taken from Huq et al. 2013)

Treatment Unit Parameter	Frequency of Testing per TG 100	Frequency of Testing per TG 142	Example Test
Unit Output			
Dosimetric constancy	At least every 3 or 4 days (for normal fractionation). Daily for few fraction treatments	Daily	Detector measurement in phantom at depth
Dose linearity with respect to number of monitor units	Commissioning or after major repairs (see part 1 on commissioning) with at least one, preferably independent, repeat verification	Annually	Detector measurement in phantom at depth, IMRT vs. normal delivery
Dosimetric constancy as a function of dose rate	Commissioning or after major repairs; at least one, preferably independent, repeat verification	Monthly	Detector measurement in phantom at depth to assess dosimetric constancy at all dose rates used.
Dosimetric constancy with respect to gantry angle	Commissioning, check after repair on bending magnet or beam alignment; at least one, preferably independent, repeat verification	Annually	Periodically perform the dosimetric consistency test with lateral beams or under the table
Stabilization for small monitor-unit settings	Commissioning or after major repairs; at least one, preferably independent, repeat verification. Also limiting the use to range of stable number of MU, check after major beam tuning		Detector measurement in phantom at depth
Beam Characterization			
Flatness and symmetry (beam profile)	Commissioning, then performed together with output constancy check using measurement device with off-axis detectors	Monthly	1-D or 2-D detector measurement in phantom at depth; *for checks, one off-axis point for each axis*
Beam energy	Commissioning, then performed with the output constancy check using flatness	Annually (monthly for electrons)	1-D or 2-D detector measurement in phantom at depth; *for checks, one off-axis point for each axis*
Collimation			
Positioning and calibration of MLC	Daily operational checks, at least weekly picket fence or similar IMRT-related tests	Weekly picket fence; monthly non-IMRT patterns and IMRT leaf position accuracy	Preferably image-based checks. Use light field with template if imaging is unavailable
Consistency of MLC with gantry orientation	Commissioning, and QA check with frequency dependent on sensitivity determined at commissioning	Monthly	Preferably image-based checks or light field (if accuracy validated)

Table 9–12, cont. Example of accelerator QA checks suggested by the TG–100 analysis for the example IMRT process (taken from Huq et al. 2013)

Treatment Unit Parameter	Frequency of Testing per TG 100	Frequency of Testing per TG 142	Example Test
Speed of MLC movement (if relevant to IMRT delivery method)	Commissioning, then routine confirmation of speed and delivery accuracy. Frequency required by risk-based analysis not clear yet	Monthly	Preferably image-based checks or light field (if accuracy validated)
Accuracy of the secondary collimators	Commissioning, then observation with the output checks.	Daily	Shadow of jaws compared with a template using the light field for large and small fields
Beam positioning			
Accuracy of gantry angle	For simple isocentric treatments, monthly is probably adequate. For off-axis or VMAT-type IMRT deliveries, weekly or daily may be necessary.	Monthly	Light field consistency with marks on wall and floor or bubble level
Accuracy of collimator angle	For complex IMRT and VMAT deliveries, weekly or even daily checks are important		Consistency with marks on floor or bubble level
Accuracy/consistency of the couch position	Insufficient information to specify, depends on type of setup used. Based on history and usage for a particular facility and whether IGRT is being used	Annual couch rotation, couch translation not addressed	Consistency of readouts with the couch positioned placing the cross hairs (gantry pointing down) sequentially on two marks on the table and with two settings on the ODI
Laser accuracy	Daily if laser setup used. Study clinical use to determine frequency if all setup is done with IGRT	Daily	Consistency marks on wall and floor
NOTE: This table is an example only and should not be used without an appropriate analysis of your own clinical process. The last column tests are examples only, not meant to be an exhaustive summary of available tests; the report of Task Group 198 may well provide such a summary in the future.			

treatment management system vendors (D'Souza et al. 2012, Margalit et al. 2011, Arnold et al. 2010, Bissonnette and Medlam 2010, Huang et al. 2005, Mutic et al. 2010).

Many delivery errors can be efficiently detected if therapists carefully monitor treatment while it is in progress. The American College of Radiology (ACR 2012) recommends staffing of two RTTs per treatment unit under "a standard schedule" and states that additional RTTs may be needed for longer hours or heavy patient loads. The newly published ASTRO document "Safety Is No Accident" states, "It is recommended that a minimum of two qualified individuals be present for any routine external beam patient treatment" (Zietman et al. 2012). These recommendations should be taken seriously by administrators. Situations where monitoring by an alert therapist can prevent machine performance from jeopardizing patient safety include verifying MLC motion during IMRT treatment, either between sMLC segments or dynamically during dMLC deliveries, acting promptly on clearly anomalous machine behavior, and notifying the physicist about all peculiar or unusual machine behaviors. It is incumbent upon the physicist to take such reports seriously, respond when called, and investigate the reported problems. Because attention to the treatment is a major function of the therapists, the console area, workflow, and department policies should be designed to minimize distractions and other pathways for attention lapses.

9.4.3.6 QM of Multimodality Imaging for Target Delineation

Specifying image sets to be used for target delineation, particularly when they are obtained outside a radiation oncology department is a serious potential source of difficult-to-detect (high D) errors in the planning process. In many centers, this process consists of the attending physician or resident reviewing the patient's imaging studies using the radiology PACS system. The desired image set is then identified and its study number is passed to dosimetrists who contact the appropriate radiologic technologist and request them to export the desired DICOM dataset into the RTP file server. There are many potential sources of error here, including propagation of an incorrect ID number, miscommunication (between therapist, dosimetrist, physician, and radiology technologist), and the possibility that the radiology technologist will export an incorrect image set. Because of the growing number and variety of MR and PET imaging studies that are used for planning, the dosimetrist and physicist cannot, on their own, verify the correctness of the secondary image sets imported into the planning system. Only if the physician notices that an incorrect dataset has been selected will the error be detected. Ways to change the process to decrease the likelihood of this failure mode include:

1. Obtain a modern PACS system that allows the physician to directly download the desired studies when they are viewed. This part of the solution requires a high-level managerial decision, but, if implemented, eliminates opportunities for miscommunication.
2. Expand the site-specific protocol for workup of a patient prior to IMRT planning (Example Checklist 1) to include the technique factors (e.g., MR pulse sequence, contrast, patient position, volume) to be used for each major clinical site and presentation. This will provide a basis for verifying the image datasets selected for planning.
3. Develop, and require the physician to complete, an on-line form that not only identifies PACS study ID, but also the date of the procedure and imaging technique desired.
4. Require the dosimetrist to verify that the imported secondary dataset is consistent with items 2 and 3.

FTA shows that QC check placed prior to importing DICOM images from the PACS server (where full technique information is available) to the RTP (where full DICOM header information is not reviewable in the model for the facility considered) reduces the probability of error. Furthermore, many of the initial planning directive error pathways can be managed by the same strategy: a dosimetrist performs QC checks of the inputs into the planning process by comparison against the treatment protocol, while the more comprehensive downstream physics checks of the plan use the same information for the final QA check. For example, to make an error in the secondary-to-primary image registration more detectable, the treatment protocol documents should specify the standard registration process to be used for the given clinical site (which image set is primary, type of registration used, e.g., manual vs. automated, which landmarks to align, etc.) This places the onus on the attending physician to request and document variances from the standard procedure where medically indicated. It should be noted that primary image set selection errors are not limited to cases where additional secondary imaging studies must be imported. It is not uncommon to have multiple CT simulation datasets (e.g., repeat exams for changing medical condition, adaptive replanning, or to correct a simulation error). Other errors that can be intercepted with this strategy include incorrect specification of goals and constraints and treatment planning approach/parameters. The treatment protocol can be implemented as a patient-specific form to be inserted in the patient's chart. Default or standard choices (e.g., DVH planning or evaluation constraints) would be printed in the form, so if the physician wants to modify these values, the default number is crossed out and the physician-specified number written by hand. This eliminates transcription errors characteristic of a form with simple blanks. On the other hand, it introduces the potential failure of using the default in error because the physician neglected to make a change (Thomadsen et al. 2003).

9.4.3.7 QM of Treatment Startup

The initial treatment session, Day 1 Treatment ("Day 1"), includes, the first day of planned changes within a single course of treatment: examples include cone-downs, field changes done in response to peer review or patient changes, and the introduction of a new treatment site concurrent with an on-going treatment. As the adaptive radiation therapy paradigm becomes more prevalent, the number of Day-1 sessions per patient is likely to increase. The major concern for the Day-1 treatment is establishing or verifying the treatment parameters that will be duplicated through the entire treatment course, since errors that are not detected at Day 1 may become systematic errors that will affect many or all of the treatments. There are many issues that must be handled within the QM program associated with Day-1 treatment, including the following:

- Parameters (e.g., couch positions, shifts from patient reference marks) and treatment accessories (e.g., bolus) may be defined or added to the plan during the first day's treatment in a way that circumvents the standard flow and checks of the treatment preparation process. The Day-1 QM must verify the correctness of these additions and assure that they are correctly continued through the treatment course.
- A QM check of the entire treatment delivery script before treatment is crucial. The "when, how, and who" for the performance of these checks depends on the details of the process used for preparation, plan download,

and Day-1 setup and verification. In all cases, though, the QM system must ensure that all parts of the plan are validated before treatment.

- It is essential to confirm that the correct patient and the correct treatment plan have been selected, endpoints addressed by the time-out process required by the Joint Commission. Although there is no specific guidance that a time out should be required for each treatment session, it is a good idea. During the time out, the patient's identity and treatment site, especially laterality, are confirmed, and any changes in patient condition that have bearing on the treatment are noted and conveyed to the physician. It is also confirmed that the correct files have been opened in the delivery system and the correct instructions in the paper or electronic record, and that prescribed changes in treatment have been addressed (with proper signatures in place). Especially at Day 1, radiation oncologist participation in the time-out process is a guard against deviations from physician intent.
- The entire Day-1 treatment process should be structured by written departmental procedures that clearly identify the parameters to be validated before treatment, and should be part of the training for all new physicians, physicists, dosimetrists, and therapists. All staff involved should understand the patient's treatment plan and associated treatment parameter tolerances. Individuals associated with each patient's treatment should be clearly identified.
- The initial imaging session and resultant marking of the patient or accessories may set the standard for the patient's position for the whole treatment course. Each facility should develop a policy describing the process for Day-1 imaging, setup, and verification of patient position and treatment isocenter(s). Patient-positioning errors need to be corrected via the appropriate image guidance strategy. For some disease sites, this refers to traditional weekly portal and orthogonal field imaging. With proper training and protocols, daily positioning corrections implemented by therapists and based on image guidance with off-line review by the radiation oncologist, can improve setup accuracy. However, if an incorrect isocenter placement remains uncorrected over much of the treatment, or if anatomy is misidentified or misinterpreted during the Day-1 procedure, a high severity treatment failure may result, indicating the need for QM procedures to mitigate this risk. One pragmatic approach defines classes of treatments with set tolerances. For example, a hypothetical protocol may allow prostate patients, as a class, up to 2 mm discrepancies between the DRR and beam image, while lung patients may be allowed up to 5 mm. Defining such classes of patient beforehand removes ambiguity and possible errors at the time of imaging.
- Depending on departmental policy, the monitor unit setting per segment (or the equivalent for other forms of IMRT) is validated prior to the first treatment by a second, independent calculation program or by measurements. Assuming that the validation methodology has been thoroughly

commissioned at or before Day 1, the medical physicist need only review that the independent check fell within department-specified tolerances.

- For both Day-1 and Day-N treatments, the human factor of "inattention" was frequently identified as a cause of failure. Treatment sessions can become repetitive exercises for therapists, and it is difficult for any individual to remain alert at all times. Training, policies, and managerial actions (e.g., sufficient staffing to allow for short breaks or rotating therapists between machines to keep them fresh) are partial solutions, but an additional layer of technical protection would be a much stronger and more effective approach. The TG recommends that manufacturers develop techniques to address such verification, such as a method of comparing records of the MLC positions in real time through the treatment to the pattern in the treatment plan.

9.4.3.8 QM of Human Factors

Several FMs are illustrative of issues that arise at the boundary of human and technological systems. The higher-ranking failures involve incorrect data being used for treatment, such as changes made but not entered correctly, or not entered at all, into the delivery system computer, or changes made inappropriately. The lower-ranked failures involve software or hardware failures. Many of the human failures have the same causes: lack of standardized procedures, inattention, inadequate training, and lack of communication. Prevention of these types of problems requires QM using both technological and human factors methods. The following list gives some recommended QM measures:

- Independent checks of all delivery system treatment plan parameters against those originally approved for use. A QA check of all this information is critical, and some kind of check has long been part of the weekly physics check, though weekly checks are clearly inadequate for some hypo-fractionated treatments. Modern treatments contain very many treatment parameters, so developing robust automated checks of this information is crucial.
- Methods to flag changes in delivery parameters and prevent further treatments until review and approval are performed. Such a feature exists in at least one modern treatment management system with regard to major delivery parameters. It should become a universal feature.
- Procedures to assure the consistency of daily treatment with the approved prescription(s) and plan(s) are crucial. Typically, this has been an important part of the weekly chart review by the medical physicist or dosimetrist and a separate weekly review by a therapist, both of whom should include checking the consistency of the daily treatment record(s) with the most current physician prescription(s). This check should also verify that all treatments have been correctly recorded in the official record, whether paper, electronic, or a combination.

- Methods to draw attention to unplanned changes, as well as expected changes that do not show up. Detecting unfulfilled change orders is often difficult if the orders are verbal or poorly documented. Especially in a combined electronic and paper environment, such issues can be relatively common if there is no established and uniform procedure. For any system—electronic, paper, or combination thereof—the process for making changes and triggering the appropriate QA checks of the change must be rigidly designed and followed. Though the common QA practice (also tied to billing and the recommendations of TG 40) of weekly paper or electronic chart checks helps detect some of these failures, this check is not adequate for many clinical scenarios.
- Dissemination of warnings about nonstandard behavior of the treatment system or involving the patient to appropriate staff for timely investigation. An anomalous condition should not be allowed to persist long enough to adversely affect any patient.
- Policies that establish electronic and procedural "permissions" so that change approvals are performed by the appropriate staff.

Lower-ranked failures include the treatment unit computer not loading the patient's file correctly (after having done so on Day 1) or file corruption. Assuming that the file corruption does not bring the treatment to a halt or trigger software messages, such failures can be extremely difficult to detect. Therapist monitoring might detect incorrect MLC movement, but many cases would not be apparent. Much of the ability to detect or prevent such problems relies on good software design. For example, use of file checksums to confirm validity of files can increase the detectability of such problems and should be encouraged. Other developments involving automatic monitoring would also address this FM.

9.4.3.9 QM of Data Transfer Integrity

Delivery data corruption is a high-severity failure mode that involves corruption of an MLC file used to control MLC leaf motions required to deliver an IMRT plan. Such a corruption might present as an unreadable file, which might prevent a failure at this step, or it could present as an empty file that might be used for treatment. The most hazardous situation involves plan revision on Day N, as in a recently reported case in which the MLC trajectories were lost during transfer of a revised plan (Bogdanich 2010). Since the MLC trajectory information was missing, open fields were treated with the IMRT MU, resulting in severe overdoses to the patient. There are also anecdotal (but undocumented) reports of incorrect leaf motions in previously treated but unchanged plans. For this failure to occur, the file must be accepted by the machine as valid, but contain an incorrect set of leaf sequences. While the likelihood of occurrence may be low, such errors clearly can occur, even though their true frequency is unknown. Because of its high-severity ranking, the QM program should address this FM. Testing all fluence patterns to confirm their correct behavior and resulting intensity distributions before any clinical use is crucial. While some institutions have substituted calculation checks for a physical plan delivery, such checks must be accompanied by strict QA of the generic performance of the MLC. Even with these, it is incumbent on the institution to verify that this combination of checks can really confirm the correctness of a delivery based on a given MLC description. In the absence of such verification methods, this task group recommends physical deliveries of any new fluence pattern be confirmed for correctness when delivered on a machine. We also note that pretreatment verification does not address potential "Day N" delivery problems. Checking the leaf pattern daily could provide protection from the most serious effects on the patient, but for a facility with even a third of the patients under IMRT treatments at any given time, this check would consume a great deal of time and be clinically unfeasible. It is crucial that the vendors provide automated tools to avoid this daily problem (e.g., use of checksums or other automated checks, automated comparison of EPID dose or dose back-calculated from MLC log files with corresponding calculations from the TPS or with pretreatment data) that might demonstrate the MLC descriptions were identical and unchanged from day to day (Luo et al. 2006, Litzenberg et al. 2002, Prabhakar et al. 2011, Mans et al. 2010, Stell et al. 2004, Sabet et al. 2012). Similar automated checks for all files related to a patient's treatment are also necessary. At present, there is no widely available procedure to prevent this potential failure.

9.4.4 Quality Management Program Components

Design and implementation of a quality management program based on all the information generated in an FMEA or FTA analysis of this scope is a large and complex task, the details of which are highly dependent on how the planning and delivery processes are implemented in the institution, as well as on the types of treatments administered. Thus a QM program should be individualized to the relevant processes, case mix, methods, and equipment used in each center, and it is not expected that one standard set of QM guidelines and methods will be appropriate for each clinic. To efficiently encompass the QM tasks that have been identified, it is clear that collecting appropriate quality assurance, quality control, and procedural tasks into a set of recommendations will help organize the QM to help increase safety and the quality of treatment for patients treated with IMRT.

The QM elements discussed above can be viewed in the context of Table 9–13, which is based on the suggestions of the ISMP (ISMP 1999). It should be noted that education, while at the bottom of the list in this table, is essential for correct planning and execution of procedures. However, even with the best training, humans fail, and relying on education to prevent all failures proves less effective than the tools ranked more highly. Redundancy, independent checks, and operational checks (periodic QA) fall in the middle of the list, yet serve important functions in radiotherapy QM. When possible, the most effective tools should be used, but resources and practicality often lead to tools from later in the table. Used judiciously, any of the tools provide value in controlling quality and preventing the propagation of failures. After performing a process for some time, reevaluation of the process itself provides a refined picture of the effectiveness of the quality measures instituted and

Table 9–13. Ranking of QM tools based on their effectiveness with examples, in part following the suggestions of the Institute for Safe Medical Practices (ISMP 1999)

1. Forcing functions and constraints • Interlock • Barriers • Computerized order entry with feedback **2. Automation and computerization** • Bar codes • Automated monitoring • Computerized verification • Computerized order entry **3. Protocols, standards, and information** • Check-off forms • Establishing protocol/clarify protocol • Alarms • Labels • Signs • Reduce similarity **4. Independent double check systems and other redundancies** • Redundant measurement • Independent review	• Operational checks • Comparison with standards • Increase monitoring • Add status check • Acceptance test **5. Rules and policies** • Priority • Establishing/clarify communication line • Staffing • Better scheduling • Mandatory pauses • Repair • PMI (preventive maintenance inspection) • Establish and perform QC and QA (hardware and software) **6. Education and Information** • Training • Experience • Instruction

can identify new hazardous steps. Input for the reevaluation comes principally from three sources:

- *Records of events, failures, and near misses.* Establishing a reporting system and database for events and capturing information from root-cause analyses can add *a-posteriori* statistical data to the *a-priori* estimates used by FMEA (Ekaette et al. 2006). Such data can also uncover problems either not recognized during the FMEA or created by the quality program. Reports of events focus not on potential failures, but failures that happened, giving much more power to actions established to prevent future failures. Near miss reports prove extremely valuable. During near misses, failures occurred, but some actions prevented the failures from compromising the patient's treatment. The actions that saved the situation from affecting the patient give insights into what actions might be effective in intercepting similar failures in future cases. Reporting systems also can be open to receiving information on hazardous situations noted by the staff before developing into failures. All three types of information can assist in refining the quality management program.
- *Quality audits.* Quality audits consist of having knowledgeable persons from outside the facility review the program. The audits include a "product" audit which, in a medical setting, would consist of reviewing cases and assessing whether all patient care was appropriate and complete, and a "process" audit that reviews the standardized procedures and evaluates whether they function well in the setting. The audits might also include measurements, either on-site or performed remotely, to assess the accuracy of treatment unit calibration or other operating parameters.

- *Quality improvement.* Input for quality improvement comes from the event database, the audits, and the QA and QC procedures. QI identifies the parts of the quality program that need changes or enhancement and the parts of the process that would benefit from redesign.

Finally, one of the general results of the FMEA and associated FTA is the clear need to define site-specific treatment planning and delivery protocols that serve as the basis for simulation, planning, and treatment delivery expectations, methods, and QM procedures. This general standardization and documentation of the methods to be used addresses many of the most common failure modes for many of the most critical steps in the planning and delivery process, and these steps are crucial to avoiding training and procedure lapses. A checklists that summarizes issues, procedures, decisions, and QM that should be defined for each clinical, site-specific, pretreatment workup protocol provides considerable protection against failures.

9.4.5 Summary and Conclusions

The creation and maintenance of a complete quality management program for clinical use of any complex technology, such as IMRT, requires detailed analysis, continual improvement, and development of increasingly more effective QM measures, plus continuous attention to both details and the overall goal of achieving safe and effective patient treatment.

In this section, we have demonstrated the use of failure modes and effects analysis and fault tree analysis methods to study a generic planning and delivery process for IMRT to illustrate how to apply these general tools to a complex radiotherapy technology and formulate a more comprehensive IMRT quality management program that more effectively promotes quality with more efficient use of available resources. The generic nature of the analyses does not make it possible to give complete guidance for any single specific clinical implementation. However, the recommendations set forth should be a guide to individual institutions as they apply these techniques to their individual processes.

The FMEA and FTA were used to determine the most likely points of failure and to construct a model QM program for one example generic IMRT treatment process. Analysis of the types and causes of failures and their relative severity (S), likelihood of occurrence (O), and lack of detectability (D) assigned by TG 100 was used to order failure modes by risk and severity. This ranked list of failure modes was then analyzed to determine, or at least identify, quality management steps that would mitigate those failure modes. Aside from the many QM recommendations produced by the analysis, a number of "key core components" for quality were identified. Their absence in the QM program significantly increases the likelihood that a large fraction of the failure modes identified will actually occur. The key core components that any safe and high-quality IMRT program must include are:

- standardized procedures,
- adequate training of staff, and
- clear lines of communication among staff.

In addition to these, other components essential for quality treatments include:

- maintenance of hardware and software resources and
- adequate staffing, physical environment, and computer resources.

Regardless of a department's treatment process or methods, it is expected that other physicists will identify various individual potential failure modes from their own experience of incidents or near misses that are not considered in this work. These issues must be included in future analyses so that the QM program for IMRT (and other techniques) continues to become more successful at preventing safety and quality problems. It is essential that all members of the radiation therapy team continue to enhance the quality of the QM program.

References

Abt Associates Inc. *The Abt Study of Medical Physicist Work Values for Radiation Oncology Physics Services: Round III.* Prepared for the American College of Medical Physics (ACMP) and the American Association of Physicists in Medicine (AAPM), March, 2008.

American College of Radiology (ACR). "Radiation Oncology Accreditation Program Requirements." Reston, VA: American College of Radiology, 2012. Available at http://www.acr.org/Quality-Safety/Accreditation/RO.

Arthur, D.W., F.A Vicini, D.A Todor, T.B Julian, M.R Lyden. (2011). "Improvements in critical dosimetric endpoints using the Contura multilumen balloon breast brachytherapy catheter to deliver accelerated partial breast irradiation: preliminary dosimetric findings of a phase IV trial." *Int Radiat Oncol Biol Phys* 79:26–33.

Bayouth, J.E. (2008). "Siemens multileaf collimator characterization and quality assurance approaches for intensity-modulated radiotherapy." *Int J Radiat Oncol Biol Phys* 71:S93–7.

Bissonnette, J-P., G. Medlam. (2010). "Trend analysis of radiation therapy incidents over seven years." *Radiother Oncol* 96:139–144.

Bogdanich, W. (2010. "Radiation offers new cures and new ways to do harm." *New York Times*, January 23, 2010.

Bogdanich, W., K. Rebelo. "Radiation Errors Reported in Missouri." *New York Times*, Feb. 25, 2010.

Calandrino, R., G.M. Cattaneo, C. Fiorino, B. Longobardi, P. Mangili, P. Signorotto. (1997). "Detection of systematic errors in external radiotherapy before treatment delivery." *Radiother Oncol* 45:271–4.

Chera, B.S., M. Jackson, L.M. Mazur, R. Adams, S. Chang, K. Deschesne, T. Cullip, L.B. Marks. (2012). "Improving quality of patient care by improving daily practice in radiation oncology." *Semin Radiat Oncol* 22:77–85.

Ciocca, M., M.C. Cantone, I. Veronese, F. Cattani, G. Pedroli, S. Molinelli, V. Vitolo, R. Oreccia. (2012). "Application of failure mode and effects analysis to intraoperative radiation therapy using mobile electron linear accelerators." *Int J Radiat Oncol Biol Phys* 82:e305–11.

Clark B.G, R.J. Brown, J. Ploquin, P. Dunscombe. (2013). "Patient safety improvements in radiation treatment through five years of incident learning." Prac Radiat Oncl 3:in press.

Clark, B.G., R.J. Brown, J.L. Ploquin, A.L. Kind, L. Grimard. (2010). "The management of radiation treatment error through incident learning." *Radiother Oncol* 95:344–349.

Constantinou, C., E.S. Sternick. (1984). "Reduction of the 'horns' observed on the beam profiles of a 6-MV linear accelerator." *Med Phys* 11:840–2.

Cunningham, J., M. Coffey, T. Knöös, O. Holmberg. (2010). "Radiation Oncology Safety Information System (Rosis) – Profiles of Participants and the First 1074 Incident Reports." *Radiother Oncol* 97:601–07.

D'Souza, N., L. Holden, S. Robson, K. Mah, L. De Prospero, C.S. Wong, E. Chow, J. Spayne. (2012). "Modern palliative radiation treatment: do complexity and workload contribute to medical errors." *Int J Radiat Oncol Biol Phys* 84:e43–e48.

Denny, D.S., D.K. Allen, N. Worthington, D. Gupta. (2012). "The use of failure mode and effect analysis in a radiation oncology setting: the Cancer Treatment Centers of America experience." *J Healthc Qual* Feb 24, 2012. [Epub ahead of print.]

Dunscombe, P.B., S. Iftody, N. Ploquin, E.U. Ekaette, R.C. Lee. (2007). "The Equivalent Uniform Dose as a severity metric for radiation treatment incidents." *Radiother Oncol* 84:64–6.

Duwe, B., D.B. Fuchs, J. Hansen-Flaschen. (2005). "Failure mode and effects analysis application to critical care medicine." *Crit Care Clin* 21:21–30.

Ekaette, E.U., R.C. Lee, D.L. Cooke, K.L. Kelly, P.B. Dunscombe. (2006). "Risk analysis in radiation treatment: application of a new taxonomic structure." *Radiother Oncol* 80:282–7.

Ford, E., K. Smith, J. Keck, K. Harris, S. Terezakis, G. Sibley. (2012). "A Method of Streamlined Failure Mode and Effect Analysis to Improve Patient Safety." *Med Phys* 39:3600 (abstr).

Ford, E.C., L. Fong de los Santos, T. Pawlicki, S. Sutlief, P. Dunscombe. (2012). "Consensus recommendations for incident learning database structures in radiation oncology." *Med Phys* 39:7272–7290.

Ford, E.C., R. Gaudette, L. Myers, B. Vanderver, L. Engineer, R. Zellars, D.Y. Song, J. Wong, T.L. Deweese. (2009). "Evaluation of Safety in a Radiation Oncology Setting Using Failure Mode and Effects Analysis." *Int J Radiat Oncol Biol Phys* 74:852–8.

Gregoire, V., P. Levendag, K.K. Ang, J. Bernier, M. Braaksma, V. Budach, C. Chao, E. Coche, J.S. Cooper, G. Cosnard, A. Eisbruch, S. El-Sayed, B. Emami, C. Grau, M. Hamoir, N. Lee, P. Maingon, K. Muller, H. Reychler. (2003). "CT-based delineation of lymph node levels and related CTVs in the node-negative neck: DAHANCA, EORTC, GORTEC, NCIC, RTOG consensus guidelines." *Radiother Oncol* 69:227–36.

Huang, G., G. Medlam, J. Lee, S. Billingsley, J.P. Bissonnette, J. Ringash, G. Kane, D.C. Hodgson. (2005). "Error in the delivery of radiation therapy: results of a quality assurance review." *Int J Radiat Oncol Biol Phys* 61:1590–5.

Hunt, M.A., G. Pastrana, H.I. Amols, A. Killen, K. Alektiar. (2012). "The impact of new technologies on radiation oncology events and trends in the past decade: an institutional experience." *Int J Radiat Oncol Biol Phys* 84:925–931.

Huq, M.S., B.A. Fraass, P.B. Dunscombe, J.P. Gibbons, Jr., G.S. Ibbott, P.M. Medin, A. Mundt, S. Mutic, J.R. Palta, B.R. Thomadsen, J.F. Williamson, E.D. Yorke. (2008). "A Method for Evaluating Quality Assurance Needs in Radiation Therapy." *Int Radiat Oncol Biol Phys* 71 (Suppl 1):S170–S173.

Huq, M.S., B.A. Fraass, P.B. Dunscombe, J.P. Gibbons, Jr., G.S. Ibbott, P.M. Medin, A. Mundt, S. Mutic, J.R. Palta, B.R. Thomadsen, J.F. Williamson, E.D. Yorke. (2013) "Application of risk analysis methods to radiation therapy quality management: Report of AAPM Task Group 100." *Med Phys* (in press).

Institute for Safe Medical Practices (ISMP). *Medication error prevention "toolbox."* 1999. Available from http://www.ismp.org/msaarticles/toolbox.html. Accessed March 18, 2013.

International Atomic Energy Agency (IAEA). Training course slides, especially set 2.6, 2.10, and 3.1. Available at https://rpop.iaea.org/RPOP/RPoP/Content/AdditionalResources/Training/1_TrainingMaterial/AccidentPreventionRadiotherapy.htm. Accessed January 27, 2013.

Israelski, E.W., W.H. Muto. (2004). " Human factors risk management as a way to improve medical device safety: a case study of the Therac 25 radiation therapy system." *JT Comm J Qual Saf* 30:689–95.

Kapanen, M., M. Tenhunen, R. Parkkinen, P. Sipila, H. Jarvinen. (2006). "The influence of output measurement time interval and tolerance on treatment dose deviation in photon external beam radiotherapy." *Phys Med Biol* 51:4857–67.

Karzmark, C.J. "Radiotherapy Treatment Simulator Design" in *Advances in Radiation Therapy Planning*, A.E. Wright and A.L. Boyer, eds. Medical physics monograph #9. Madison, WI: Medical Physics Publishing, 1982.

Kaus, M.R., K.K. Brock, V. Pekar, L.A. Dawson, A.M. Nichol, D.A. Jaffray. (2007). "Assessment of a model-based deformable image registration approach for radiation therapy planning." *Int J Radiat Oncol Biol Phys* 68:572–80.

Keall, P.J., G.S. Mageras, J.M. Balter, R.S. Emery, K.M. Forster, S.B. Jiang, J.M. Kapatoes, D.A. Low, M.J. Murphy, B.R. Murray, C.R. Ramsey, M.B. Van Herk, S.S. Vedam, J.W.

Wong, E. Yorke. (2006). "The management of respiratory motion in radiation oncology report of AAPM Task Group 76." *Med Phys* 33:3874–900.
Klein, E.E. (2010). "A grid to facilitate physics staffing justification." *J Appl Clin Med Phys* 11:2987.
Klein, E.E., J. Hanley, J. Bayouth, F.F. Yin, W. Simon, S. Dresser, C. Serago, F. Aguirre, L. Ma, B. Arjomandy, C. Liu, C. Sandin, T. Holmes. (2009). "Task Group 142 report: quality assurance of medical accelerators." *Med. Phys.* 36:4197–4212.
Kubo, H.D., G.P. Glasgow, T.D. Pethel, B.R. Thomadsen, J.F. Williamson. (1998). "High Dose-Rate Brachytherapy Treatment Delivery: Report of the Aapm Radiation Therapy Committee Task Group No. 59." *Med Phys* 25:375–403.
Kuettel, M., ed. *Manual for ACRO Accreditation.* American College of Radiation Oncology (ACRO). Bethesda, MD: ACRO, 2013. Available at http://acro.org/Accreditation/ACROManualWeb%209-21-12.pdf.
Kutcher, G.J., L. Coia, M. Gillin, W.F. Hanson, S. Leibel, R.J. Morton, J.R. Palta, J.A. Purdy, L.E. Reinstein, G.K. Svensson. (1994). "Comprehensive QA for radiation oncology: report of AAPM Radiation Therapy Committee Task Group 40." *Med Phys* 21:581–618.
Layfield, L.J., G.M. Anderson. (2010). "Specimen Labeling Errors in Surgical Pathology An 18-Month Experience." *Am J Clinl Pathol* 134: 466–470.
Lee, W.R., M. Roach, 3rd, J. Michalski, B. Moran, D. Beyer. (2002). "Interobserver variability leads to significant differences in quantifiers of prostate implant adequacy." *Int J Radiat Oncol Biol Phys* 54:457–61.
Litzenberg, D.W., J.M. Moran, B.A. Fraass. (2002). "Verification of dynamic and segmental IMRT delivery by dynamic log file analysis." *J Appl Clin Med Phys* 3:63–72.
Liu, C., T.A. Simon, C. Fox, J. Li, J.R. Palta. (2008). "Multileaf collimator characteristics and reliability requirements for IMRT Elekta system." *Int J Radiat Oncol Biol Phys* 71:S89–92.
LoSasso, T. "IMRT delivery system QA." In *Intensity-Modulated Radiation Therapy*, J.R. Palta and T.R. Mackie, eds. Madison, WI: Medical Physics Publishing, 2003.
LoSasso, T., C.-S. Chui, C.C. Ling. (2001). "Comprehensive quality assurance for the delivery of intensity modulated radiotherapy with a multileaf collimator used in the dynamic mode." *Med Phys* 28:2209–2219.
Luo, W., J. Li, R.A. Price, Jr., L. Chen, J. Yang, J. Fan, Z. Chen, S. McNeeley, X. Xu, C.M. Ma. (2006). "Monte Carlo based IMRT dose verification using MLC log files and R/V outputs." *Med Phys* 33:2557–64.
Macklis, R.M., T. Meier, M.S. Weinhous. (1998). "Error rates in clinical radiotherapy." *J Clin Oncol* 16:551–556.
Mans, A., M. Wendling, L.N. McDermott, J.J. Sonke, R. Tielenburg, R. Vijlbrief, B. Mijnheer, M. van Herk, J.C. Stroom. (2010). "Catching errors with in vivo EPID dosimetry." *Med Phys* 37:2638–44.
Margalit, D.N., Y.H. Chen, P.J. Catalano, K. Heckman, T. Vivenzio, K. Nissen, L.D. Wolfsberger, R.A. Cormack, P. Mauch, A.K. Ng. (2011). "Technological advancements and error rates in radiation therapy delivery." *Int J Radiat Oncol Biol Phys* 81:e673–9.
Marks, L.B., K.L. Light, J.L. Hubbs, D.L. Georgas, E.L. Jones, M.C. Wright, C.G. Willett, F.F Yin. (2007). "The impact of advanced technologies on treatment deviations in radiation treatment delivery." *Int J Radiat Oncol Biol Phys* 69:1579–1886.
Marwick, C., E.Watts, J. Evans, P. Davey. (2007). "Quality of care in sepsis management: development and testing of measures for improvement." *J Antimicrob Chemother* 60: 694–697.
Moran, J.M., M. Dempsey, A. Eisbruch, B.A. Fraass, J.M. Galvin, G.S. Ibbott, L.B. Marks. (2011). "Safety considerations for IMRT: Executive summary." *Pract Radia Oncol* 1:190–195.
Mu, G., E. Ludlum, P. Xia. (2008). "Impact of MLC leaf position errors on simple and complex IMRT plans for head and neck cancer." *Phys Med Biol* 53:77–88.
Mutic, S., J.R. Palta, E.K. Butker, I.J. Das, M.S. Huq, L.N. Loo, B.J. Salter, C.H. McCollough, J. Van Dyk. (2003). "Quality Assurance for Computed-Tomography Simulators and the Computed-Tomography-Simulation Process: Report of the AAPM Radiation Therapy Committee Task Group No. 66." *Med Phys* 30:2762–92.
Mutic, S., R.S. Brame, S. Oddiraju, P. Parikh, M.A. Westfall, M.L. Hopkins, A.D. Medina, J.C. Danieley, J.M. Michalski, I.M. Naqa, D.A Low, B. Wu. (2010). "Event (Error and

near-Miss) Reporting and Learning System for Process Improvement in Radiation Oncology." *Med Phys* 37: 5027–36.

Nath, R., L.L. Anderson, J.A. Meli, A.J. Olch, J.A. Stitt, J.F. Williamson. (1997). "Code of Practice for Brachytherapy Physics: Report of the Aapm Radiation Therapy Committee Task Group No. 56. American Association of Physicists in Medicine." *Med Phys* 24:1557–98.

National Patient Safety Foundation (NPSF). Patient Safety Dictionary. http://www.npsf.org/for-healthcare-professionals/resource-center/definitions-and-hot-topics/patient-safety-dictionary_Accessed March 18, 2013.

Novak, P., E.G. Moros, W.L. Straube, R.J. Myerson. (2005). "Treatment delivery software for a new clinical grade ultrasound system for thermoradiotherapy." *Med Phys* 32:3246–56.

Paté-Cornell, M.E., L.M. Lakats, D.M. Murphy, D.M.Gaba. (1997). "Anesthesia patient risk: a quantitative approach to organizational factors and risk management options." *Risk Anal* 17:511–23.

Pawlicki, T., M. Whitaker, A.L. Boyer. (2005). "Statistical process control for radiotherapy quality assurance." *Med Phys* 32:2777–86.

Perkins, C., D. Balma, R. Garcia, Members of the Consensus Group, Susan G. Komen for the Cure. (2007). "Why current breast pathology practices must be evaluated. A Susan G. Komen for the Cure white paper: June 2006." *Breast J* 13:443–47.

Perks, J.R., S. Stanic, R.L. Stern, B. Henk, M.S. Nelson, R.D. Harse, M. Mathai, J.A. Purdy, R.K. Valicenti, A.D. Siefkin, A.M. Chen. (2012). "Failure Mode and Effect Analysis for Delivery of Lung Stereotactic Body Radiation Therapy." *Int J Radiat Oncol Biol Phys* 83:1324–29.

Prabhakar, R., J. Cramb, T. Kron. (2011). "A feasibility study of using couch-based real time dosimetric device in external beam radiotherapy." *Med Phys* 38:6539–52.

Raab S.S., D.M. Grzybicki. (2010). "Quality in Cancer Diagnosis." *CA Cancer J Clin.* 60:139–165.

Raab, S.S., D.M. Grzybicki, J.E. Janosky, R.J. Zarbo, F.A. Meier, C. Jensen, S.J. Geyer. (2005). "Clinical impact and frequency of anatomic pathology errors in cancer diagnoses." *Cancer* 104:2205–13.

Radiation Oncology Safety Information System (ROSIS). http://www.rosis.info/ Accessed January 27, 2013.

Rangel, A., P. Dunscombe. (2009). "Tolerances on MLC leaf position accuracy for IMRT delivery with a dynamic MLC." *Med Phys* 36:3304–9.

Royal College of Radiologists, Society and College of Radiographers, Institute of Physics and Engineering in Medicine, National Patient Safety Agency, British Institute of Radiology. *Towards Safer Radiotherapy.* London: The College of Radiologists, 2008. Available at https://www.rcr.ac.uk/docs/oncology/pdf/Towards_saferRT_final.pdf.

Sabet, M., P. Rowshanfarzad, P. Vial, F.W. Menk, P.B. Greer. (2012). "Transit dosimetry in IMRT with an a-Si EPID in direct detection configuration." *Phys Med Biol* 57:N295–306.

Santanam, L., C. Hurkmans, S. Mutic, C. van Vliet-Vroegindeweij, S. Brame, W. Straube, J. Galvin, P. Tripuraneni, J. Michalski, W. Bosch. (2012). "Standardizing naming conventions in radiation oncology." *Int J Radiat Oncol Biol Phys* 83:1344–49.

Sawant, A., S. Dieterich, M. Svatos, P. Keall. (2010). "Failure mode and effect analysis-based quality assurance for dynamic MLC tracking systems." *Med Phys* 37:6466–79.

Sheridan-Leos, N., L. Schulmeister, S. Hartranft. (2006)."Failure mode and effect analysis: a technique to prevent chemotherapy errors." *Clin j Oncol Nurs* 10:393–8.

Smith, B.D., D.W. Arthur, T.A. Buchholz, B.G. Haffty, C.A. Hahn, P.H. Hardenbergh, T.B. Julian, L.B. Marks, D.A. Todor, F.A. Vicini, T.J. Whelan, J. White, J.Y. Wo, J.R. Harris. (2009). "Accelerated Partial Breast Irradiation Consensus Statement from the American Society for Radiation Oncology (ASTRO)." *Int J Radiat Oncol Biol Phys* 74: 987–1001.

Stamatis, D.H. *Failure Mode and Effect Analysis: FMEA from Theory to Execution.* Milwaukee, WI: ASQC Quality Press, 1995.

Staradub, V.L., K.A. Messenger, N.J. Hao, E.L. Wiley, M. Morrow. (2002). "Changes in breast cancer therapy because of pathology second opinions." *An Surg Oncol* 9:982–987.

Stell, A.M., J.G. Li, O.A. Zeidan, J.F. Dempsey. (2004). "An extensive log-file analysis of step-and-shoot intensity modulated radiation therapy segment delivery errors." *Med Phys* 31:1593–602.

Terezakis, S.A., P. Pronovost, K. Harris, T. DeWeese, E. Ford. (2011). "Safety strategies in an academic radiation oncology department and recommendations for action." *JT Comm J Qual Patient Saf* 37:291–299.

Thomadsen, B., S.-W. Lin, P. Laemmrich, T. Waller, A. Cheng, B. Caldwell, R. Rankin, J. Stitt. (2003). "Analysis of treatment delivery errors in brachytherapy using formal risk analysis techniques." *Int J Radiat Oncol Biol Phys* 57:1492–508.

Wetterneck, T.B., K.A. Skibinski, T.L. Roberts, S.M. Kleppin, M.E. Schroeder, M. Enloe, S.S. Rough, A.S. Hundt, P. Carayon. (2006). "Using failure mode and effects analysis to plan implementation of smart i.v. pump technology." *Am J Health Syst Pharm* 63:1528–38.

Williamson, J.F., P.B Dunscombe, M.B. Sharpe, B.R. Thomadsen, J.A. Purdy, J.A. Deye. (2008). "Quality assurance needs for modern image-based radiotherapy: recommendations from 2007 interorganizational symposium on 'quality assurance of radiation therapy: challenges of advanced technology.'" *Int J Radiat Oncol Biol Phys* 71 (1 Suppl):S2–12.

Zannis, V., P. Beitsch, F. Vicini, C. Quiet, A. Keleher, D. Garcia, H. Snider, M. Gittleman, H. Kuerer, E. Whitacre, P. Whitworth, R. Fine, B. Haffty, A. Stolier, J. Mabie. (2005). "Descriptions and outcomes of insertion techniques of a breast brachytherapy balloon catheter in 1403 patients enrolled in, the American Society of Breast Surgeons MammoSite breast brachytherapy registry trial." *Am J* Sug 190:530–538.

Zietman, A.L., J.R. Palta, M.L. Steinberg, eds. *Safety is No Accident.* Fairfax, VA: American Society for Radiation Oncology, 2012. Available at https://www.astro.org/uploadedFiles/Main_Site/Clinical_Practice/Patient_Safety/Blue_Book/SafetyisnoAccident.pdf.

Exercise 1

Work Systems Design

Barrett Caldwell and Jennifer L. Johnson

Objectives:

After this exercise, the participant will be able to:

1. Analyze a situation using work system design concepts
2. Evaluate the elements of the safety culture in the example
3. Create recommendations for a safer work system design

Scenario:

- The physician was not fully present for the CT simulation (á la "drive by sim"). The CT scan did not cover enough of the anatomy (superiorly.)
- The dosimetrist completing the treatment planning may or may not have asked either the physician or physics about the scan. The dosimetrist completed treatment planning on the inadequate scan by just extended the axial slices superiorly to get the volume. A representative generated DRR is included below.

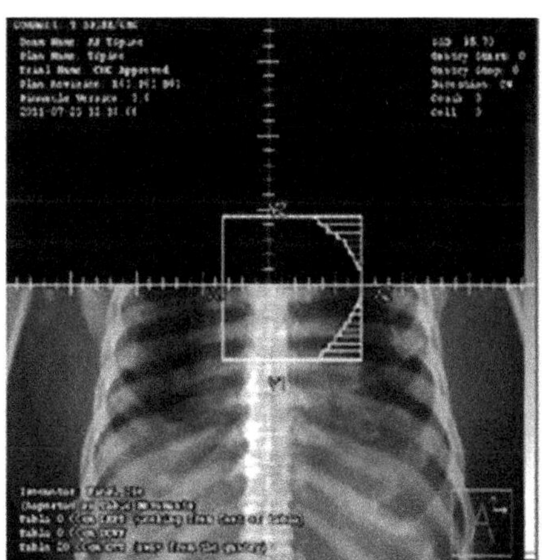

- The physician reviewed and approved the treatment plan.
- The physicist reviewed and approved the treatment plan but did not communicate any issues to the therapists.

- The therapists did not receive the plan with enough time to review before the patient was brought into the room. With the patient on the table, the therapists discovered the treatment field DRRs did not have sufficient anatomy in the treatment field to compare with daily IGRT images. Therapists called physics for assistance.
- A different, covering physicist determined that doing an on-treatment couch simulation would "fix" the treatment field DRR problem. The physician was unavailable. The therapists, uncomfortable with the plan, stopped the process and did not proceed with the treatment.

Tasks:

1. Identify a work system design concept in each phase (delineated by the bullets) of the Work Systems Design example scenario.
2. Appraise the safety culture of the organization in the example scenario and justify the judgment (e.g., is it strong? weak? both? neither?).
3. Consider and discuss how the safety culture and work culture can affect individual performance and the use of technology in an organization.
4. Create three suggestions integrating both safety culture and work system design concepts for a safer work system design of the simulation-to-treatment process for this organization.

Exercise 2

Taxonomy

Bruce Thomadsen

Objectives:

After this exercise, participants will be able to:

1. Classify the nature of failures according to several taxonomies
2. Understand the type of information that forms the focus of the taxonomies considered

Scenario:

- A patient was treated for cervical cancer with high dose-rate, intracavitary brachytherapy. It was very difficult to place the applicator due to disease eroding the cervical os, and the insertion took much longer than usual.
- When the patient arrived at the MR suite, because of the delay in the case, the staff at the unit was in the middle of a very involved case, and the brachytherapy patient had to wait.
- After the imaging, one of the facility's two physicists performed the treatment plan. Due to the complexity of the case, several trial plans were generated. The plan that covered the high-risk CTV best would deliver excessive doses to the rectum. The approved plan brought the dose off the rectum but sacrificed the HR-CTV coverage.
- Again because of the delays in the case, the facility's other physicist had to go to a different clinic to plan a stereotactic radiosurgery case and was not available to perform the checks on the case.
- The planning physicist erroneously uploads the plan that treated the entire HR-CTV to the treatment unit.
- Because the case ran so late and the second physicist was not present, the physics check of the loaded program was omitted and the treatment was initiated and completed.
- At the beginning of the next fraction, the error in the first fraction was noticed.

Exercises:

1. Classify the failures in this case using the AAPM Consensus Taxonomy (Table 3–4) and the other taxonomy from Chapter 3 (Table 3–3, Figures 3–3a, b, and c, and Figure 3–5).
2. Distinguish between the classifications established using a Labyrinth and pinball approach.
3. Determine what sort of information each of the taxonomies return.

Exercise 3

Process Mapping

Derek Brown and Eric Ford

Objective:

The goal of this exercise is to develop a process map for IMRT treatment planning from the time the dosimetrist receives the final region-of-interest contours from the physician to the time the plan is ready to be treated.

Exercise:

Below is a rough guide that can be used as an outline for this exercise, as well as some tips for creating useful process maps.

Step 1: Decide which process to map. The scale of the process is an important concern here. Mapping the entire external beam radiation oncology process, for example, is a large project that could take many weeks. We'll use these process maps in the FTA and FMEA exercises—this should be considered when deciding the scope and granularity of the map.

Step 2: Form a group and identify a team leader. Normally we would include a representative from all professional groups involved, but in the context of this exercise, this is not possible.

Step 3: Create an initial process map. It is often useful to make a first draft that does not attempt to capture the entire process in detail, but rather the workflow at a more general level.

Step 4: Iterative mapping. Refine the process map, adding levels of detail as necessary.

Step 5: As a large group we'll compare the process maps created by each smaller groups.

Step 6: Use the process map as the basis of the FTA and FMEA exercises.

Tips for Creating Useful Process Maps

1. It is often useful to look at processes from the patient's perspective.
2. For clinical processes, a **multi-professional team** is necessary for the development of a valid map.
3. The number of subprocesses identified should be the **smallest number** to meet the objective.
4. The users of the map should have the **same understanding** of the meaning of the sub-processes.

5. Choose the right level of detail. A map that is too general loses its utility, while one that is too detailed becomes unmanageable, and staff members can lose the big picture.
6. Don't get hung up on fancy graphics. There is value in the **process of creating the map.**

Example process maps

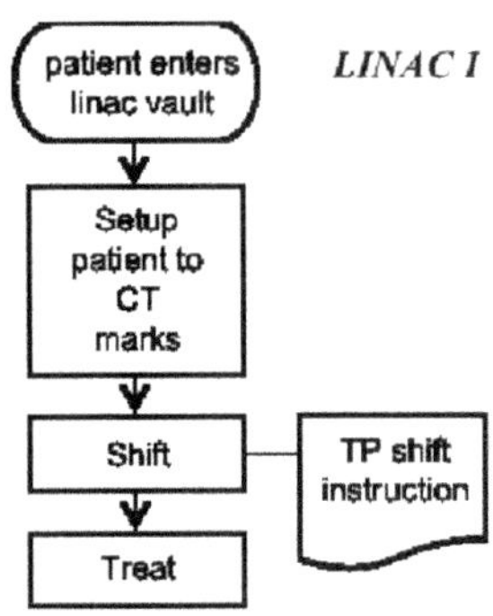

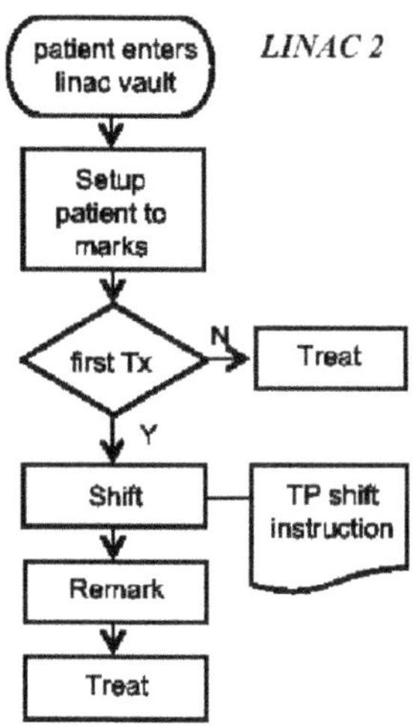

Exercise 4

Process Failure Mode and Effects Analysis

Frank Rath

Objectives:

After this exercise, the participants should be able to:

1. Perform a basic Failure Mode and Effects Analysis
2. Identify risks or hazards for a given process.

Overview:

Participants working in small teams will complete a PMFEA for a step(s) identified in the process tree segment for Intensity Modulated Radiation Therapy below. "Evaluate Plan" will be used to generate FMEA and FTA examples in this and following workshops.

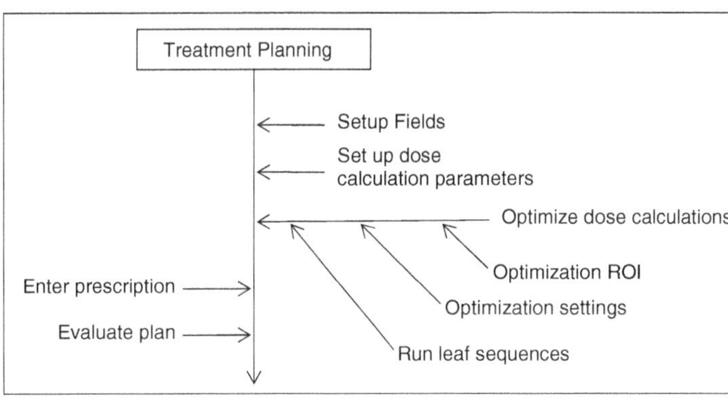

Treatment planning segment from a process tree describing the IMRT process

Steps:

1. Form your team. Teams familiar with the process being analyzed always produce a higher quality PFMEA than an individual.
2. Select one of the steps from the treatment planning process tree segment and use the form handed out (reproduced below) to perform a PFMEA on that step.

Performing the PFMEA

1. List the process step your team selected.

Process Step	Potential Failure Mode	Potential Cause of Failure Mode	Effects of Potential Failure Mode	Current Controls	Occurrence - Cause	Detectability of Failure Mode	Severity of Effect from Failure Mode	RPN	Corrective Action	Resulting RPN

FMEA form

2. Identify ways in which the process step can fail (failure modes). List at least four. In order to minimize confusion, your team should use a consistent process for identifying failure modes, such as always defining failure modes in terms of specific process failures. For example, for delivering a specific prescribed dose of radiation, failure modes should include dose delivered to the wrong location, too little radiation delivered, too much radiation delivered, etc.
3. For one of the failure modes you identify, list several causes that could result in that failure mode. It is important that you list causes that could occur and not limit the analysis to causes of failure modes that your team thinks are likely to occur. Typical causes of failure modes include, but are not limited to, the following:
    - Lack of formal and written procedures, work instructions, or work methods
    - Inadequate training
    - Insufficient time to complete a task due to other tasks requiring attention
    - Equipment or software malfunction
    - Stressful work environments leading to mistakes
4. Identify the potential effects that could result when the failure mode occurs. It is important to identify the worst possible outcome of a failure mode. Your team should not consider how likely an effect is to occur. Very serious effects could occur as a result of many failure modes in radiation therapy.
5. List all process controls currently in place and being used. There are three categories of process controls.
    - Controls that reduce the likelihood of specific causes of failure modes occurring. Examples include but are not limited to:

    - Operator training or certification
    - Written procedures and work instructions
    - Process checklists
    - Statistical Process Control (SPC)
- Controls that detect failure modes prior to serious effects resulting. All types of in-process inspections are the most commonly used detection controls. Examples include, but are not limited to, the following:
    - Peer review of process decisions
    - Downstream process checks
- Controls that will moderate the severity of effects that could result from a failure mode. This category of control is typically difficult to execute in radiation therapy. The time between a failure mode occurring and the resulting, potentially very serious, effects is very short, and damage is often inevitable once a failure mode occurs.

6. Judge the effectiveness of the current controls by:
    - Defining the likelihood that a specific cause of a failure mode will occur
    - Defining the probability that the current controls will detect the failure mode before any serious effects result
    - Specifying the seriousness of the effects resulting from the failure mode
    - Using the table on the next page, assign values for occurrence of a cause, detection of a failure mode, and seriousness of effects
7. Calculate the Risk Priority Number (RPN) by multiplying the Occurrence, Detection, and Severity rankings.
8. Identify and list new process controls that will:
    - Improve the likelihood of preventing specific causes of failure modes from occurring
    - Improve the probability of detecting a failure mode before any serious effects occur
9. Estimate the improvements resulting from the recommended actions in terms of:
    - Reducing the occurrence of the cause
    - Improving the detection of the failure mode
    - Calculating the new RPN by multiplying the estimated occurrence detection rankings resulting from the recommended actions and carryover severity rankings

FMEA ranking scales for Occurrence, Detection and Severity

Rank	Occurrence	Detection	Severity
	Probability that the cause will occur and lead to the failure mode	Probability that the failure mode will be detected before resulting in the end effect	Seriousness of the end effect when it occurs
1	Remote probability	Always	No effect
2	Low probability	High likelihood	Minor effect
3	Low probability	High likelihood	Minor effect
4	Moderate probability	Moderate likelihood	Moderate effect
5	Moderate probability	Moderate likelihood	Moderate effect
6	Moderate probability	Moderate likelihood	Moderate effect
7	High probability	Low likelihood	Serious effect
8	High probability	Low likelihood	Serious effect
9	Very high probability	Very low likelihood	Injury
10	100% probable	Never	Death

Example PFMEA for the process step – "Import images into RTP system data base" from IMRT

Step	Potential Failure Modes	Potential Causes of Failure	Potential Effects of Failure	O	S	D	RPN	Comments
Input images into RTP system database	Wrong patient's images	Miscommunication User error	Wrong dose distribution Wrong volume	3	9	5	135	
	Wrong imaging study (correct patient) Viz. Wrong phase of 4D CT selected for planning Wrong MR for target volume delineation	Ignorance of available imaging studies Ambiguous labeling of image sets Inadequate training Miscommunication User error	Wrong dose distribution Wrong volume	7	8	7	392	
	File(s) corrupted	Network problem	Lost images Wrong dose distribution	4 3	3 9	2 4	24 108	File probably would not open

The table above shows an example PFMEA for the process step "Import images into RTP system data base" from IMRT. The PFMEA is completed only through the baseline or current situation and controls and resulting RPN. It does not include recommended actions or the estimated effects that the recommended actions will have on the new RPN.

The table on the next two pages shows the actual PFMEA for the process step "Evaluate Plan" from TG 100.

PFMEA from TG 100 for "Evaluate Plan" process step from IMRT

Rank #	Step #	Major Process	Step	Potential Failure Modes	Potential Causes of Failure	Potential Effects of Failure	AVE O	AVE S	AVE D	AVE RPN
91	125	8 - Treatment Planning	11. Evaluate leaf sequences	2. Incorrect evaluation (i.e., OK when not, etc.)	1. Poor training 2. Inattention. 3. Poorly designed evaluation tools. 4. Software error.	Very wrong dose	3.78	7.89	5.89	179.33
Examples of causes and failures: Similar to the last one: How do you evaluate the leaf sequence? I do not know how to evaluate the leaf sequence?										
120	124	8 - Treatment Planning	11. Evaluate leaf sequences	1. Unintentional modification of sequences	1. Software error 2. User error	Very wrong dose	3.44	7.89	5.78	157.22
Examples of causes and failures: In order to understand whether the leaf sequence is right or wrong, one will have to generate i) the fluence profile using the final leaf sequences generated by the sequencer and compare it with the desired fluence profile; ii) one will have to generate the dose distribution and then determine whether the distribution is the desired one or not. If it is not then it could be due to many reasons, one of which is an incorrect leaf sequence. But the question is how does one know whether the leaf sequence is correct or not?										
100	126	8 - Treatment Planning	12. Transfer sequencer results to TPS for final dose calc (if needed)	TPS (Dose calculation algorithm) gets wrong intensities.	Inadequate commissioning 1. Incomplete or failed transfer. 2. Software error 3. Bad data or data handling 4. bad interface definition	Very wrong dose	4.00	7.67	5.22	170.78
Examples of causes and failures: Everything stated in the cause column can cause error in the final dose calculation										
10	127	8 - Treatment Planning	13. Evaluate plan (DVH, isodose, dose tables, etc.)	1. Inadequate evaluation	1. Not enough time/effort spent 2. Inadequate training 3. Poor evaluation strategy 4. Incorrect final prescription	Wrong dose Wrong dose distribution	5.63	7.00	7.13	303.00
Examples of causes and failures: One needs to look at DVH and dose distribution on a slice by slice basis. But because of poor training or not spending enough time to evaluate the DVH or isodose distribution, the result can be an inadequate evaluation of the plan. This may result in the effects outlined in the effect column.										
28	128	8 - Treatment Planning	14. Evaluate plan (DVH, isodose, dose tables, etc.)	Incorrect DVHs, IDLS, etc	1.Software error; 2. User selects wrong parameters for DVH 3. Wrong or inappropriate structure(s) for DVH(s) 4. Poor evaluation tools 5. Wrong normalization point selected Inadequate commissioning	Very wrong dose	5.00	8.11	5.44	230.67
Examples of causes and failures: Everything stated in the cause column may result in an incorrect evaluation of DVH or IDLs, etc.										

PFMEA from TG 100 for "Evaluate Plan" process step from IMRT

172	132	8 - Treatment Planning	15. Evaluate delivery system limitations	3. Patient and delivery system collision	1. Incorrect delivery system definitionintheTPS; 2. Lack of machine limitations in TPS 3. Failure to recognize the impact of patient immobilization and/or positioning aids on collision free zone 4. Poor knowledge of patient location + positioning on table	Non-radiation related injury a) Treatment delivery is aborted (O=5,S=3,D=1), b) Patient is hurt (O=5,S=10,D=3) a) Treatment delivery is aborted (O=6,S=6,D=5), b) Patient is hurt (S=5,S=9,D=5) a) Treatment delivery is aborted (O=4,S=3,D=6), b) Patient is hurt (O=4,S=8,D=3) a) Treatment delivery is aborted (O=4,S=9,D=3), b) Patient is hurt (O=4,S=9,D=3) a) Treatment delivery is aborted (O=3,S=2,D=1), b) Patient is hurt (O=3,S=8,D=2) a) Treatment delivery is aborted (O=5,S=3,D=8), b) Patient is hurt (O=5,S=10,D=8)	4.33	5.78	4.11	106.33
Examples of causes and failures: Explanation of 1 and 2. Certain gantry and table angles will result in collision but these angles are not defined in the TPS. So, TPS will allow this but in reality this will cause collisions. 3. For large patient or a patient in an immobilization system there will be collisions at certain gantry angles. This is due to failure to recognize the limitations of large patients or immobilization devices.										
206	129	8 - Treatment Planning	15. Evaluate delivery system limitations	1. Treatment volume inappropriate for the delivery system	1. Failure to recognize the limitations of the delivery system 2. Inadequate training 3. Inadequate commissioning	Inconvenience (staff and patients) 1. Wasted time and effort 2. Patient treatment delayed	3.71	3.43	2.71	53.00
Examples of causes and failures: –										
208	131	8 - Treatment Planning	15. Evaluate delivery system limitations	2. Undeliverable treatment plan	1. Incorrect delivery system definitionintheTPS; 2. Lack of machine limitations in TPS 3. Poor knowledge of patient location + positioning on table Inadequate commissioning	1. Wasted time and effort 2. Patient treatment delayed	4.11	3.00	2.78	49.00
Examples of causes and failures: For Varian machine the collimator cannot rotate full 360 degrees. But the plan asks for a rotation of the collimator in the forbidden region. This will be an example of nos. 1 and 2 causes identified in the cause column. You want to treat the leg but you scanned the patient head in first. This may result in placement of the patient on the table that does not allow access to the target for treatment.										

Exercise 5

Fault Tree Analysis

Saiful Huq and Bruce Thomadsen

Objectives:

After this exercise, the participants should be able to:

1. Construct a fault tree from an FMEA
2. Perform basic Fault Tree Analysis (FTA). The determination of the QM program from this FT comes with Exercise 7

Exercise Overview:

In this exercise your team will complete an FTA for the step "Evaluate Plan" from the FMEA constructed in Exercise 4.

Steps:

1. Form your team. Teams composed of members who were involved in the development of the process tree and performing the FMEA for various steps of the process tree tend to produce a higher-quality fault tree than a single individual.
2. Perform the FTA:
   - The failure mode to the far left would be "Treatment Planning Failure." There will be many more paths that will lead to that than the one you will construct, but you will not consider those now.
   - The next box to the right will begin the portion of the tree that you will develop, and contain "Failure in Evaluating the Plan." When multiple potential failures could lead to the failure independently, connect the causes on the right with the failure to the left with an OR gate. If multiple causes have to happen simultaneously, connect them to the failure with an AND gate.
   - Working to the right, from each box, continue to add boxes for potential causes that could directly lead to each failure.
   - Remember to stop a pathway when it reaches the end of the facility's control.

- As you work on the fault tree, you may uncover potential failure modes that were overlooked on the FMEA. Add these to your tree.
- Add the RPN and severity scores to the branches of the tree.

From your fault tree you will now be able to see how failures propagate and can potentially cause harm to the patient.

Exercise 6a

Checklist Development

Luis Fong de los Santos, Jean-Pierre Bissonnette, Ellen Yorke, and Jatinder Palta

Objectives:

After this exercise participants will:

1. Have an understanding of the goals, development, and implementation process of checklists as error mitigation tools
2. Be able to identify those processes and areas in the radiotherapy workflow where checklists can provide meaningful results

Scenarios:

Scenario 1. Your department decided to treat on-call palliative cases. The patients will be treated using an isocentric clinical setup with parallel opposed fields. No simulation or planning system will be used for the procedure. The linac used for this procedure only has an EPID as an image verification tool. The group agreed that the following individuals will be present during the procedure: two therapists and an on-call radiation oncologist. The group decided that a checklist will help to mitigate the potential of an error during the treatment, and physics has been asked to help on the development of the checklist.

Scenario 2. You are coming to a brand new center. As part of the implementation of new procedures, you decided to create a checklist for your plan check process for IMRT plans. The department setup is as follow:

- 1 CT simulator
- Treatment planning system from vendor A
- Record and verify system from vendor B
- 2 linacs from vendor C
- 2 physicists (including you)

Tasks:

As part of the exercise, each small group will:

1. Identify the type of checklist you think should be developed for the two scenarios.
2. Identify the part of the process where you plan to use the checklist (i.e., before, during, or after the specific process).
3. Develop the two checklists.

   - Describe the checklist format
   - Briefly describe the implementation process
4. Describe the method to evaluate the performance of the checklists.

Questions:

After completing the above exercises, answer these questions not as a group, but individually:

1. What were the major differences between the checklists used for scenarios 1 and 2?
2. What were the major challenges that you faced during the development of the checklist?
3. What would be the major challenges of implementing checklists in your practice?
4. Provide three more clinical examples where you think a checklist would help your practice.

Exercise 6b

Procedure Drafting

Jean-Pierre Bissonnette, Luis Fong de los Santos, Ellen Yorke, and Jatinder Palta

Objectives:

After this exercise, the participant will master the basics of writing an effective procedure:

1. The procedure must have clear objectives
2. The conformity policy, including how to deal with variances, must be well defined
3. The life cycle of the procedure must also be determined

The process for establishing a useful procedure is also of importance. Two hypothetical tasks are outlined below. Your task is to write a clear procedure that is intended for technical staff.

Task 1:

Technical staff must capture data from a Panel A and calculate the sum of both columns, write the result at the bottom of a form, then date and sign the form. It is imperative that data is recorded in the following order: Column 1 from top to bottom, and Column 2 from top to bottom.

Once the procedure has been written, an independent party will review it and provide feedback.

Panel A

Voltage 1	32 V	Current 1	1.5 A
Voltage 1	28 V	Current 1	2.5 A
Voltage 1	12 V	Current 1	3.5 A
Voltage 1	38 V	Current 1	4.5 A

Task 2:

Technical staff must capture data from a Panel B and calculate the sum of both columns and subsequently divide each sum by 2. Staff must write the result at the bottom of a form, then date and sign the form. It is imperative that data is recorded in the following order: Row 1 from left to right, and each following row, in the same order.

Once the procedure has been written, an independent party will review it and provide feedback.

Panel B

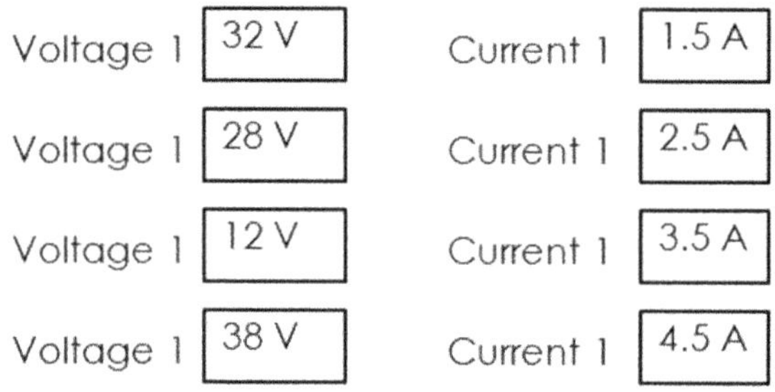

Reference:

Wieringa, D., C. Moore, and V. Barnes. Procedure Writing: Principles and Practice, 2nd ed. Columbus, OH: Battelle Press, 1998.

Exercise 7

Quality Management Design

Bruce Thomadsen

Objectives:

After this exercise, the participants should be able to:

1. Understand how to address potential failures and causes of failure identified during FMEA and FTA
2. Understand how to establish a quality management program

Overview:

In this exercise your team will address the potential failure modes and causes identified for the procedural step "Evaluate Plan" from your FMEA in Exercise 4 and your FTA in Exercise 5. You also may need the process map from Exercise 3.

Steps:

1. Scan through the potential causes for failures on the right of the FTA. Identify those causes that might indicate inadequate resource allocation to perform the task. These should be addressed by recommendations to increase support. Be specific as to what resources would be requested.
2. Identify those causes that result from the lack of any of the key core components (training, procedures and policies, and communications). Even if these are low-scoring causes, they may potentially indicate a general problem in the facility that needs to be addressed. List recommended actions to mitigate the causes.
3. Again, scan through the potential failure modes and causes, paying particular attention to those with the higher RPN or severity values. Is there any redesign of the process that would eliminate these potential failures or reduce their RPN values? (Nothing will reduce the severity.) If a redesign looks appropriate, would that lead to new potential failures or increase the RPN values for those previously identified?
4. For the remaining potential failure modes and causes, begin with the box with the highest-ranking PRN value.
   - Would thorough commissioning eliminate this potential failure mode?
   - If not, at this point it must be addressed through quality management: quality assurance, quality control, or a combination of the two.

- Most likely, quality control would be associated with the particular step, entering the fault tree as a "Failure in QC" in parallel with the cause and joined through an AND gate going to the resulting potential failure mode.
- Quality assurance would work down stream, after the cause. It is efficient to design QA such that it may cover several causes or potential failures modes.

5. Add the QM steps to your fault tree:
   - Specify the recommended tool and methodology. Note the strength of the tool according to the ranking of the Institute for Safe Medical Practice, and
   - For QA steps, estimate how frequently the tests should be performed.
6. Continue the exercise as in step 4 for the box with the next highest-ranking PRN value. Continue addressing boxes until the RPN values are so low that it would not be worth using resources to prevent their effects. However, make sure all potential failures with severity values of four or greater are addressed.

Exercise 8

Reporting Systems

Steven Sutlief and Sasa Mutic

Objectives:

After this exercise, the participant will be able to:

1. Design a reporting system based on the intended objectives, setting, and resources available
2. Assess the limitations of a reporting system based on its design constraints
3. Apply a reporting system design to sample incidents, and evaluate how well the intended objectives have been met

Introduction:

In Part 1 of this exercise, you will be tasked with designing an incident reporting system for your radiation therapy department. Reporting to the hospital patient safety office or to outside regulatory agencies will be handled separately. Given your knowledge of the staff culture, buy-in, and management support, you will make a number of design decisions to develop the system most suited for your department.

In Parts 2 and 3 you will consider two incident scenarios in light of your incident reporting system.

PART 1:

Item 1a: Intended use of the reporting system

Characterize your department by circling one or the other in each of the following pairs of choices:

Alternative A	Alternative B
Management is supportive of the quality improvement program and a positive functional relationship is established	Either management is indifferent to the quality improvement program or the staff fears that management involvement will lead to more harm than good
Physicians are generally distrustful of reporting their incidents or of the reporting of incidents involving their patients (that more harm than good will come of it)	Physicians are generally supportive of reporting their incidents or the reporting of incidents involving their patients
Other staff are generally distrustful of reporting their incidents (fearing that more harm than good will result)	Other staff are generally supportive of reporting their incidents
Facility has its own reporting system to which radiation oncology must also report	No facility-wide (beyond radiation oncology) reporting system exists
Subject to regulatory reporting criteria	Not subject to regulatory reporting criteria

Item 1b: Characteristics of the reporting system

Design your system by circling one or the other in each of the following pairs of choices:

Alternative A	Alternative B
Anonymous reporting	Reporter identified
Anonymous patient	Patient identified
Electronic submission	Paper-based submission
Voluntary reporting	Mandatory reporting (which means a reportability threshold must be established)
Events and close calls (near misses)	Events only
Response to individual submissions only	Response to trends as well as individual submissions
Only QI committee has access to the reports	All staff have access to the reports
Prompt reporting required	No requirement on reporting timeliness
The department's quality improvement team will review reports as they come	Reports are reviewed only at scheduled intervals during formal department quality improvement team meetings
Incident reports (not pertaining to negligence or wrongful intent) cannot be used as the basis for disciplinary action	Incident reports can be used as the basis for disciplinary action
The department's quality improvement team will meet regularly to ensure follow-up of corrective actions	The department's quality improvement team will not meet regularly
Formal RCA is used within the department (team selection, investigation, and a written report)	Only informal methods of RCA are used in conjunction with the department's incident learning system

Item 1c: Data elements

Select which data elements will be collected by circling items in the following list. This list is drawn from Appendix E of the consensus recommendations for incident learning database structures (Ford 2012). Based on recommendations of the consensus document, a preceding "r" indicates required items, "s" indicates suggested items, and "o" indicates optional items. For the sake of the exercise, make your selection of what is most appropriate for your department (per items 1a and 1b) without regard for whether the element has been denoted as required, recommended, or optional.

Item 1d: Department committees

How will the department's quality improvement committee be run? When it meets, what will the goals be?

Reporter's Form

[r]Date of incident(s)
[r]Time of incident(s)
[s]Date of report
[o]Date of discovery
[r]Incident type (Near miss or actual event)
[r]Person affected
[r]Number of incorrect fractions
[r]Incident description
[s]Where found
[s]How discovered
[s]Patient's first name
[s]Patient's last name
[s]Patient's medical record number
[s]Treatment modality
[o]Person reporting name
[o]Person reporting's role
[o]Action taken by reporter

Responder's Form

[o]Clinical action scale *(A, B, C, D - meaning determined by clinic)*
[s]Intervention (*treatment stopped, treatment modified, added additional treatment fractions,)*
[s]Safety barriers *(pull-down list of barriers already in place)*
[o]Corrective action
[o]Preventive action
[o]Learning actions
[o]Incident closed

Analyst's Form

[s]Location* *(city and state/province)*
[s]Facility profile *(number of linacs, patients on treatment, etc.)*
[s]Patient's age at the time of the incident
[s]Patient's race (pull-down list)
[s]Gender
[s]Patient notification
[s]Treating radiation oncologist
[s]Treating radiation oncologist notified
[s]Referring physician(s) notified
[s]Disease being treated
[o]Disease stage
[s]Intended treatment site
[o]Treatment intent (*curative, palliative, unknown, not applicable)*
[s]Total prescribed dose
[s]Dose per fraction
[s]Course *(primary phase, boost phase, modification during treatment, other (specify), not applicable)* *Treatment technique *(simple, 3D conformal, IMRT, etc.)*
[o]Image guidance *(kV or MV radiographs, IGRT: kV cone-beam CT, IGRT: MV cone-beam CT, etc.)*
[o]Person discovering name
[o]Person discovering's role (*attending radiation oncologist, resident radiation oncologist, etc.)*
[s]Treatment unit manufacturer(s)
[s]Treatment unit model(s)
[s]Treatment planning system manufacturer
[s]Treatment planning system model
[s]Record and verify system manufacturer
[s]Record and verify system model
[s]Third-party ancillary device manufacturer
[s]Third-party ancillary device model
[o]Number of patients on-treatment
[o]Number of radiation therapists present
[o]Number of other staff present
[o]Staff involved *(names and roles)*
[o]Error type *(wrong patient, wrong site, wrong laterality, wrong dose, wrong modality or energy, other)*
[o]Dosimetric severity (pull-down *list)*
[o]Reportable event
[r]Medical severity *(pull-down list)*
[r]Severity assessment type *(estimated severity, actual observed severity, not applicable)*
[s]When severity assessed (*before 24 hours, after 24 hours but before 3 days)*
[o]Incident analysis (from evaluator)
[s]Where originated
[o]Attached files *(e.g., screenshots, pdf reports, etc.)*
Causes

PART 2

Scenario:

- A patient in supine position was treated with an AP field, but from the posterior direction.
- Both the radiation oncologist and the medical physicist had approved the plan and fields.
- Even though it is from the wrong angle and is effectively a mirror image of the portal shape, the prescription for dose to a point has still been met.

Please address the following questions:

1. Is it an error?
2. How would you report it?
3. How would the department quality improvement team handle it?

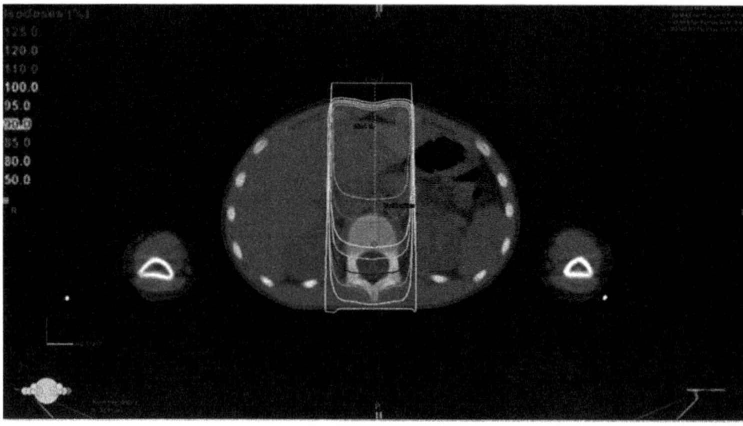

PART 3

Scenario:

- The radiation oncologist reviews a patient verification field image and recommends a 6 mm inferior shift.
- Similar shifts, ranging from 5 mm to 10 mm, are required later for some, but not all, of other patients starting treatment.
- It is later suspected that the erroneous superior-inferior positions occur for those patients who have received couch coordinate settings on one linear accelerator but then received filming and treatment on a different one. During follow up investigation it is found that the graticule on the second machine is off by 7 to 8 mm.

Please address the following questions:

1. Is it an error?
2. How would you report it?
3. How would the department quality improvement team handle it?

Exercise 9

Root Cause Analysis

Steven Sutlief and Barrett Caldwell
(Adapted from a case study provided by Stephen Breen)

Objectives:

After this exercise, the participant will be able to:

1. Analyze a radiation therapy incident for factual attributes, deviation from expected outcome, and causal factors.
2. Organize an RCA investigation.
3. Formulate corrective actions and follow up which address root causes rather than superficial conditions.
4. Understand and recognize the impact of performance shaping factors and RCA investigation processes on incident descriptions and possible corrective actions.

Introduction:

You are responsible for the quality improvement committee in the radiation oncology department. During a patient's course of treatment, erythema develops in an unexpected surface region. You are tasked with identifying the causes which led to this situation and implementing corrective actions which mitigate the possibility of the situation arising again.

PART 1

You are given the following case details:

- Patient undergoes external beam radiotherapy of the thyroid
- 50 Gy in 25 fractions
- Treatment progresses without unusual observations
- An RTT notices reddening on the shoulders of the patient

What are your immediate impressions? In particular, please address the following questions:

1. For the case described what kind of dose would you expect the shoulders to receive for this type of case?
2. For what approximate dose would you expect skin erythema to manifest?

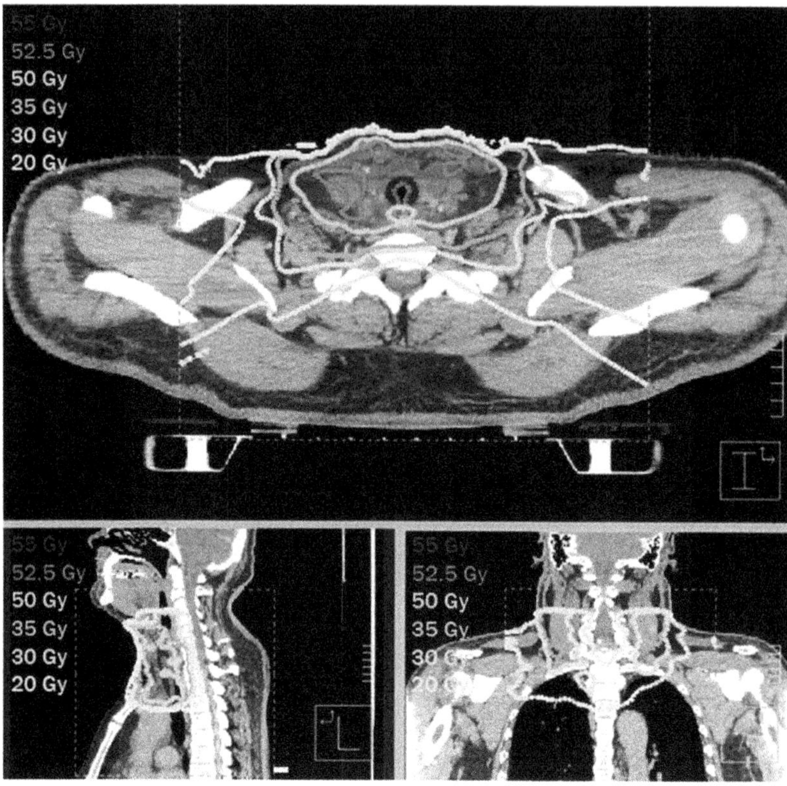

Exercise 9 Figure 1: Initial treatment planning images. (See CD for color images.)

PART 2

Your job is to gather information, assess the problem, and identify causes. Please address the following questions:

1. In selecting members of the investigation team, what specializations and background do you want represented on the team?
2. What materials do you want to gather?
3. Who do you want to talk to and what do you want to ask them?

PART 3

In the course of your investigation, you obtain the treatment planning images shown in Figure 1. You also are able to reconstruct the following timeline:

Day 0	Planning CT
Day 12	First fraction
Day 35	Erythema on shoulders observed by experienced staff

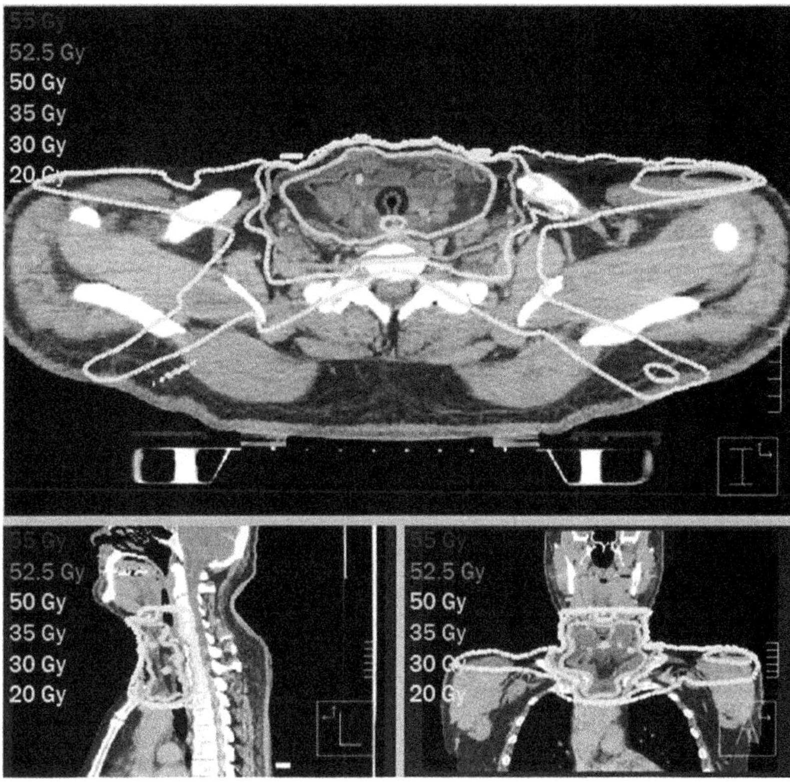

Exercise 9 Figure 2: Revised treatment planning images. (See CD for color images.)

Please address the following questions:

1. What are you looking for in the images?
2. What is your assessment of the information you have been given?
3. What is your next step?

PART 4

Based on your request for a recomputation of the isodoses, you receive the revised treatment planning images shown in Figure 2.

1. What are you looking for in the images?
2. What is your assessment of the information you have been given?
3. What is your next step?

PART 5

After all relevant materials have been gathered and all relevant parties have been talked to, the next step is to analyze the information and identify causal attributes. An appropriate methodology to use here is the five whys: the first cause you come up with will likely involve a deeper causal precursor; to get to the deeper causation, ask why the purported causal factor arose and then repeat the questioning process a handful of times. Using the causal taxonomy shown in the appendix to Chapter 6, identify the appropriate causal node(s) and describe why they characterize elements of the situation.

1. Which causal assignments are essential and which contributed to make the error possible?
2. Are the causal assignments limited to a particular specialist?
3. What safety barriers needed to fail in order for this error to occur?

PART 6

For the basic cause(s) you have identified, generate appropriate corrective actions to mitigate the problem from occurring again.

1. What corrective actions did you come up with?
2. Do the changes occur at the sharp end or the blunt end of the process? (The sharp end is where people are in direct contact with safety clinical processes, while the blunt end is the organization behind the activities at the sharp end.)
3. Can the corrective actions be monitored?

Exercise 10

Error-proofing

Peter Dunscombe and Eric Ford

Objectives:

In this exercise the participant will:

1. Gain experience applying the principles of error-proofing
2. Understand how error mitigation strategies apply to a realistic clinical scenario

Exercise:

This exercise will explore a practical application of the principles of error-proofing laid out in chapters 5 and 7. As you explore the case scenario presented below and consider error-mitigation strategies, it will be useful to recall the "hierarchy of safety interventions" (see Chapter 7, Table 7–1) and also the three categories of error-proofing, namely: preventing errors from happening, making errors more detectable, and implementing a fail-safe design.

Scenario:

The events and timeline of the case are as follows:

- A patient is referred to the radiation oncology department for emergent treatment for spinal cord compression.
- The patient is admitted into the hospital and has a consult with a radiation oncology resident at 9:00 a.m. Friday.
- Due to scheduling miscommunication within the in-patient department, the patient does not arrive at radiation oncology until 4:00 p.m.
- CT simulation is performed. The isocenter is set by the attending physician who is present. Per standard procedure for emergent cases, the physician places beams in the treatment planning system and DRRs are generated and sent to the treatment unit computer. Localization marks are established (ink marks with adhesive tape).
- A plan is formulated to treat the patient off-hours by on-call staff. The total treatment will be 30 Gy delivered in 10 fractions.
- Upon further discussion with each other, the attending and resident physician decide to move the isocenter so as to accommodate half-beam blocking. This is thought to be useful if future field-matching is

required. They move the isocenter in the treatment planning system and create new treatment beams. At 6:00 p.m. they turn over the plan to the on-call dosimetrist to finalize the work.

- The dosimetrist is not aware that isocenter has been moved from its location at simulation.
- The resident physician calls the on-call therapist and says that treatment will start in half an hour. He mentions that the isocenter must be moved.
- The dosimetrist finalizes the plan and sends the new treatment fields to the treatment unit computer. The plan uses the new fields and isocenter defined by the physician. The old DRRs generated at the time of simulation remain in the treatment unit computer.
- There is no physics check per standard policy of only requiring a check before the third fraction is treated.
- At 6:30 p.m. the patient is brought into the treatment room by the on-call treating therapist ("Therapist B").
- Since Therapist B has been informed by the resident that there is an isocenter shift, she pulls up the shift documentation from the planning system printout and begins to set up the patient accordingly.
- A second therapist ("Therapist A" who performed the CT simulation) enters the room to assist with treatment. He informs therapist B that the isocenter was already set at the time of simulation and, therefore, no shifts are required. Therapist B voices some concern but eventually defers to therapist A, judging that if there is a shift, it will be visualized on portal images.
- The two therapists set up the patient to the marks established at the time of simulation with no shifts.
- They acquire portal images and compare them to the DRRs generated at the time of simulation. They judge the localization to be very good. The attending physician is paged to the treatment machine and approves the images.
- Treatment is completed using the new fields. The patient is released to go back to the floor.
- The following day is a weekend, and the patient receives two more treatments with the same beams.
- On Monday morning, a physicist checks the plan and discovers the error. The treatment beams on the plan use a different isocenter from those that were imaged. The patient has received three treatments to a location 5 cm inferior to that intended.

In summary, an error was made in not shifting the patient to the new isocenter and multiple errors were made in communications between staff. These errors went undetected because of another error in not generating new DRRs when the isocenter was moved.

In the wake of this incident, the department leadership elects to conduct a root-cause analysis. The department has had three similar incidents in the last two years. In addition to the factual patterns above, the following contributing factors are identified: time pressure to treat the patient, inexperience on the part of some staff (in particular, a new resident), incomplete understanding of workflow by the physician, a reluctance to question authority (therapist B feeling reluctant to question the more senior therapist A), and failures of communication.

Assignment:

You are tasked by the department leadership to develop one or more interventions or changes in practice to prevent such an incident from occurring again. You are free to suggest any changes and have assurances from the leadership that all reasonable changes will be implemented. You must, however, work within the confines of a relatively limited budget and staffing levels.

To structure your response, please fill out the following table. For each proposed intervention, provide a numerical score for expected effectiveness:

0 = not effective at all
5 = extremely effective)

and perceived feasibility:

0 = extremely difficult
5 = very easy

Finally rank your proposed interventions and decide which you will present for implementation. Discuss how you will proceed with implementation.

Proposed Error-mitigation Strategies

Proposed Intervention	Staff Involved	Expected Effectiveness	Perceived Feasibility	Rank

Exercise 11

The Use of Control Charts

(monitoring daily linear accelerator output)

Todd Pawlicki

Objectives:

After this exercise, the participant should be able to:

1. Create an individuals and a moving range control chart
2. Understand how to use the charts to characterize process performance for quality improvement

This exercise is loosely based on the paper by Pawlicki, T, M. Whitaker, A.L. Boyer. (2005). "Statistical process control for radiotherapy quality assurance." *Med Phys* 32:2777–86.

Scenario:

You are working in a department with two dual-energy linear accelerators (linacs). The morning daily output QC device for one of the linacs has just returned from repair. You have accepted and commissioned the device, i.e., ensured that it is working properly, created baseline output values for the different linac energies, and revisited the policy and procedures for morning output QC. The device reports the percent difference from the baseline. You have also educated the therapists on how to use the device, including how to set up the device and how to obtain an output measurement using the appropriate amount of build-up and device settings for different energies. The device is set up with 100 cm SSD to the top of the device and an appropriate build-up for the energy selected.

You check the output readings from the device at least once per week. As part of the policy and procedure you set up, the therapists are required to stop and call you if a reading is greater than ±5% from baseline. No patient should be treated until you resolve the reason for a discrepancy greater than ±5%. If any output reading is greater than ±3% from baseline, then the therapists can continue, but they should notify you of the deviation.

May 31 was the first day that the repaired device was used by the therapists. You have made a point to be at the linac every morning from May 31 to June 13 to help the therapists carefully set up the device and answer any questions. From June 14 onward, you let the therapists do the morning QA on their own. On June 21, the therapists rotated from one linac to the other as part of their routine policy and procedure. The data for the second beam energy of the hypothetical daily linac

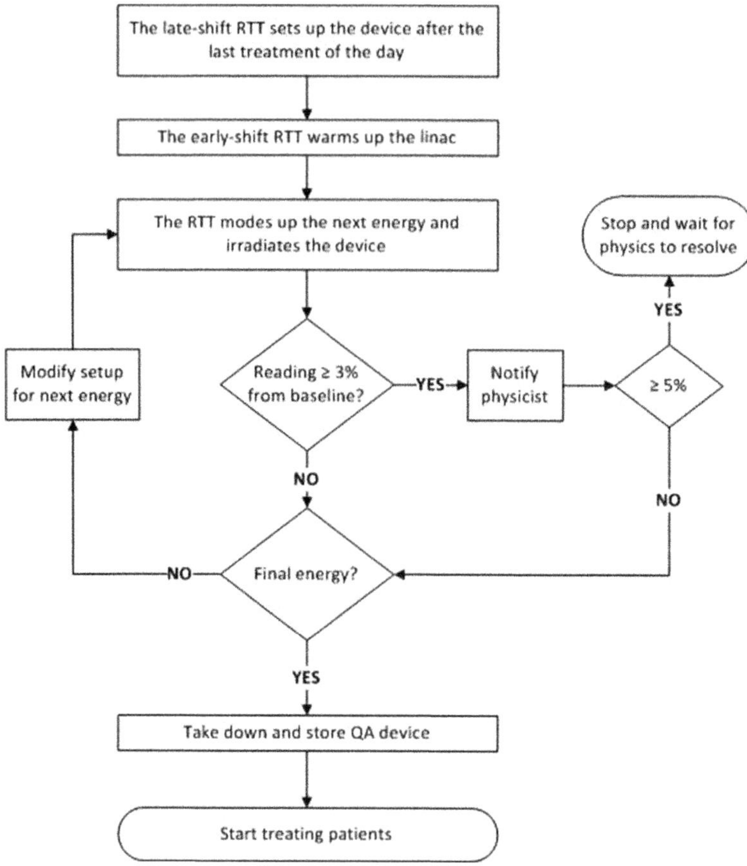

Process map for this exercise

output verification process are given below in Table 1. A flowchart (process map) is also provided below.

Your task is to identify if/when the process has changed. Then, characterize the process performance and discuss what actions might be taken.

Exercise 9 Table 1: Output QC data (percent deviation from baseline) for the second energy that was checked during morning QA. Remember, in real-life, you would get these data one day at a time.

j	Date	Data	
1	Fri, May 31	-0.32	
2	Mon, June 3	0.73	
3	Tue, June 4	-0.19	
4	Wed, June 5	0.24	
5	Thur, June 6	1.04	
6	Fri, June 7	1.16	
7	Mon, June 10	0.63	
8	Tue, June 11	-0.79	
9	Wed, June 12	-1.28	
10	Thur, June 13	0.03	
11	Fri, June 14	-1.69	
12	Mon, June 17	0.44	
13	Tue, June 18	-0.91	
14	Wed, June 19	-1.16	
15	Thur, June 20	-0.44	
16	Fri, June 21	2.02	
17	Mon, June 24	2.45	
18	Tue, June 25	0.70	
19	Wed, June 26	-0.61	
20	Thur, June 27	0.42	

Before proceeding, please consider the data in Table 1 and make some notes about the following questions.

- Would you ever be notified by the therapists (see flowchart on the preceding page)?
- How would you characterize this process?
- Is this process is acceptable? Why or why not?

Steps:

1. Calculate the data average over a period where the process was carefully monitored (keep paragraph 3 of the narrative on page 1 in mind).
   - Draw a center line on the individuals chart (see next page).

   $\bar{x} = \frac{1}{N}\sum x =$ ________

2. Write mR in the Column 4 heading of Table 1.
   - Calculate the moving range (mR) and enter the first value in row 2 of the table.

   $\left|x_j - x_{j-1}\right| =$ fill in Table 1

3. Calculate the average moving range (for the same points in Step 1).
   - Draw the average moving range on the moving range chart (see next page).

   $\bar{x} = \frac{1}{N}\sum x =$ ________

4. Calculate the upper limit for the average chart.
   - Draw the upper limit on the individuals chart.

   $\bar{x} + 2.66 \cdot m\bar{R} =$ ________

5. Calculate the lower limit for the average chart.
   - Draw the lower limit on the individuals chart.

   $\bar{x} - 2.66 \cdot m\bar{R} =$ ________

6. Calculate the upper limit for the moving range chart.
   - Draw the upper limit on the moving range chart.

   $3.268 \cdot m\bar{R} =$ ________

7. Plot the data and moving ranges on the individuals and moving range chart, respectively.
   - Use different symbols for the points used to calculate the chart limit and the later points.
   - Remember to start at point 2 on the moving range chart.
8. To answer the question on how best to characterize this process, see Table 2 below.
   - Is this process in control or out of control?
   - What is the action require for this process?

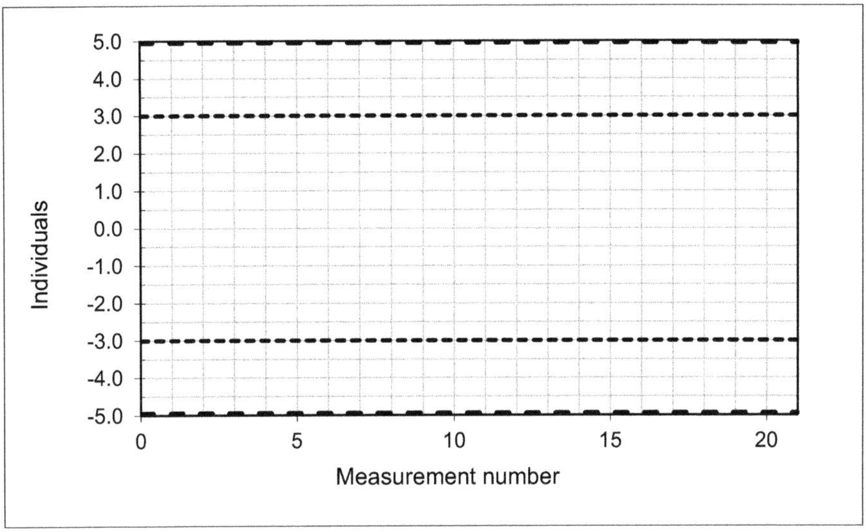

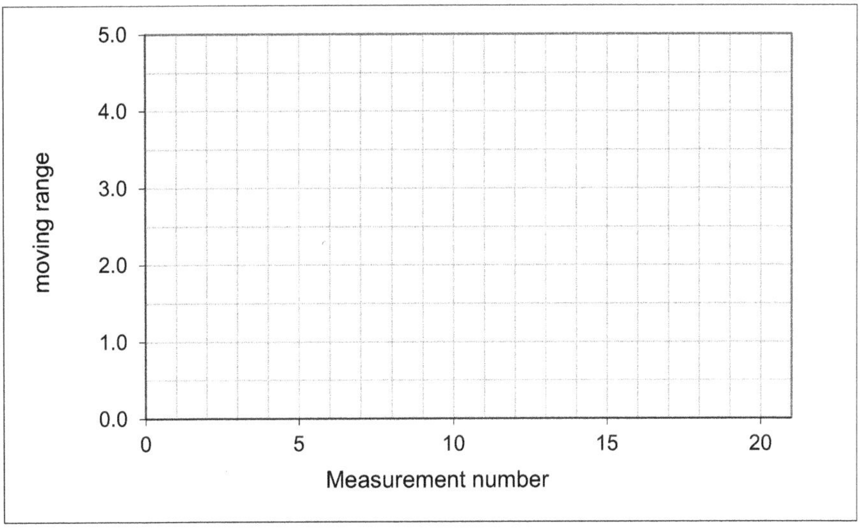

Control charts. Note, the dashed lines on the average chart are the conventional clinical tolerance limits (±3) and action limits (±5).

Exercise 9 Table 2: A schema to characterize process performance.

Process Characteristics	Action Required
In-control Control limits are within tolerance or action limits	Continue monitoring process indefinitely
In-control Control limits are outside tolerance or action limits	Re-engineer process or widen tolerance or action limits
Out-of-control Control limits within tolerance or action limits	Analyze process and remove sources contributing to process variation[1]
Out-of-control Control limits outside tolerance or action limits	Analyze process and re-commission and/or re-engineer process[2]

An in-control process is one where all data points are within the control limits. An out-of-control process is one where any data point is outside the control limits.
Note: this characterization scheme is adapted from Figure 1.8 of Wheeler, D.J. and D.S. Chambers. *Understanding Statistical Process Control,* 2nd edn. Knoxville: SPC Press, 1992.

[1] Other quality tools can be helpful here, for example, when a point is out-of-control, then perform a mini Root-Cause-Analysis to find out why. Process mapping and Value Stream Mapping can also be useful.
[2] In this case, a mini FMEA can be helpful because you are looking wide and deep at the process. Fault Tree analysis may also be helpful for this analysis.

Other questions to think about when analyzing this exercise:

1. What are all the components where variation could enter the process of daily linac QA (some parts of an FMEA could be useful to answer this question)? What are some reasons this process might be out-of-control?
2. Do you need tolerance limits (±3%) and action limits (±5%)? Would it be better to just use action limits and control limits as the tolerance limits? Why or why not?
3. Would you expect to need separate control chart limits for the other linacs in your department? What about different control charts for each energy?